Cost-Benefit Studies of Natural Resource Management in Southeast Asia

David James • Herminia A. Francisco
Editors

Cost-Benefit Studies of Natural Resource Management in Southeast Asia

Springer

Editors
David James
Independent Consultant
Whale Beach, NSW, Australia

Herminia A. Francisco
Economy and Environment Program
for Southeast Asia (EEPSEA), WorldFish
Laguna, Philippines

ISBN 978-981-287-392-7 ISBN 978-981-287-393-4 (eBook)
DOI 10.1007/978-981-287-393-4

Library of Congress Control Number: 2015935671

Springer Singapore Heidelberg New York Dordrecht London

Printed on acid-free paper

Springer Science+Business Media Singapore Pte Ltd. is part of Springer Science+Business Media (www.springer.com)

Foreword

Cost-benefit analysis (CBA) as a technique to help decision makers make investments, assess regulations, and assess difficult public policy tradeoffs has been around for over 50 years. What makes another book about cost-benefit unique and worth reading? What does a reader have to learn? This book is different because it focuses on pragmatic issues in Southeast Asia examined by researchers from the region. As sponsor of the research, the Environment and Economy Program of Southeast Asia (EEPSEA) mentors each researcher to help create studies that can be used in the classroom and help inform decision makers. EEPSEA researchers participate in workshops and training courses where they get valuable feedback on their work and hands-on experience working with international experts. The studies in this book should thus appeal to a wide audience who want to see how economic techniques are applied in practical settings and also learn about critical environmental and resource issues in Southeast Asia. A variety of economic tools are used: benefit cost analysis as well as cost effectiveness and multi-criteria analysis. In this way they "triangulate" on the benefit and cost estimates using multiple methods. Many of the studies examine policies rather than facilities or projects as is common in traditional benefit cost analysis. One area of particular importance is the estimation of the benefit side of cost-benefit analysis. Benefits create measurement problems for the analyst, and in many studies are ignored or measured very crudely. The result is an over-emphasis on the cost side amongst policy makers rather than a fulsome look at the total picture. This one-sided approach can lead to decisions that are not in the best interests of society.

Part I, Chap. 1, introduces the purpose of the text, the regional setting, and the role of EEPSEA. Chapter 2 provides the historical context for CBA and covers the core concepts and conceptual bases fundamental to a CBA study. Part II focuses on studies in the natural resource areas of agriculture, forestry, and fisheries. The first two chapters in part 2 compare the cost effectiveness of biofuels to conventional fossil fuels in reducing greenhouse gases (GHGs) and other pollutants. In Chap. 3, Wang explores under what circumstances biodiesel from a plant (*Jatropha curcas*) would be competitive with fossil fuels using a lifecycle analysis methodology.

A financial analysis shows that biodiesel is not competitive. But if yields of the plant are improved and/or carbon emissions from fossil fuels are priced in a more fulsome social cost and benefit accounting framework, biodiesel becomes more competitive. The study illustrates the value of undertaking a lifecycle approach as well as considering the environmental costs and benefits of alternative fuels. Thanh in Chap. 4 also compares the cost effectiveness of biofuels – ethanol and biodiesel – to fossil fuels in a lifecycle analysis for Vietnam and finds that improvements in biofuel productivity are needed to make biofuel as cost efficient as fossil fuels. Ethanol, however, can compete favorably with fossil fuels if used in a fuel efficient vehicle.

Turning to agriculture, Launio and co-authors in Chap. 5 also use cost effectiveness to assess methods to alter farm management practices. Rice straw, a by-product of rice production, is typically treated as waste by farmers, and when burned emits methane (a GHG) and nitrogen oxide. The authors use production functions to estimate the magnitude of the problem and assess the cost effectiveness of alternative government policies and farm management practices to reduce carbon emissions. Ranking the options, they find that on-farm practices of incorporating straw stubble into the soil, followed by composting, are the most cost effective.

Economic models of forest management traditionally focused only on maximizing the net value of timber harvests. In recent years, researchers have incorporated the benefits of sequestering carbon in timber into a theoretical model of optimal forest harvests. Nhung's paper in Chap. 6 finds that land holders harvest timber at a younger age than what the theoretical model yields when carbon sequestration is included. The challenge is to find policies that provide incentives for land owners to delay harvests. Payments to farmers to increase the rotation interval – a lump-sum payment of carbon benefits at the beginning of the rotation – and a planting cost subsidy are explored.

How can land be used to benefit both the land holders and the environment? Yem and co-authors in Chap. 7 compute the net social benefits of land managed as large scale rubber plantations compared to smallholder rubber plantations in Cambodia. Their study shows the importance of incorporating the social component – the impact of land conversion on the local population. Conversion from crop production (maize, soybean, cassava, and cashew) to smallholder rubber plantation provides the largest benefit to farmers involved in those conversion schemes. Removing the natural forest and converting the land to large-scale rubber plantation ranks last among their options.

The final chapter in Part II explores policies to reduce water pollution from inland aquaculture practices in Vietnam. Lang and co-authors compute the cost effectiveness of different waste management approaches and add an important component to their technical analysis – consultations with fish farmers on the practicality of different policies. The authors recommend three policy directions: the establishment of fish planned zones to help create concentrated wastewater treatment systems and reduce treatment costs; set and enforce emission standards;

and set up an environmental fund to provide long-term loans with preferential interest rates to fish farmers to enable them to build treatment systems.

Part III contains two papers on river basin management – developments of river basin resources. Project evaluation of such development projects typically entails evaluation of benefits and costs with and without the project. In Chap. 9, Thoradeniya examines a reservoir development project in this framework, but also includes stakeholder involvement and a range of nonmarket values. Limited information was available for the environmental valuation and demand analysis – but the researcher found various ways to approximate the values. The most novel aspect of this study is that the study incorporated stakeholders and communities in a validation/calibration approach to assess valuation and benefit cost analysis. Gunaratne also examines a resource development project and evaluates options for a sand mining project, but in addition to benefit cost analysis, he employs multi-criteria analysis to evaluate the options. As input into the analysis, the preferences of workers are incorporated (via choice modelling) and expert opinion is also employed. Both of these project evaluation examples use multiple methods, stakeholder input, and an array of other sources to inform project selection.

Part IV of the volume focuses on economic analysis of protected areas. Economic analysis of protected areas has been hampered by the lack of information on various benefit categories (non-timber forest products, recreation/tourism, threatened species benefits, etc.). The authors of these chapters recognize these limitations and either collect primary data or employ benefit transfer techniques. What is also admirable is that in many cases the papers augment the formal benefit cost analysis with stakeholder interviews as a form of "calibration" or as a method to include political feasibility or distributional aspects of the issue within the problem analysis. In Chap. 11, Baylatry et al. evaluate multiple outputs (market and non-market) including soil erosion, forest carbon, tourism and other economic outputs. In some cases they rely on benefit transfer while in others they collect primary data. Hou et al. assess community forestry policies in Cambodia in Chap. 12. Their analysis also includes multiple categories of outputs such as timber, water and other services and they obtain primary data in many of the cases. In Chap. 14, Roongtawanreongsri et al. examine the possible options for a suburban forest area of southern Thailand. In this case the range of benefits examined in assessing land use options includes timber, water supply, carbon, flood control and biodiversity. This study also discusses some rather unusual benefit categories that have not been addressed in much of the literature. While there are clearly limitations in the availability of information on biological and economic factors like biodiversity and flood control, this study, like most in this volume, uses various sources of data or benefit transfer techniques. This study also raises questions about benefits capture (is there a way to capture or monetize the nonmarket values) and the fact that the benefit cost analysis is only the first step in policy development. The discussion of the establishment of a payment for ecosystem services program to actually implement the protection program illustrates the role that benefit cost analysis can play in aiding policy design.

Jian et al. employ cost effectiveness analysis rather than full benefit cost analysis in their assessment of wetland conservation in Chap. 14. They also use multi-criteria analysis to compare the policy options. This interesting study recognizes the distributional aspects of benefit cost analysis or cost effectiveness analysis by engaging with stakeholders and searching for a policy option that results in a relatively efficient yet equitable outcome.

The last three papers in the volume in Part V examine adaptation to climate change and the evaluation of projects that may protect against adverse effects of climate change. Arias et al. in Chap. 15 examine the benefits and costs of construction of coastal infrastructure to protect against the impacts of sea level rise in the Philippines. A major component of the analysis is the assessment of areas affected. A set of advanced techniques are used to forecast the impacts. Values of properties as well as environmental values (measured using benefits transfer) were incorporated into the analysis. Danh examines a similar issue when investigating the benefits and costs of a sea dike. This study rigorously incorporated risk analysis and sensitivity analysis to identify the variability of benefits and costs over time. The timing of the construction of the dikes is also an issue in this case, suggesting the possibility of a real options analysis in future research. In contrast to the previous studies, Arias et al., in Chap. 17, use the demand for early warning services as valued by individuals affected to determine the benefits of climate adaptation projects. This set of three papers illustrates the wide range of methods, components of value, and a variety of technical approaches to sensitivity analysis to evaluate options for mitigating the effects of climate change. While all three studies find options that appear to be the best to use to combat the effects of climate change, such as building sea dikes, they all also recognize the sensitivity and uncertainty of these outcomes.

The papers in this volume illustrate the challenges in conducting benefit cost analysis in areas involving environmental values (e.g. nonmarket values), uncertainty due to a lack of information, or an inherently variable system (e.g. climate change). The researchers employ a number of different methods to address these challenges, including stakeholder input, methods that attempt to triangulate the benefit cost analysis, or other novel approaches. While there are many uncertainties about the accuracy of the value estimates and others aspects of these studies, as there often are with any benefit cost analysis, the papers in this volume are creative, attempt to use multiple methods and employ sensitivity analysis in various ways to do the best with limited data. The result is a set of informative examples of benefit cost analysis involving environmental challenges in Southeast Asia.

University of Alberta
Edmonton, AB, Canada
Vic Adamowicz

Simon Fraser University
Burnaby, BC, Canada
Nancy Olewiler

Preface

As a regional organization tasked with developing the capacity of Southeast Asian researchers in environmental economics research to inform policy formulation in the region, the Economy and Environment Program for Southeast Asia (EEPSEA) has supported a number of researchers over the years through its research grant program. Among the various methods used by EEPSEA researchers in conducting policy relevant studies is cost-benefit analysis (CBA). For this book, we selected 15 studies that highlight how CBA was used to understand and/or evaluate solutions to natural resource and environmental problems. The aim is to provide researchers with reference material that demonstrates how they can use CBA to support improved natural resource and environmental management in their own countries.

The studies were carried out in a developing country setting where data availability – and sometimes expertise to implement such research – has a bearing on the quality of the analysis. Hence, we were guided by two main goals in selecting cases for this book. Firstly, we wish to demonstrate that, despite such problems, researchers can still generate meaningful results that can help decision makers. Secondly, we wish to provide teachers in the region with useful discussion and case materials showing how they can improve the application of CBA to achieve a better environment. This latter objective is consistent with EEPSEA's mandate and with what our donors – the International Development Research Centre (IDRC), the Swedish International Development Cooperation Agency (Sida) – and our host organization, WorldFish, expect us to deliver as part of their development support for researchers in Southeast Asia and China.

I want to thank the people whose tireless efforts helped produce this book: our lead editor, Dr. David James, who also guided most of the researchers in this collection during the conduct of their studies; Dr. Canesio Predo for co-writing a chapter in this book and for his logistical support to the project; and Dr. Noor Aini Zakaria, Julienne Bariuan-Elca and Mia Mercado for their logistical support in communicating directly with the authors and in getting the book into its present form. Finally, we want to thank the various researchers for their contributions in transforming their studies into versions appropriate for inclusion in the book and for agreeing that their studies could be used as teaching cases to train future researchers in this field.

Southeast Asia is endowed with rich natural and environmental resources. About half of the world's terrestrial and marine biodiversity resources are found in the region. Unfortunately, many of the region's resources are in varying states of degradation and depletion, clearly indicating the need to improve their management. To do so requires the use of various tools that different disciplines have to offer. In the case of economics, the field offers several tools capable of determining how to manage these valuable limited resources more efficiently or cost-effectively – CBA being one of them. As such, I hope this book provides useful insights into the importance of these CBA studies as a means of identifying best courses of management action.

Los Banos Laguna, Philippines Herminia A. Francisco

Contents

Part I Guide to Cost-Benefit Analysis and Case Studies

1 **Introduction** 3
Herminia A. Francisco and David James

2 **Principles and Practice of Cost–Benefit Analysis** 11
David James and Canesio Predo

Part II Agriculture, Forestry and Fisheries

3 **A Cost-Benefit Analysis of *Jatropha* Biodiesel Production in China** 49
Zanxin Wang

4 **Comparing the Social Costs of Biofuels and Fossil Fuels: A Case Study of Vietnam** 69
Loan Le Thanh

5 **Economic Analysis of Rice Straw Management Alternatives and Understanding Farmers' Choices** 93
Cheryll C. Launio, Constancio A. Asis, Rowena G. Manalili, and Evelyn F. Javier

6 **Optimal Forest Management for Carbon Sequestration: A Case Study of *Eucalyptus urophylla* and *Acacia mangium* in Yen Bai Province, Vietnam** 113
Nghiem Thi Hong Nhung

7 **Rubber Plantation Development in Cambodia: A Cost-Benefit Analysis** 137
Dararath Yem, Neth Top, and Vuthy Lic

8 **Environmental Consequences of and Pollution Control Options for Pond "Tra" Fish Production in Thotnot District, Can Tho City, Vietnam** 157
Thi Lang Vo, Quang Vinh Ky, and Thi Thanh Truc Ngo

Part III River Basin Management

9 **Application of an Educated Trade-Off Analysis Framework for the Ma Oya River Basin Development Project, Sri Lanka** 179
Bhadranie Thoradeniya

10 **Policy Options for Sustainable River Sand Mining in Sri Lanka** 201
L.H.P. Gunaratne

Part IV Protected Areas

11 **A Cost-Benefit Analysis of Ecotourism: A Case Study of Xe Pian National Protected Area, Champasak Province, Lao PDR** 227
Malabou Baylatry, Sivannakone Malivarn, and Philayvanh Viravouth

12 **A Cost and Benefit Analysis of the Community Forestry Project in Chumkiri District, Kampot Province, Cambodia** 251
Kalyan Hou, Sothunvathanak Meas, Maredi Im, Chanthy Ros, and Han Keam

13 **Economic Benefit of Management Options for a Suburban Forest (Kho Hong Hill) in South Thailand** 275
Saowalak Roongtawanreongsri, Prakart Sawangchote, Sara Bumrungsri, and Chaisri Suksaroj

14 **Cost-Effectiveness of Policy Options for Sustainable Wetland Conservation: A Case Study of Qixinghe Wetland, China** 299
Jian Wu, Xiaoxia Wang, Kunyu Niu, and Shushan Li

Part V Adaptation to Climate Change

15 **Economic Vulnerability and Possible Adaptation to Coastal Erosion in San Fernando City, Philippines** 325
Jaimie Kim B. Arias, Moises A. Dorado, and Rowena Dorado

16 Adaptation to Sea Level Rise in the Vietnamese Mekong River Delta: Should a Sea Dike Be Built? 345
Vo Thanh Danh

17 Exploring the Potentials of a Technology-Based Flood Early Warning System in Laguna, Philippines: A Cost-Benefit Analysis 369
Jaimie Kim B. Arias, Rowena A. Dorado, Maria Emilinda T. Mendoza, and Vicente G. Ballaran Jr

Index 389

Contributors

Jaimie Kim B. Arias Department of Economics, College of Economics and Management, UPLB, Laguna, Philippines

Constancio A. Asis Soils and Plant Physiology Division, Philippine Rice Research Institute (PhilRice), Muñoz, Nueva Ecija, Philippines

Vicente G. Ballaran Jr Agrometeorology and Farm Structures Division, Institute of Agricultural Engineering, College of Engineering and Agro-industrial Technology, UPLB, Laguna, Philippines

Malabou Baylatry Greater Mekong Subregion Division, The Lao National Mekong Committee Secretariat, Ministry of Natural Resources and Environment, Vientiane Capital, Lao PDR

Sara Bumrungsri Department of Biology, Faculty of Science, Prince of Songkla University, Hat Yai, Songkhla, Thailand

Vo Thanh Danh School of Economic and Business Administration, Can Tho University, Can Tho City, Vietnam

Moises A. Dorado Agrometeorology and Farm Structures Division, Institute of Agricultural Engineering, College of Engineering and Agro-industrial Technology, University of the Philippines Los Baños (UPLB), Laguna, Philippines

Rowena A. Dorado Department of Economics, College of Economics and Management, UPLB, Laguna, Philippines

Herminia A. Francisco Economy and Environment Program for Southeast Asia, WorldFish, Los Banos Laguna, Philippines

Ngo Thi Lam Giang Tat Thanh Institute of Agrobiology, Nguyen Tat Thanh University, Ho Chi Minh City, Vietnam

L.H.P. Gunaratne Department of Agricultural Economics and Business Management, Faculty of Agriculture, University of Peradeniya, Peradeniya, Sri Lanka

Kalyan Hou The Center for People and Forest (RECOFTC), Forestry Administration, Phnom Penh, Cambodia

Maredi Im The Center for People and Forest (RECOFTC), Forestry Administration, Phnom Penh, Cambodia

David James EEPSEA Resource Person and Independent Consultant, Whale Beach, NSW, Australia

Evelyn F. Javier Soils and Plant Physiology Division, Philippine Rice Research Institute (PhilRice), Muñoz, Nueva Ecija, Philippines

Han Keam National Committee for Sub-national Democratic Development (NCDD), Kampong Thom Province, Cambodia

Quang Vinh Ky Climate Change Coordination Office of Can Tho City, Can Tho City, Vietnam

Cheryll C. Launio Socioeconomics Division, Philippine Rice Research Institute (PhilRice), Muñoz, Nueva Ecija, Philippines

Shushan Li School of Environment & Natural Resources, Renmin University of China, Beijing, People's Republic of China

Vuthy Lic Environment and Natural Resources Unit, Cambodia Development Research Institute (CDRI), Phnom Penh, Cambodia

Sivannakone Malivarn Greater Mekong Subregion Division, The Lao National Mekong Committee Secretariat, Ministry of Natural Resources and Environment, Vientiane Capital, Lao PDR

Rowena G. Manalili Socioeconomics Division, Philippine Rice Research Institute (PhilRice), Muñoz, Nueva Ecija, Philippines

Sothunvathanak Meas The Learning Institute, Phnom Penh, Cambodia

Maria Emilinda T. Mendoza Department of Social Development Services, UPLB, Laguna, Philippines

Thi Thanh Truc Ngo Department of Agricultural, Natural Resources and Environmental Economics, Can Tho University, Can Tho City, Vietnam

Nghiem Thi Hong Nhung Department of Public Health, University of Otago, Wellington, New Zealand

Kunyu Niu School of Environment & Natural Resources, Renmin University of China, Beijing, People's Republic of China

Canesio Predo College of Forestry and Natural Resources, University of the Philippines Los Baños, Laguna, Philippines

Saowalak Roongtawanreongsri Environmental Economics Research Unit, Faculty of Environmental Mangement, Prince of Songkla University, Hat Yai, Songkhla, Thailand

Chanty Ros National Committee for Sub-national Democratic Development (NCDD), Kampong Thom Province, Cambodia

Prakart Sawangchote Department of Biology, Faculty of Science, Prince of Songkla University, Hat Yai, Songkhla, Thailand

Chaisri Suksaroj Department of Civil Engineering, Faculty of Engineering, Prince of Songkla University, Hat Yai, Songkhla, Thailand

Loan Le Thanh Faculty of Economics, Nong Lam University, Ho Chi Minh City, Vietnam

Bhadranie Thoradeniya Division of Civil Engineering Technology, Institute of Technology, University of Moratuwa, Moratuwa, Sri Lanka

Neth Top Environment and Natural Resources Unit, Cambodia Development Research Institute (CDRI), Phnom Penh, Cambodia

Philayvanh Viravouth Greater Mekong Subregion Division, The Lao National Mekong Committee Secretariat, Ministry of Natural Resources and Environment, Vientiane Capital, Lao PDR

Thi Lang Vo School of Economics & Business Administration, Can Tho University, Can Tho City, Vietnam

Xiaoxia Wang School of Environment & Natural Resources, Renmin University of China, Beijing, People's Republic of China

Zanxin Wang School of Development Studies, Yunnan University, Kunming, Yunnan Province, People's Republic of China

Jian Wu School of Environment & Natural Resources, Renmin University of China, Beijing, People's Republic of China

Dararath Yem Environment and Natural Resources Unit, Cambodia Development Research Institute (CDRI), Phnom Penh, Cambodia

Part I
Guide to Cost-Benefit Analysis and Case Studies

Chapter 1
Introduction

Herminia A. Francisco and David James

Abstract Southeast Asia (SEA) is home to 600 million (Brown, Southeast Asia: region on the rise, Inbound logistics, January. www.inboundlogistics.com, 2013) living in about 20 % of its 4.5 million km^2 land area (ESCAP, Review of the state of the environment in Asia and the Pacific. Report to the 5th ministerial conference on environment and development, Bangkok, 2005). The region comprises mainland SEA (Cambodia, Lao PDR, Myanmar, Thailand, and Vietnam) and maritime SEA (Brunei, East Timor, Indonesia, Malaysia, the Philippines, and Singapore). The population growth rate declined from 2.26 % per annum (1980–1985) to only 1.33 % per annum (2000–2005), but it is still more than 2 % for the Philippines, Malaysia, Brunei, and East Timor (Jones, The population of Southeast Asia, ARI working paper #196, Asia Research Institute, National University of Singapore. www.nus.ari.edu.sg/pub/wps.htm, 2013). With a large population base, the region's growing population will place continuing pressure on its natural and environmental resources.

Keywords Southeast Asia • Environment • Natural resources • EEPSEA • Valuation

Overview of Environmental Problems in Southeast Asia

Southeast Asia (SEA) is home to 600 million (Brown 2013) living in about 20 % of its 4.5 million km^2 land area (ESCAP 2005). The region comprises mainland SEA (Cambodia, Lao PDR, Myanmar, Thailand, and Vietnam) and maritime SEA (Brunei, East Timor, Indonesia, Malaysia, the Philippines, and Singapore). The population growth rate declined from 2.26 % per annum (1980–1985) to only 1.33 % per annum (2000–2005), but it is still more than 2 % for the Philippines,

H.A. Francisco (✉)
Economy and Environment Program for Southeast Asia, WorldFish,
Los Banos Laguna, Philippines
e-mail: H.Francisco@cgiar.org; hermi@eepsea.net

D. James
EEPSEA Resource Person and Independent Consultant, Whale Beach, NSW, Australia
e-mail: ecoservices@iprimus.com.au

D. James, H.A. Francisco (eds.), *Cost-Benefit Studies of Natural Resource Management in Southeast Asia*, DOI 10.1007/978-981-287-393-4_1

Malaysia, Brunei, and East Timor (Jones 2013). With a large population base, the region's growing population will place continuing pressure on its natural and environmental resources.

Rapid economic growth is a major driver of the loss and degradation of the region's natural and environmental resources. This growth (about 5 % per year in the last decade) is due in part to the greater economic integration of SEA with the fast-growing economy of China and the rising world demand for its primary products. This trade dependency is quite high as indicated by a trade-to-gross domestic product (GDP) ratio of more than 150 % (Brown 2013). An even higher growth rate is thus expected with ASEAN economic integration in 2015. Poverty levels have improved overall, but disparity in income levels persists widely across countries. Myanmar's per capita GDP is the lowest in the region at USD849, while Singapore's per capita GDP is USD49, 936 (Das 2013). Furthermore, the number of poor people remains high for many countries (i.e., 33 % in Myanmar, 2007; 31 % in Cambodia, 2007). Many of the poor have high dependence on natural resources for subsistence and livelihoods.

How fast is the region losing its natural resources? From 1990 to 2005, the region's forest resources went down from 55 % of the land area to 45 % (203 million hectares). This represents an annual forest loss rate of 1.35 % (2.75 million hectares) compared to the world average of 0.2 % (FAO 2006a). The forest loss is higher for countries like the Philippines, Cambodia, and Indonesia (FAO 2006b). It is worth noting that the SEA region's forest, which averages 45 % of land area, is still higher than the broader Asian average of 18 % and the global average of 30 % (UNEP 2004). However, the United Nations Economic and Social Commission for Asia and the Pacific (ESCAP) 2005 Review of the State of Environment in Asia and the Pacific pointed out that because the region has a big population, its biological and land resources on a per capita basis are lower than the global averages of 60 % and 80 %, respectively. This big population is projected to increase by 19 % (113 million people) from 2010 to 2030 (Jones 2013).

Forests are an important source of wood and rich biodiversity resources. About half of the world's terrestrial and marine biodiversity resources are found in SEA. The mega diversity countries of Malaysia, the Philippines, and Indonesia are home to 80 % of the global biodiversity resources (UNEP 2004). The loss of forest resources, particularly those due to land use changes, contributes to greenhouse gas (GHG) emissions. Hooijer et al. (2006) estimated that 12 million hectares of the 27 million hectares of peatland forest in SEA was deforested from 1995 to 2005. Indonesia is responsible for most of the GHG emissions, resulting mainly from the rapid degradation of its peat swamp forest.

Excessive logging, land use conversion to agriculture (both small scale and for plantation crops like oil palm and rubber), and expansion of settlement areas contribute to the rapid loss of forest resources in the region. Rola and Coxhead (2005) have pointed out that government-sanctioned timber extraction, motivated by the need to consolidate political power, has played a major role in the rapid depletion of forest resources in the region.

The region is also endowed with rich coastal resources, comprising mangroves, coral reefs, and sea beds that support a rich variety of marine animal and plant species. These ecosystems support extensive fisheries, making SEA a major producer and consumer of fish products globally. The region has 35 % of the world's total mangrove resources (4.9 million hectares), 60 % of which are found in Indonesia. The region's mangroves are more developed and are the most diverse in the world (Giesen et al. 2006; FAO 2004; Wilkie and Fortuna 2003). In terms of coral reef resources, the coral triangle, home to about a third of the world's coral reefs, is found between Indonesia, Malaysia, and the Philippines. This marine area is home to 600 species of reef-building coral and 3,000 species of reef fish. It supports more than 100 million people who depend on marine-based industries (Kool et al. 2011).

Mangroves and coral reefs are under serious threat despite their huge ecological value. Perhaps, this in itself is the main problem. Their biggest contributions in terms of their ecological services (e.g., fish habitat, nursery and spawning grounds, and protection against shoreline erosion and storms, among others) are often unrecognized and indirect. Thus, their economic values are not fully appreciated, leading to their rapid loss. From 1980 to 2005, 20 % of the total preexisting mangrove area in 1980 (3.6 million hectares) was lost, representing a loss of 1,000 km^2 per year (FAO 2007). Rapid mangrove conversion to fishponds, excessive harvesting of wood products, excessive fishing coupled with the use of destructive harvesting practices, land reclamation, pollution, and sedimentation have all contributed to the loss of mangrove resources and to the destruction of coral reefs. In addition to these, climate change poses an even bigger threat to the coral reef ecosystem through increasing acidification and coral bleaching. About 40 % of the region's reefs are already lost, 45 % are under threat, and 15 % are under low threat (Hoegh-Guldberg et al. 2009).

SEA's environmental resources are under threat also from its growing population and expanding urbanization and rapid industrialization that often accompany rapid economic growth. These conditions bring about a host of problems related to water pollution, air pollution, growing volume of solid waste, and congestion. An Asian Development Bank Institute (ADBI) working paper showed that, based on data for 2000–2004, the concentrations of major pollutants (e.g., PM10, SO_2, and NOx) in major Asian cities have already exceeded the standards set by WHO guidelines (Howes and Wyrwoll 2012). These cities include Metro Manila, Bangkok, Jakarta, Kuala Lumpur, Ho Chi Minh, and Hanoi.

Water supply and water quality issues continue to place pressure on water resources. With a growing population, the region expects the water demand for agriculture, industrial use, and urban usage to grow significantly. Likewise, water pollution is already a major concern for most urban centers in the region. Water pollutants include pathogens, organic matter, nutrients, heavy metals and toxic chemicals, sediments and suspended solids, silt, and salts. The large populations and rising affluence in Asian cities are already creating a big challenge in solid waste management. In 1999, urban areas in Asia generated 2.7 million m^3 of solid waste. This has been expected to grow to 5.2 million m^3 in 2025 (World Bank 1999).

Role for Environmental Economics Research in SEA

The previous section describes major problems confronting SEA countries in terms of their natural and environmental resources. The population is big and is still growing. Poverty levels have improved but income disparity has not. Vested interest groups, often with favorable political connections, control a large portion of the region's natural resources, while those with very limited economic means are left to depend on a very small portion of the natural resource pie. This situation and the availability of jobs in urban centers have driven growing urbanization, resulting in pollution problems in cities. It is clear that such a pattern of economic growth, which depletes the region's natural capital base and damages the environment, is not sustainable or desirable. The issue of sustainability compels us to look into the rate of resource depletion and the associated costs. The issue of desirability requires us to take into account the cost of using limited natural resources to society, particularly costs arising from resource depletion and degradation and environmental pollution. Environmental economics, coupled with institutional analysis and behavioral economics studies, has much to offer in terms of helping the region to address both the sustainability and desirability issues relating to natural and environmental resource use.

The Economy and Environment Program for Southeast Asia (EEPSEA) was established in May 1993 to support research and training in environmental and resource economics. Member countries are Cambodia, China, Indonesia, Lao PDR, Malaysia, Myanmar, the Philippines, Thailand, and Vietnam. Sri Lanka, while part of South Asia, was a member of EEPSEA for many years. The Program is supported by the International Development Research Centre (IDRC), the Swedish International Development Cooperation Agency (Sida), WorldFish, and the Canadian International Development Agency (CIDA).

EEPSEA's main objective is to enhance local capacity to undertake economic analysis of environmental problems and policies. A networking approach is adopted, involving training courses, meetings, technical support, and access to literature and opportunities for comparative research. The Program supports various kinds of research projects on priority national environmental and natural resource management concerns in SEA.

Since its inception, EEPSEA has supported more than 300 research grants. The distribution of research grants by subject matter is shown in Fig. 1.1. From 1993 to 2006, forest resource issues dominated the research focus of EEPSEA-supported projects. This was followed by pollution control issues and by research on water issues and marine and coastal resources. From 1997 to the present, research on climate change adaptation (and a few, on mitigation) became the dominant theme of EEPSEA-funded studies. The dominant themes of EEPSEA research continually evolve to reflect current main issues. In the last 5 years, the pressing issues have been climate change, pollution, marine and coastal resources, forest management, and water resource issues.

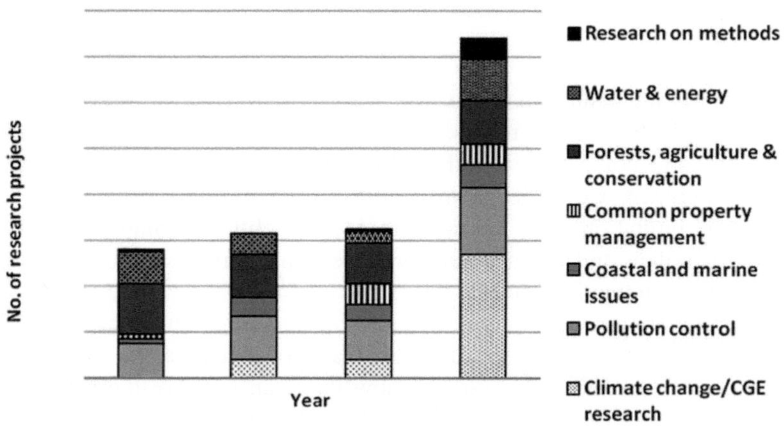

Fig. 1.1 Research focus of EEPSEA-supported projects over the years

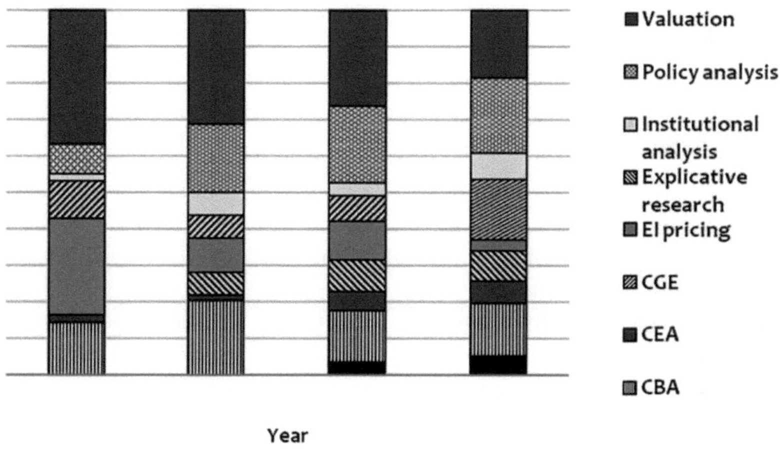

Fig. 1.2 Topical focus of EEPSEA-funded studies over the years

In terms of topical focus, valuation research conducted by EEPSEA is declining in favor of a growing interest in policy analysis and cost-benefit analysis (CBA) studies (Fig. 1.2). While the derivation of economic values associated with natural resources and the environment has always been a major focus of EEPSEA research, there is now much greater emphasis on applying such values in the evaluation of policies or projects. Indeed, suggestions for policy or management responses often accompany the technical analyses that are undertaken. The coverage of the case studies appearing in this volume attests to the practical role that evaluation techniques such as CBA can play in managing natural resources and environment within the SEA region.

Aims of the Book

This book has two main aims. The first is to present a selection of case studies undertaken by EEPSEA researchers, highlighting the wide range of environmental issues challenging policymakers in the SEA, as well as the practical advantages that CBA can provide by way of decision support. The studies demonstrate that, despite some of the limitations that researchers face, useful information can still be provided for policy and management purposes. In many cases, the research has led to important policy initiatives.

The second aim has been to produce a set of analytical guidelines and demonstration case studies that can be used by teachers in the region in educational programs and future research, showing how they can improve the application of CBA to achieve a better environment. Already, the programs conducted by EEPSEA with donor support over the years have helped to enhance the careers of teachers and researchers in the region, strengthening their positions in universities and research institutions and creating new opportunities for them to act as expert advisers or participants in government and in the broader community.

Outline of the Book

The book contains studies from most of the countries in Southeast Asia, including China, Vietnam, the Philippines, Cambodia, Thailand, and Lao PDR. Two studies are included from Sri Lanka.

Common themes in environmental and natural resource management are represented in separate parts of the book. Each of the case studies selected comprises a shortened version of a fuller report submitted to EEPSEA on completion of a research project. Most of the full studies can be accessed on EEPSEA's website: www.eepsea.org.

Part I of the book is introductory. It includes the present chapter. Chapter 2 presents and discusses general principles of CBA, including the analytical techniques and criteria that support economic evaluations of natural resource and environmental policies and projects, referring to case studies in the book to demonstrate practical applications of the relevant concepts and techniques. Difficulties of particular importance in a developing country context are highlighted. Chapter 2 is not intended to be a comprehensive treatise on CBA, as many excellent texts on CBA are already available. Instead, only a general guide is presented. Key references for more detailed study are cited.

Part II focuses on agriculture, forestry, and fisheries. It includes studies on biofuel production, agricultural practices, forestry, rubber plantation development, and pollution associated with aquaculture in a riverine environment.

Part III deals with river basin management. The two studies in this section relate to sand mining in Sri Lanka, the damage that is caused, and possible policy interventions designed to ameliorate adverse environmental impacts.

Part IV contains studies on the management of protected areas. One of the studies demonstrates the economic advantages of properly managing protected areas to promote ecotourism; another examines the trade-offs involved in community forestry projects, while a third study highlights the broader community benefits of conserving biodiversity in a forested area, adopting an approach based on the concept of ecosystem services. This part of the book also contains a study addressing the problem of how to manage competing uses of water resources for farming and conservation of a wetland area in China.

Part V deals with adaptive responses to climate change. The focus is on coastal areas, which are particularly vulnerable to sea level rise, storm surge, saline intrusion, and flooding. The options examined include construction of sea dikes, other protective structures, planned retreat, and early warning systems for flood events.

The compendium of case studies contained in this volume indicates the wide range of natural resource and environmental management problems that can be evaluated from an economic perspective using CBA.

References

Brown J (2013) Southeast Asia: region on the rise, Inbound logistics. www.inboundlogistics.com

Das SB (2013) Institute of Southeast Asian Studies: enhancing regional and sub-regional cooperation and connectivity in ASEA, ISEAS working paper #3. www.iseas.edu.sg

ESCAP (2005) Review of the state of the environment in Asia and the Pacific. Report to the 5th ministerial conference on environment and development, Bangkok

FAO (2004) Forest resources assessment working paper 063. FAO, Rome

FAO (2006a) Report of the expert consultation on Global Forest Resources Assessment: towards FRA 2010, 12–16 June 2006, Kotka/Rome

FAO (2006b) Global forest resource assessment 2005 – progress towards sustainable forest management, Forestry paper 147, FAO, Rome

FAO (2007) The world's mangroves, 1980–2005, FAO forestry paper. FAO, Rome

Giesen WS, Wulffraat M, Zieren M, Scholten L (2006) Mangrove guidebook for Southeast Asia. FAO and Wetlands International, Bangkok

Hoegh-Guldberg O et al (2009) The coral triangle and climate change: ecosystem, people, and societies at risk. WWF, Brisbane, 276 pp

Hooijer A, Silvius M, Wösten H, Page S (2006) PEAT-CO2, assessment of CO2 emissions from drained peatlands in SE Asia, Delft Hydraulics report Q3943. International Wetlands

Howes S, Wyrwoll P (2012) Asia's wicked environmental problems, ADBI working paper series #348, Asian Development Bank Institute. www.adbi.org/workingpaper/2012/02/28/5009.asia.wicked.environmental.problems

Jones GW (2013) The population of Southeast Asia, ARI working paper #196, Asia Research Institute, National University of Singapore. www.nus.ari.edu.sg/pub/wps.htm

Kool JT, Paris CB, Barber PH, Coreten RK (2011) Connectivity and the development of the population genetic structure in Indo West Pacific Coral Reef – Communities. Glob Ecol Biogeogr 20:695–706

Rola AC, Coxhead I (2005) Economic development and environmental management in the uplands of Southeast Asia: challenges for policy and institutional development. Agric Econ Int Assoc Agric Econ 32:243–256

UNEP (2004) Environmental indicators Southeast Asia. Regional Resource Centre for Asia and the Pacific, United Nations Environment Programme, Pathumthani, Thailand

Wilkie ML, Fortuna S (2003) Status and trends in mangrove area. FAO, Rome

World Bank (1999) What a waste: solid waste management in Asia. Urban Development Sector Unit, East Asia and Pacific Region, The International Bank for Reconstruction and Development/The World Bank, Washington, DC

Chapter 2
Principles and Practice of Cost–Benefit Analysis

David James and Canesio Predo

Abstract Cost–benefit analysis (CBA), or benefit–cost analysis (BCA) as it is often called, has a long history as a methodology for assessing the economic efficiency with which resources are used to support human wellbeing. Its theoretical origins lie in the foundations of welfare economics established by economists such as Hicks, Kaldor, Scitovsky and Little. This review of principles and practice focuses on the advantages and limitations of CBA and related methods in a developing country context. General guidelines are presented.

Keywords Cost–benefit analysis • Willingness to pay • Economic value • Externalities • Cost-effectiveness analysis

Background to CBA

Evolution of CBA

Cost–benefit analysis (CBA), or benefit–cost analysis (BCA) as it is often called, has a long history as a methodology for assessing the economic efficiency with which resources are used to support human wellbeing. Its theoretical origins lie in the foundations of welfare economics established by economists such as Hicks (1939), Kaldor (1939), Scitovsky (1941) and Little (1957).

CBA originated in the US Flood Control Act of 1936 as a means of assessing projects involving public sector investments in the public interest. The Act stipulated that flood control projects would be desirable if 'the benefits to whomsoever they may accrue are in excess of the estimated costs'. The first practical guidelines on how to conduct CBA appeared in 1950 when the US Federal Inter-Agency River

D. James (✉)
EEPSEA Resource Person and Independent Consultant, Whale Beach, NSW, Australia
e-mail: ecoservices@iprimus.com.au

C. Predo
College of Forestry and Natural Resources, University of the Philippines Los Baños, Laguna, Philippines
e-mail: cdpredo@yahoo.com

D. James, H.A. Francisco (eds.), *Cost-Benefit Studies of Natural Resource Management in Southeast Asia*, DOI 10.1007/978-981-287-393-4_2

Basin Committee released the *Proposed Practices for Economic Analysis of River Basin Projects*, popularly referred to as the *Green Book* for project appraisal. Pioneering texts and guidebooks on CBA were subsequently published by economists such as Gittinger (1982), Little and Mirrlees (1974), Pearce (1971), Dasgupta and Pearce (1972), Marglin et al. (1972), Sugden and Williams (1979), Pearce and Nash (1981) and Ray (1984). Early texts on project appraisal in developing countries paid particular attention to shadow pricing, adjusting for scarcity of foreign exchange, inflation, taxes and subsidies, unemployment, unpaid labour and inequalities in income and wealth.

The incorporation of environmental values in CBA occurred somewhat later, with the emergence of environmental economics as a recognised branch of the economics discipline. Freeman (1979) described approaches to valuing environmental benefits, while other authors emphasised the place of CBA and valuation techniques in environmental and natural resource management (Hufschmidt et al. 1983). Case studies for Southeast Asian countries appeared soon after in Dixon and Hufschmidt (1986).

Most contemporary texts and guidelines for CBA highlight the importance of environmental costs and benefits, with an explanation of how to estimate and incorporate them in an economic analysis (Zerbe and Diverly 1994; Hanley and Spash 1995; Bateman et al. 2005; Hanley and Barbier 2009). CBAs are now commonly performed in evaluations of policies, programmes and projects, environmental impact assessments and the management of natural resources and environment more generally (ADB 1997; James 1994; Pearce et al. 1994, 2002; UK Treasury 1997; US EPA 2000). The case studies presented in this volume are just some examples of applications in Southeast Asian countries. Nevertheless, they highlight conceptual and practical difficulties that beset practitioners, especially in a developing country context.

Conceptual Basis of CBA

Wellbeing and Economic Efficiency

CBA is applied principally to evaluate the economic efficiency of different options that are capable of achieving some predetermined policy or management objective. Where the public interest is involved, CBA takes the form of a *social* benefit–cost analysis, concentrating on the wellbeing of the community as a whole. Wellbeing is defined in terms of the utility experienced in the consumption of goods and services produced by the economic system or that are otherwise made available, such as those provided by natural systems and the environment.

CBA focuses on only certain aspects of wellbeing, disregarding other indicators of wellbeing such as social relations, equity and personal security. Fundamental assumptions of CBA are that individual preferences count regarding the use of resources and that the wellbeing of the community comprises an aggregation of the wellbeing of its members. It is assumed that utility itself cannot be measured directly, and interpersonal comparisons of utility are debarred in the analysis.

According to the Pareto *potential economic welfare criterion*, an efficient use of resources is achieved when it is not possible to make some individuals better off without making others worse off. Application of the criterion implies that an economically efficient outcome is achieved when net benefits (total benefits minus total costs) are maximised. Following some change in economic circumstances, the gainers should in principle (but not necessarily in practice) be able to compensate the losers without being made worse off.

The key indicator of utility in CBA is the willingness to pay (WTP) by individuals or the community for positive increases in wellbeing or for the avoidance of losses. In CBA a benefit can be a cost avoided and a cost can be a benefit forgone. Although in theory it is possible to use the willingness to accept (WTA) compensation to indicate a decrease in wellbeing, it is not generally favoured as a basis for evaluation. Such measures may be subject to various kinds of bias, such as exaggerated claims for compensation, leading to inappropriate estimates of the welfare changes involved.

Willingness to Pay and Market Values

The willingness to pay for any good or service by an individual is assumed to be a reflection of his/her underlying utility function. In general, as larger quantities of a good or service are consumed, total utility will increase but at a diminishing rate. The individual's marginal utility accordingly declines. Where the good or service is provided through a market mechanism, the price that the individual is prepared to pay decreases as the quantity consumed increases. This leads to the concept of an *individual demand function* or, when represented graphically, an individual demand curve for the good or service in question.

For the community as a whole, the market demand function for the good or service comprises an aggregation of the demand functions of all individuals in the community. The market demand curve typically slopes downward to the right, expressing the prices that the community is prepared to pay as increasing quantities of the good or service are consumed. The area under the demand curve for a given quantity consumed defines *total benefits* measured in monetary units. Where the WTP exceeds the actual market price, the expenditure that could have been extracted is defined as *consumers' surplus*.

Market demand functions (described as *Marshallian* demand functions) are based on the premise of constant income. Hicks (1943) postulated four kinds of consumers' surplus, in which utility is assumed constant. In reality, only market demand functions can be observed and estimated empirically. However, as argued by Willig (1976), the difference between the Hicksian and Marshallian versions is considered to be so small that Marshallian demand functions suffice for most applications in CBA. Even so, the difficulties of obtaining sound econometric representations of market demand functions should not be underestimated.

Costs in CBA are defined as *opportunity costs*. They measure the benefits forgone by using available resources to provide a good or service rather than using them elsewhere or in some other way to support wellbeing. In general, the total cost of producing the good or service rises at an increasing rate as the quantity produced increases. The marginal cost of production (the extra cost of producing an additional unit of output) typically increases as the volume of output expands.

In a market situation, producers need to charge a price equal to the marginal cost of production to cover their costs of production and reach an acceptable level of profit. The supply curve for the good or service thus increases to the right as larger quantities are offered for sale. The area under the supply curve, for the particular quantity produced and offered for sale, indicates the *total cost* of producing a given quantity of the good or service. Where the market price prevailing in the market exceeds the price that producers are prepared to accept, the additional returns received by producers in excess of their costs of production are defined and measured as *producers' surplus*. Mutual acceptance of a price and quantity by both buyers and sellers in the market leads to a market equilibrium. Summing consumers' surplus and producers' surplus yields an estimate of the *net benefits* received.

Expressed in diagrammatic form, this occurs where the market demand and supply curves intersect. Figure 2.1 illustrates this condition.

In Fig. 2.1 the horizontal axis measures the quantity of a good X traded in the market, while the vertical axis measures its price. The market demand curve is shown as AEB and the supply curve as SET. Intersection of the curves (point E) reveals the equilibrium price (OP_1) and quantity (OQ_1) that clear the market. The total WTP for good X (the total benefit experienced by consumers) is measured as the area $OAEQ_1$, and the total cost of supplying it is $OSEQ_1$. Consumers' surplus is measured as the area AEP_1 and producers' surplus as the area SEP_1. The sum of these two areas represents the net benefits of producing and consuming OQ_1 of good X in the market.

It is instructive to note that most of the studies presented in this volume focus on producers' surplus under the assumption of fixed market prices (i.e. horizontal demand curves). None of the studies explicitly estimates a downward-sloping market demand curve, and the assessments of changes in producers' surplus are all based on the assumption of fixed prices. The study by Bhadranie Thoradeniya (Chap. 7), however, relies on an imputed downward-sloping demand curve and the concept of consumers' surplus when applying an individual travel cost model to estimate the recreation benefits of a river basin in Sri Lanka.

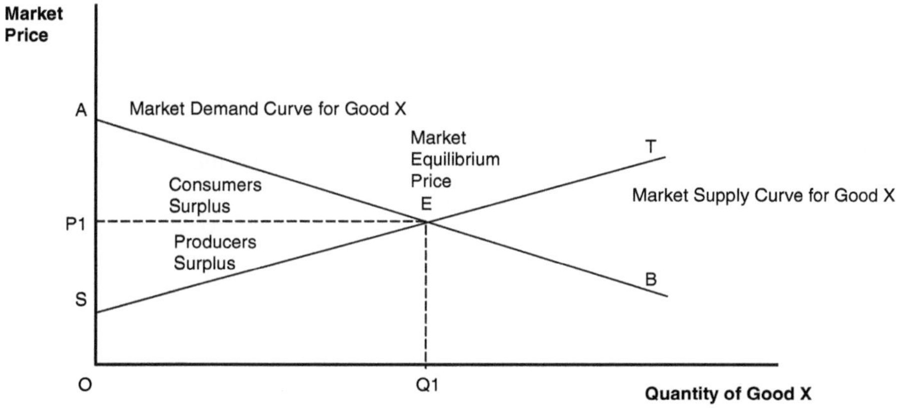

Fig. 2.1 Market supply and demand

Values of the Environment and Natural Resources

Total Economic Value

Ecosystems and natural environments provide goods and services that are essential for human existence, as well as supporting other aspects of wellbeing. Pearce and Turner (1989) introduced the concept of the total economic value (TEV) of the environment, decomposing it into various categories of use and non-use values. Such values have commonly been defined and estimated for natural forests. Valuation methodologies for timber, non-timber forest products (NTFPs) and non-use values are discussed by Bann (1998) among others. A schema for forest-related economic values is presented in the study of rubber plantation development in Cambodia by Yem Dararath et al. (Chap. 7), while Kalyan Hou et al. (Chap. 12) provide estimates of forest-based benefits potentially achievable from community forestry schemes. In the study by Saowalak Roontawanreongsri et al. (Chap. 13), the benefits of forests are categorised and valued in terms of ecosystem services. Similar studies have been performed for other natural resource systems.

It is important, in determining the value of any natural resource system, to avoid double-counting the benefits. For example, when valuing the ecosystem services provided by native forests, while the forest may contain a potential stock of timber as well as non-timber forest products and other attributes of cultural value, it would be erroneous to add all these values together, because timber-felling would be in conflict with other forest uses and values. The lesson to be drawn here is that natural resource and environmental values should be estimated only with respect to a particular policy or resource management scenario.

Environmental Externalities

Many goods and services provided by the environment are used or experienced without passing through markets. In many cases they are provided free or at prices that do not reflect their full opportunity cost or value to the community. An important task in conducting a CBA is to incorporate implicit values of the benefits that are generated by goods and services provided by the environment.

Environmental effects occurring off-site or indirectly as a consequence of human activity are defined as *environmental externalities.* In a CBA, where changes in the environment are favourable, they are specified and valued as *external environmental benefits*, while adverse environmental impacts are valued as *environmental damage costs.* External environmental costs are typically not taken into account in the evaluations, plans and activities of private producers, resulting in a divergence between the private and social costs of production, establishing a strong case for appropriate policy interventions by government. The general aim is to internalise the externalities and induce producers to modify their behaviour to improve community wellbeing.

The studies by Zanxin Wang et al. (Chap. 3), Loan Le Thanh (Chap. 4) and Cheryl Launio et al. (Chap. 5) deal explicitly with emissions of CO_2e as an economic externality, while the forest modelling carried out in the study by Nghiem Thi Hong Nhung (Chap. 6) treats reductions in CO_2e as an important co-benefit of timber production. Damage costs associated with sand mining in river basins are assessed by Bhadranie Thoradeniya (Chap. 9). Vo Thi Lang et al. (Chap. 8) estimate the damage costs resulting from water pollution and its effects on aquaculture. Jaimie Kim B. Arias et al. (Chap. 15) provide estimates of damage costs resulting from storms in San Fernando City, Philippines, while Vo Thanh Danh (Chap. 16) includes damage costs in his study of flooding and salinisation predicted to result from sea level rise in the Mekong Delta, Vietnam.

Conduct of a CBA

Steps in a CBA

Steps in conducting a CBA for a proposed project are defined in the studies by Malabou Baylatry et al. (Chap. 11) and Jaimie Kim B. Arias et al. (Chap. 17) following Boardman et al. (2010). They can be summarised as follows:

- Define the boundaries and time horizon of the project.
- Define the referent groups.
- Select the portfolio of project options.
- Catalogue the potential physical impacts of the project.
- Define what would happen without the project.

- Quantify and predict the outputs and impacts of the project over its expected life.
- Monetise the outputs and impacts.
- Calculate NPV, BCR and IRR for each option, assuming a given discount rate.
- Describe the distribution of costs and benefits.
- Perform sensitivity analysis.
- Make recommendations to decision-makers.

Construction of BAU Scenario

When evaluating options, NPV, BCR and IRR (defined below) should be assessed relative to a base case or business as usual (BAU) scenario. What counts are the incremental changes associated with any option compared with the base case. To construct a BAU scenario, it is necessary to determine the general characteristics of the scenario, currently and in the future, including its geographic boundaries, attributes and functions of the surrounding environment and natural resource systems, the population or community involved, market trends, relevant economic activities, possible impacts of human activity on the environment, the time horizon for analysis and institutional factors including industry structure and government policies and regulations.

A BAU scenario does not imply 'do-nothing' as it may include policies and activities that would be in place independently from the options under consideration. Rather, it should incorporate the best assessments of what can be expected over the planning period in the absence of the foreshadowed options. Nevertheless, the general aim of constructing the BAU scenario is to conduct the analysis 'with and without' proposed possible actions, noting that the 'with and without' situations should not be interpreted as 'before and after' chronologically.

Incremental benefits and costs of each option relative to the BAU scenario should be estimated for each future year up to the planning horizon. This involves constructing a scenario for each option, modelling all the changes that can be expected to take place and valuing them in monetary terms. Incremental benefits are calculated as the expected change in benefits relative to the BAU. A similar approach is taken to calculate incremental costs.

The studies by Yem Dararath et al. (Chap. 7), Kalyan Hou et al. (Chap. 12) and Jamie Kim B. Arias et al. Ch (15) all contain clearly defined baseline or BAU scenarios, against which feasible options are compared.

Criteria for Evaluating Options

Several criteria, defined below, are used in CBA to determine on economic efficiency grounds whether or not to undertake a particular investment policy, programme or project. It is convenient to discuss these, in what follows, at the project scale.

Net Present Value (NPV)

This criterion measures the present value of the net benefits of the development project. The formula for calculating NPV is:

$$\text{NPV} = \Sigma(B_t - C_t)/(1 + r)^t \ldots\ldots t = 1..n$$

where B_t is the benefit at time period t and C_t is the cost at period t. The planning horizon or terminal year is n.

For the project to be acceptable on economic grounds, the NPV should be positive. A positive NPV means that the option produces greater net economic benefits, assessed in terms of present values, in comparison with the BAU scenario. Where there are mutually exclusive options, the option with the highest NPV is preferred. Any proposed options that have negative NPVs must be rejected as economically undesirable. In some appraisals, the best option is the BAU scenario itself. In that case, the BAU scenario will have a zero value for its NPV, while all other options will have negative NPVs.

Benefit–Cost Ratio (BCR)

This is the ratio of the present value of benefits to the present value of costs. The formula for calculating the BCR is:

$$BCR = \Sigma B_t/(1 + r)^t/\Sigma C_t/(1 + r)^t \ldots\ldots t = 1..n$$

If the BCR of a project exceeds 1, the present value of benefits is greater than the present value of costs; thus the project is acceptable in terms of economic efficiency. If the BCR is less than 1, the project should be rejected. The BCR should not be used to rank mutually exclusive options, however, as it can lead to rankings that are inconsistent with those obtained using NPV as the ranking criterion.

Internal Rate of Return (IRR)

The IRR is the rate of discount that equates the present value of benefits with the present value of costs. IRR appears as the 'unknown' i in the following equation:

$$\Sigma(B_t - C_t)/(1+i)^t = 0 \ldots\ldots t = 1..n$$

This equation cannot be solved explicitly for the value of i. The only way to determine i is to postulate an initial value and solve for i on a trial-and-error basis – that is, by simulating different values of i until a solution is reached. For some options it may not be possible to calculate an IRR. This can occur when the time path for net benefits of the option fails to change sign (−ve to + ve or vice versa). In some cases, where the time path changes sign more than once, multiple solutions may be obtained for the IRR.

The IRR is typically used by finance departments to compare the internal financial productivity of a project with the official interest rate or cost of funds, to see whether the project is desirable as a financial investment. The IRR should not be used to rank mutually exclusive options, as it also can result in a ranking that is inconsistent with a ranking based on NPV.

Internal Economic Rate of Return (IERR)

A distinction is sometimes drawn between an economic IRR and financial IRR. The only difference is that for an IERR, all values for benefits and costs comprise economic rather than financial values. Application of the IERR criterion follows the same rules as for the IRR.

NPV per Unit of Investment (NPVI)

NPVI is calculated as the ratio of the present value of all positive future NPVs of the project to the present value of all negative NPVs. This criterion is applied where there is only a fixed supply of investible funds. The option with the highest NPVI is the preferred option on economic efficiency grounds.

Frameworks for Conducting CBA

Spreadsheet models offer the most effective framework for compiling and analysing all information involved in a CBA. A guidebook demonstrating how to construct and apply a CBA spreadsheet model using Microsoft Excel is available on the EEPSEA website (Predo and James 2006). It is frequently used in EEPSEA training courses and in CBA studies by EEPSEA researchers. Similar guidelines are provided by Campbell and Brown (2003) among others. A major advantage of using a spreadsheet model is that sensitivity analysis (exploring *what if* scenarios by varying the data input or model structure) can be readily carried out as a means of exploring different policy options and identifying critical aspects of the analysis.

Sensitivity Analysis

The economic desirability of a project depends on the values that are estimated for the various categories of benefits and costs, as well as the discount rate that is adopted. The results of an economic analysis should always be subjected to sensitivity analysis to assess their robustness and the factors that could change any initial ranking of options. Sensitivity analysis is especially useful where uncertainty prevails for particular benefits or costs. The problem of uncertainty and how to handle it in CBA is dealt with later in this chapter.

Applications of CBA Evaluation Criteria in Case Studies

For most of the studies in this volume, complete CBAs are conducted, making use of the above evaluation formulae. The CBA of an ecotourism project for Xe Pian National Protected Area in Lao PDR by Malabou Baylatry et al. (Chap. 11) shows how to set up an appropriate spreadsheet model, specify the base case and calculate the incremental NPV, BCR and IRR for the project, using the format of the EEPSEA CBA guide. Incremental analysis is carried out in the studies of land use options for rubber plantations by Yem Dararath et al. (Chap. 7) and for community forestry by Kalyan Hou et al. (Chap. 12). Other studies also apply the relevant formulae for the options involved and compare the results.

Economic vs Financial Analysis

Differences Between Economic and Financial Analysis

The economic analysis conducted in a CBA differs from a financial analysis. As noted, a CBA focuses on the welfare of a community as a whole. Rather than maximising net social benefits as in a CBA, a financial analysis attempts to determine how individuals, households or commercial enterprises can maximise their own net financial returns or minimise the costs of conducting their activities. Indeed, a potential conflict between private and public interests is often what initiates a social CBA.

An important reason for complementing a social CBA with a financial analysis is that financial factors are key determinants of the *decisions and behaviour* of individuals, households and enterprises. The aim of public policy is often to implement an effective system of incentives (financial, informative or regulatory) that aligns private financial interests more closely with the public interest. Monetary incentives have a powerful influence on behaviour.

In a financial analysis, the usual assessment method applied is discounted cash flow (DCF) accounting. The same formulae are used as in a CBA to evaluate the financial viability of an investment or proposed activity by a private entity. Differences are discussed below.

Costs and prices in a CBA are always expressed in 'real' or constant dollar terms (i.e. excluding price and cost inflation), whereas a DCF may or may not incorporate inflationary effects. Instead of a social discount rate, a market interest rate is applied in a financial analysis. The market rate typically includes an inflationary component.

Some of the prices in a CBA can take the form of *shadow prices*, representing the true opportunity costs of resources used by the community or the true value of benefits that are received. Conversion of actual market prices to shadow prices may be required to indicate a scarcity of foreign exchange, the existence of surplus labour (thus affecting the real wage rate), environmental externalities and price distortions for goods and services provided by the environment and natural resource systems. Subsidies and taxes play a role in a DCF, but they are treated as transfer payments in a CBA and are excluded from the analysis.

Depreciation is handled differently in a financial analysis compared with a CBA. In a financial analysis, depreciation is usually represented as a sequence of write-downs of capital asset values, acting as annually recurring costs and a tax offset. In CBA, resources in the form of capital expenditure are identified as they are committed, and allowance is made for a decline in the physical condition of capital assets. Use of the assets is encapsulated in the discount rate, reflecting their opportunity cost. The residual value of assets in a CBA is counted as a benefit at the end of the planning period. It is estimated in real terms, based on what the owners are able to sell them for – hence the price that others are willing to pay to acquire them – or it may refer to the value of the assets if they are retained by their owners for continued use beyond the planning horizon.

Several of the studies in this volume conduct financial analyses alongside the corresponding economic analyses. This occurs in the studies of biofuel production by Zanxin Wang et al. (Chap. 3) and Loan Le Thanh (Chap. 4) as well as the studies on forestry management by Nghiem Thi Hong Nhung (Chap. 6), Tra fish production by Vo Thi Lang et al. (Chap. 8) and river sand mining by Gunaratne (Chap. 10). Financial costs and returns figure prominently in the study of rubber plantations by Yem Dararath et al. (Chap. 7). In several of the studies, it is suggested that the gap between private and social benefits could be bridged by government subsidies, but other interventions such as direct regulations might also be used.

Distributional Effects

Inequalities in Income and WTP

Application of the Pareto potential welfare criterion, focusing on the community as a whole, presumes that the prevailing distribution of income is acceptable. Extreme inequalities in income, commonly occurring in developing countries, inevitably influence the willingness to pay by different groups within the community and their wellbeing. Concerns must then be addressed regarding the equity or fairness of possible outcomes of policies, programmes and projects. Local communities relying on low-income or subsistence activities such as fishing or cropping may be seriously disadvantaged if the results of economic analysis suggest that they should be displaced to make way for more profitable activities such as urban development, agribusiness projects or industrial production. The willingness to pay by such groups to maintain their livelihoods will be constrained by their low incomes and ability to pay.

In such situations, it behoves the analyst to identify the impacts on different groups within the community, so that decision-makers can deal with equity issues alongside economic efficiency considerations when formulating plans of action. Some economists suggest that the relevant trade-offs can be formalised by conducting *weighted* CBA in which differential weights are attached to the benefits and costs associated with different groups within the community (Harberger 1978; Scarborough and Bennett 2012). Either way, policy decisions that forgo economic efficiency to achieve a more equitable distribution of benefits and costs inevitably require subjective judgments about the wellbeing of different groups affected.

Effects on fiscal revenues and expenditures of government agencies similarly may strongly influence policy decisions. One example of an incidence analysis in this volume is the study of an ecotourism project by Malabou Baylatry et al. (Chap. 11) which identifies the costs and benefits for local communities and local government resulting from the project.

Secondary economic impacts of projects, programmes and policies, such as the spillover effects of development on jobs and incomes in a regional or national economy, have distributional implications. Such impacts are sometimes erroneously interpreted as social benefits or costs. Input–output (I-O) models or computable general-equilibrium (CGE) models are capable of predicting such effects. Again, however, while this information may be of importance from a policy perspective, secondary impacts should be excluded from a CBA.

Cost-Effectiveness Analysis

Approaches to Cost-Effectiveness Analysis

Cost-effectiveness analysis (CEA) is often applied where the monetary values of benefits are difficult to obtain due to limitations of time, information or research resources. In a CEA, the expected outcomes of an option are defined in terms of a particular policy objective, an indicator, standard or performance target. The general aim is to achieve the desired outcome at minimum economic cost. Where the benefits of a project or policy are the same for all options, there is little advantage in estimating monetary benefits, as the most desirable outcome depends only on the comparative costs. For example, to select the most economically efficient means of generating a predetermined volume of biodiesel, only the least-cost technology needs to be identified.

An alternative formulation in CEA is to maximise the outcome where a limited budget or bundle of resources is available for implementation. Where a mix of outcomes is possible, there may be difficulties in identifying the main objective and any co-benefits associated with each option. In such cases, subjective weights must be applied to determine the optimal mix of outcomes and the allocation of limited funds.

CEA can be conducted through manual search procedures (setting up tables of options and their costs), simulation modelling (exploration of options and costs through *what if* scenarios) and mathematical programming models (e.g. choosing the least-cost option of meeting fixed targets by means of linear programming or other kinds of optimisation models). Several of the studies in this volume rely on CEA instead of CBA. Typically, the authors derive their results based on spreadsheet simulation modelling. They include the study by Loan Le Thanh (Chap. 4) of biofuel production in Vietnam, the assessment of least-cost options for pollution control in Tra fish production in Vietnam by Vo Thi Lang et al. (Chap. 8) and options for water saving to protect Qixinghe Wetland in China, evaluated by Wu Jian et al. (Chap. 14).

Cost Trade-Off Analysis

When environmental quality targets cannot easily be established, subjective judgments about the required level of environmental protection can be assisted by cost trade-off analysis. Cost trade-off analysis helps to determine an acceptable level of environmental mitigation or protection, depending on the cost at each level. Usually the total costs and marginal costs of environment protection increase sharply as higher levels of control are approached. The shape of the cost curve often suggests a logical cut-off point for environmental mitigation or protection. Environmental targets may be established as part of the cost trade-off analysis.

Threshold Value Analysis

In cases where economic development and environmental preservation alternatives are mutually exclusive, especially where irreversible environmental impacts are predicted as a consequence of development, and/or where there are difficulties in estimating the nonmarket values of development proposals, the *threshold value approach* can serve as a useful way of considering resource use options. A relevant example is deciding on the use of a wild river either for construction of a reservoir for hydropower production or its preservation for recreational use and wilderness values.

Threshold value analysis is based on the concept of opportunity cost. The opportunity cost of the preservation option consists of the value of net benefits forgone for the development alternative. The environmental benefits are thus not valued directly, but a reference value is provided against which the relative value of the environment may be assessed subjectively. The threshold value indicates the price that the community must be prepared to pay to justify the preservation option. This value can be measured as a capitalised value or as an annually recurring value. A more sophisticated approach, pioneered by Krutilla and Fisher (1985), allows for differential growth rates in development benefits and calculates an initial year's cost that would have be borne to justify the preservation option.

Threshold values may also be applied in benefit–cost analyses where the NPV and rank order of options are sensitive to environmental values. Calculations can be made to determine the threshold values of environmental benefits that would be required to change the NPVs and the rank order of options under consideration. With information on threshold values, decision-makers are obliged to judge subjectively the relevant trade-offs and identify the option that might be considered most acceptable by the community. In the present volume, however, none of the studies adopts the threshold value approach.

Discounting in CBA

The Discount Rate

Discount rates are used in economic analysis as a means of comparing present and future values of benefits and costs. The general formula translating the future value of a benefit or cost to its present value is

$$\mathrm{PV} = \mathrm{FV}_t/(1 + r)^t$$

where PV is present value, FV_t is the value at some point of time in the future, r is the rate of discount and t is the specified future year. The expression $1/(1+r)^t$ is called a *discount factor*. It reduces the future value to a smaller present value. The

Table 2.1 Discount factors and present values of $100

Discount factors						
	Planning horizon (years)					
r	**0**	**10**	**20**	**30**	**50**	**100**
3 %	1.000	0.744	0.554	0.412	0.228	0.052
7 %	1.000	0.508	0.258	0.131	0.034	0.001
12 %	1.000	0.322	0.104	0.033	0.003	0.000
Present values of $100						
	Planning horizon (years)					
r	**0**	**10**	**20**	**30**	**50**	**100**
3 %	100.00	74.41	55.37	41.20	22.81	5.20
7 %	100.00	50.83	25.84	13.14	3.39	0.12
12 %	100.00	32.20	10.37	3.34	0.35	0.00

process of applying a discount rate is known as *discounting*. The higher the discount rate, the smaller will be the present value of any future benefit or cost. Some simple calculations demonstrate the effects of different discount rates and time horizons on present values, as shown in Table 2.1.

The table reveals that with a discount rate of 7 %, the discount factor at year 10 is calculated as $1/(1+0.07)^{10}$ or 0.5083. This means that $100 worth of goods and services in year 10 would be valued at only $50.83 at the present time. The $100 at year 30 would have a present value of only $13.14. With a 12 % discount rate, $100 at years 10 and 30 would have present values of $32.20 and $3.34, respectively. Beyond 30 years, unless a very low discount rate was adopted, the present value of $100 would be negligible.

Economic Rationale for Discounting

Economists give two reasons for justifying the use of a discount rate. The first is known as the *social rate of time preference*. Given the choice of consuming a given bundle of goods or services now or the same bundle in the future, people tend to place more importance on consuming now. People do not live forever, so they prefer to enjoy life now rather than later; immediate satiation of wants takes precedence over deferred satiation. Another explanation is that there may be uncertainty about the future availability of the same bundle of goods and services, and because consumption in the present is more certain, it is more highly valued than consumption in the future.

In general, the social rate of time preference adopted by policymakers when prioritising public sector activities and investments is lower than the private rate of time preference. Society is longer-lived than individuals, and risk can be spread over a larger number of people, compared with individual risk-bearing. Yet another consideration is that, as a consequence of rising living standards, higher incomes will be available in the future to support consumption, so the value of consumption relative to income is higher at present than in the future.

The second justification of a discount rate is the concept of the *opportunity cost of capital*. Capital is defined as any resource or bundle of resources (real or financial) that is capable of generating income in the future. Suppose a given bundle of resources is invested now and left to grow at a compound rate r until sometime (year *t*) in the future. The concept is similar to placing a sum of money in a bank and watching it grow at compound interest. The opportunity forgone by consuming a given bundle of resources now is the benefit forgone that could otherwise be obtained by investing the same bundle of resources in some other alternative.

Discount rates of 3 and 6 % are used in the study of sea dikes in Vietnam by Vo Thanh Danh (Chap. 16). In the optimisation model of forest management presented by Nghiem Thi Hong Nhung (Chap. 6), discount rates of 1–8 % are simulated. The discount rate is varied in the study of biodiesel production in China by Zanxin Wang et al. (Chap. 3) indicating that the financial viability and choice of production method depend critically on the discount rate that is applied. In the same study, cost and benefit components are also subjected to sensitivity analysis.

The choice of a time horizon in a CBA is closely associated with selection of a discount rate. Many analysts choose a convenient time horizon that coincides with the expected life of a project. Others simply make an arbitrary choice such as 25 years. The effects of a project may extend well into the future, in which case an appropriate value should be included in the terminal year of the assessment, such as the residual value of the project.

The discount rate is usually predetermined by the finance department or monetary authority that oversees public sector investments by government agencies. Where an official rate is not prescribed, the usual practice is to apply the rate of interest on long-term (10-year) government bonds. This rate should be the real rate of interest, namely, the market rate adjusted for the rate of general price inflation in the national economy. In developing countries, the opportunity cost of capital is considered to be very high, so it is not unusual to encounter high rates of discount that the monetary authorities prescribe in economic and financial appraisals. What often counts in practice is whether variation of the discount rate in sensitivity analysis significantly changes any ranking of policy options.

Some Implications of Discounting

As noted, the effect of discounting is to downplay the importance of future benefits and costs. This creates difficulties in applying CBA where large magnitudes of benefits and costs are predicted for the distant future. Relevant examples are assessments of the damage costs of greenhouse gas emissions under a BAU scenario (Stern 2007; Garnaut 2008), the costs of decommissioning nuclear power plants or the benefits that accrue from hydropower schemes. Unless a low discount rate is applied, such benefits and costs will have only a minor effect on calculations of NPV. The UK Treasury (2003) has recognised the need for

Table 2.2 Schedule of declining discount rates (UK Treasury p98 2003)

Period of years	0–30	31–75	76–125	126–200	201–300	301+
Discount rate	3.5 %	3.0 %	2.5 %	2.0 %	1.5 %	1.0 %

considering lower discount rates for economic assessments with long time horizons (i.e. exceeding 30 years) and has recommended a schedule of declining discount rates, reproduced as Table 2.2.

Serious ethical concerns must be addressed where long-term effects – either in the base case or in policy options – have strong implications for the welfare of future generations. Lowering the discount rate to increase the present value of long-term costs and benefits may not be an effective or politically acceptable means of addressing long-term inequities in the inter-temporal distribution of costs and benefits. In reality, economic efficiency analysis might simply be overruled by ethical judgments and policy decisions.

Discounting can also raise difficulties in the management of natural resources. In forestry, for example, trees may take many years to reach maturity and a state suitable for timber harvesting. The net returns obtainable from a harvest planned 50 or more years into the future rarely match the returns that can be made by investing in other projects with higher productivity and more immediate economic rewards. The study by Nghiem Thi Nhung (Chap. 6) deals with fast-growing species of eucalypts and acacias, determining the optimal rotation age by means of the Faustmann formula. The optimal rotation age for *Eucalyptus urophylla* was found to be only 9 or 10 years, and for *Acacia mangium*, it was 13 years. Higher discount rates shorten the rotation age. Including values for carbon sequestration increases NPVs but also shortens rotation age. The study notes that using short-rotation small-size wood is suitable for manufacturing timber products such as chipboard, medium-density fibreboard or paper, whereas longer-rotation wood can be used for construction, wood processing and exports.

For biologically renewable resource systems, such as wild fisheries, it is well documented in the literature that unless the rate of growth of net economic returns exceeds the discount rate, economic efficiency analysis may imply that the relevant populations be driven to extinction (Clark 2010; Fisher 1981). In the mining industry, adoption of a high discount rate may lead to early exhaustion of the resource, with few alternatives to generate ongoing income. Wherever such extreme solutions are implied by the mechanical application of CBA, the possibility should be considered of introducing additional constraints in the analysis, such that economic efficiency is optimised subject to a minimum standard or target that decision-makers consider should be met in the broader interests of the community, now and in the future.

Risk Assessment in CBA

Risk Assessment

Risk assessment involves two main tasks: risk analysis and risk evaluation (Aven 2008). Risk analysis seeks answers to questions such as: (1) How does alternative *I* compare with alternative *II*? (2) Is the risk too high? (3) Is there a need to implement risk-reducing measures? In this chapter, the focus is on risk analysis to account for risk and uncertainty in the benefit–cost analysis framework.

Incorporating Risk and Uncertainty in Benefit–Cost Analysis

The terms 'risk' and 'uncertainty' are commonly used interchangeably, but a distinction is drawn for economic analysis. With risk, the probability distributions for variables are known. With uncertainty, the probability distributions are not known (Dixon et al. 1989; US Department of Transportation 2003).

Typically, benefit–cost analysis is carried out in a deterministic manner. However, the analyst is usually faced with a number of risks and uncertainties when evaluating an investment or project, as the parameter values and assumptions of any economic model are uncertain and subject to change. In many cases the problem of risk and uncertainty is ignored (Dixon et al. 1989). The common approaches to account for risk and uncertainty in CBA are (1) sensitivity analysis, (2) expected values (certainty equivalents) of scenarios and (3) risk analysis through Monte Carlo simulation. Only the third method, simulation, offers a practical methodology for analysing the overall risk of a project (Treasury Board of Canada Secretariat 1998). The following sections discuss each of these methods.

Sensitivity Analysis

The traditional means by which risk can be evaluated is sensitivity analysis. This approach can be used to account for the uncertainty in the model to quantify the impacts of policy changes and uncertain variables such as price and climate on the estimates of the net benefits of a project. Sensitivity analysis helps to test the robustness of the model results, establish critical values and discover thresholds or breakeven values around which the initially preferred option may change, thereby identifying sensitive or important variables (Pannell 1997). As a special case, involving irreversible impacts, Krutilla and Fisher (1985) applied the concept of threshold values when evaluating development versus preservation options for Hells Canyon in the USA.

In a typical sensitivity analysis, the value of an input variable or key outcome variable identified as a significant potential source of uncertainty is altered, either within some percentage of the initial value or over a range of reasonable values, while all other input values are held constant. Changes in the results of analysis are duly noted. This sensitivity process is repeated for other input variables for which risk has been identified. The input variables may then be ranked according to the effect of their variability on CBA results (ADB 1997; US Department of Transportation 2003).

A more systematic way of conducting sensitivity analysis is calculating the sensitivity indicator and associated switching value (ADB 1997). The sensitivity indicator compares the percentage change in a variable with a percentage change in a measure of project worth, usually the NPV. A switching value identifies the percentage change in a variable for the NPV to become zero, the economic internal rate of return to fall to cut-off rate and the project decision to change.

The CBA studies in this volume typically include sensitivity analysis to test the robustness of the results. For example, in the study by Bayani-Arias (Chap. 15) on coastal erosion, when comparing the present value of adaptation options, variations in the scenario of the impacts of coastal erosion are described as low, average and high. Discount rates were applied in the analysis ranging from 1 to 15 %. In the study by Vo Thanh Danh (Chap. 16), the likelihood of an extreme storm event and sea level rise was assessed, and sensitivity analysis showed that the expected NPVs of dike options were very sensitive to changes in the discount rate. If the salinity-protected area comprises 50 % of the total land area, however, the CBA results are not significantly altered.

Expected Values Approach

A straightforward and commonly used approach to valuing projects is to calculate the project's expected value. The expected value is the sum of the product of the probability of each possible state of the world and the value of the project in that state of the world.

The expected value approach is an extension of sensitivity analysis through two commonly used decision analysis techniques called payoff matrices and decision tree analysis (Pearce and Nash 1981; Dixon et al. 1989). The approach can be used to assist the decision-maker in making the best decision, but only after allowing for the decision-maker's own attitude towards risk in defining acceptable planning and management strategies. Payoff matrices are usually applied to rank alternative strategies, actions or options that are mutually exclusive.

If an option has two possible outcomes, low = \$20 and high = \$200, with probabilities of 25 and 75 %, respectively, then the expected value of the option is (\$20 × 0.25) + (\$200 × 0.75) = \$5 + \$150 = \$155. If the decision-maker has a completely rational attitude to risk, then he/she should be indifferent between investing in the option and accepting \$155 as the certainty equivalent.

For multiple but mutually exclusive options, each of which has a different outcome depending on the level of an independent 'driver' and its probability of

Table 2.3 Payoff matrix of net benefits ($′000/year) under different adaptation options and rainfall conditions

	Rainfall		
Option	Poor (0.25)	Average (0.50)	Good (0.25)
No adaptation	28	60	340
Partial adaptation	40	100	160
Full adaptation	20	140	200

occurrence, expected values can be calculated from a representative payoff matrix. For example, where options for adaptation to flooding are defined as no adaptation, partial adaptation and full adaptation and where the outcomes or benefits are affected by poor, average or good rainfall events, a payoff matrix of net benefits can be constructed, as shown in Table 2.3.

The expected payoff for each adaptation option can be calculated as:

No adaptation: (28*0.25) + (60*0.50) + (340*0.25) = $122
Partial adaptation: (40*0.25) + (100*0.50) + (160*0.25) = $100
Full adaptation: (20*0.25) + (140*0.50) + (200*0.25) = $125

In this example, the full adaptation option has the highest expected value and is thus the preferred alternative. The challenge in this approach is assigning the probabilities of events. Where probabilities are not known, arbitrary or subjective judgments must be made about the outcomes in the payoff matrix. Another drawback with the approach is that it assumes that the decision-maker is risk neutral. It does not allow for the effects of specific events, such as extreme events resulting from climate change, on human welfare (Dixon et al. 1989; Treasury Board of Canada Secretariat 1998).

Risk Analysis Using Monte Carlo Simulation

Risk analysis is any method – qualitative and/or quantitative – for assessing the impacts of risk on decision situations. The goal of any of these methods is to help the decision-maker choose a course of action, enabling a better understanding of the possible outcomes that could occur.

Quantitative risk analysis seeks to determine the outcomes of a decision situation as a probability distribution. Ideally, all CBA should be approached as a risk analysis because there is always some uncertainty in the data. In general, quantitative risk analysis involves four steps (Palisade Corporation 2009; Treasury Board of Canada Secretariat 1998):

1. Develop the basic model that will calculate NPV. This model is sometimes called the deterministic model because it uses a single deterministic value for each variable.

2. Identify and link the uncertain variables in the model to information about their maximum and minimum values (range) and about the probabilities of various values within those ranges. More specifically this involves linking the uncertain variables in the model by specifying the appropriate probability distribution function using the information available or derived from expert opinion.
3. Analyse the model with simulation. Run the model many times (a) to obtain a large number of NPVs to determine the range and probabilities of all possible outcomes for the results and (b) to determine the frequency with which various NPVs occur in the results, and, on this basis, predict the likely range of the NPV and the probabilities of various NPVs within that range.
4. Using the decision rules, interpret the results to identify the best alternative investment or, if there is only one, to decide whether it is likely to be a good investment based on the results provided and personal preferences relating to risk.

It is important to note that risk analysis is not a substitute for careful and detailed development of tables of costs, benefits and parameters. Setting up a sound deterministic model before thinking about risk is extremely important.

Adjusting for the Covariance of Related Risk Variables

When multiple uncertain inputs or output variables are considered in the model, some risk variables might be correlated. For example, if the NPV of the CBA model is based on the assumption of a high value for ‘total corn production’ and a high value for ‘average price of corn’, then the NPV may be outside the plausible range in the real world. A high production of corn normally results in a low corn price and vice versa. For the outcome of the analysis to be realistic, it must take these correlations into account. The analyst should consider adjusting for the covariance of related risk variables. Failure to take covariances into account can lead to large errors in judging risk. For example, in his pioneering study of risk analysis in project appraisal, Pouliquen (1970) cited a project for which the risk of failure was 15 % when labour productivity and port capacity were treated as independent variables but about 40 % when their positive correlation was taken into account.

Interpreting the Results of Risk Analysis

Risk analysis produces a list of NPVs, one for each run of the CBA model, which can be analysed statistically and graphically to see the probabilities of various outcomes. Two types of graphs that show the probability distribution of the NPV are (1) probability-density graph, which shows the individual probability of each NPV, and (2) cumulative-distribution graph, which shows how probable it is that the NPV will be lower than a particular value. Both types of graphs are useful for communicating with the decision-maker.

The decision rules for risk analysis in CBA in situations where there is significant uncertainty are as follows (Palisade Corporation 2009; Treasury Board of Canada Secretariat 1998; Richardson et al. 2006):

1. If the lowest possible NPV is greater than zero, accept the project.
2. If the highest possible NPV is less than zero, reject the project.
3. If the maximum NPV is higher than zero and the minimum is lower, calculate the expected NPV (ENPV). If the ENPV is greater than zero, accept the project but examine the risk of loss.
4. If the cumulative probability distribution curves for two mutually exclusive projects do not intersect, choose the option whose probability distribution is farther to the right.
5. If the cumulative probability distribution curves for two mutually exclusive projects intersect, be guided by the ENPV. If the ENPVs are similar, consider the risk profile of each alternative. Alternatively, conduct a second-degree stochastic dominance analysis using a certain risk-aversion coefficient to assess the more risk-efficient project.

Studies by Vo Thanh Danh (Chap. 16) and Predo and Francisco (2008) (not in this volume) are examples where risk analysis using Monte Carlo simulation in a CBA framework is applied. Vo Thanh Danh examined the uncertainty of cost variables using a uniform distribution and the benefits from avoided flood and storm damage using the expected value approach by obtaining the product of total damage avoided and probability of events. Although a Monte Carlo simulation was applied in the analysis, the study generated only the expected net present value (ENPV) but not the distribution of ENPV. The study by Predo and Francisco (2008) on land use alternatives identified the price of timber as the key uncertain variable and used time-series data to obtain the probability distribution function of prices. Their risk analysis provided not only mean estimates but also the entire distribution of the NPV estimates. The NPV of various land use alternatives can lie within a wide range of values. For example, the ENPV of one of the options ranged from a low value of PhP $-29{,}577\ \mathrm{ha}^{-1}$ to a maximum of PhP $31.4\ \mathrm{M\ ha}^{-1}$. While timber-based systems obtained the highest NPV, they seemed to be the most risky options, as indicated by the high coefficient variations ranging from 164 to 205 %.

Advantages and Limitations of Risk Analysis

Advantages of risk analysis documented in the literature include the following:

- It can rescue a deterministic benefit–cost analysis that has run into difficulties because of unresolved uncertainties in important variables.
- It can help bridge the communications gap between the analyst and the decision-maker. A range of possible outcomes, with probabilities attached, is inherently more plausible to a decision-maker than a single deterministic NPV. Risk analysis provides more and better information to guide the decision.

- It identifies where action to decrease risk might have the most effect.
- It aids the reformulation of projects to better suit the preferences of the investor, including preferences for risk.
- It induces careful thought about the risk variables and uses information that is available on ranges and probabilities to enrich the benefit–cost data. It facilitates the thorough use of experts.

However, risk analysis in CBA is not a panacea. It has various limitations, such as the following:

- The problem of correlated variables, if not properly contained, can result in misleading conclusions.
- The use of ranges and probabilities in the input variables makes the uncertainty visible, thereby making some managers uncomfortable.
- If the deterministic benefit–cost model is not sound, a risk analysis might obscure this by adding a layer of probabilistic calculations, thereby creating a spurious impression of accuracy.
- There is difficulty identifying the appropriate probability distribution for long-term uncertain events.

Sources of Probability Data

One of the major tasks in conducting a risk analysis is identifying the sources of probability data needed to specify the range of values each variable can take (minimum to maximum) and the probability distribution of the values in that range. There are two possible sources: historical data and expert judgment.

If historical data are available, the analyst can use the maximum and minimum values that occurred in the past as an appropriate range for the values of the variable. With the help of readily available commercial computer software packages for risk analysis, the probability distribution can easily be derived by fitting the data to an appropriate functional form. In the absence of historical data, or if historical data is not enough to underpin an estimate of the range and probabilities of a particular variable, then it may still be possible to rely on expert judgment, even though such judgments are typically subjective. For example, farmers are often able to provide expert judgment on the minimum, most likely, and maximum crop yield during El Niño, La Niña and neutral seasons based on their knowledge and long-term experience in farming. Similarly, the agronomist can provide expert judgment on the mostly likely yield of crops in any particular season.

A problem faced by SE Asia (and other parts of the world) is that, with the onset of climate change, it is anticipated that the probability distributions for a number of key driving variables such as temperature, wind, rainfall, floods and storm surge will shift and change in form. While the performance of global climate change models is continually improving, it will become increasingly necessary to rely on projected modelling results rather than historical data. It also begs the question as to whether a risk-neutral approach in economic appraisals is prudent, given that any shifts in probability distributions will have magnified effects in the tails of the distributions and the associated extreme events.

Techniques for Valuing the Environment

Valuation Methods

Over the last few decades, economists have devised a wide range of economic techniques to value the environment, some focusing specifically on developing countries (Freeman 1993; Abelson 1996; Pearce et al. 2002; Haab and McConnell 2002; Hanley and Barbier 2009; Glover and Jessup 2006; Glover 2010). The techniques belong broadly to three main categories: market-based techniques, surrogate market techniques and stated preference models. Each technique is described briefly below, including where it has been applied in the studies appearing in this volume.

Market-Based Techniques

Productivity Changes Method

This approach can be used when an environmental change leads to changes in production levels, costs or prices. Dose–response functions are usually needed to estimate physical changes in production as a consequence of changes in environmental conditions. Where changes occur only in output levels and/or costs, changes in producers' surplus can be estimated. If the change in productivity also results in a change in market equilibrium price, changes in consumers' surplus also need to be considered. The productivity changes technique is applied in a large number of the case studies in this volume: Zanxin Wang et al. (Chap. 3) for biodiesel production, Nghiem Thi Hong Nhung (Chap. 6) for forestry, Yem Dararath et al. (Chap. 7) for rubber production, Vo Thi Lang et al. (Chap. 8) for fish production, Bhadrani (Chap. 9) for tourism and industry, Gunaratne (Chap. 10) for sand mining, Kalyan et al. (Chap. 12) for forest products, Roongtawanreongsri et al. (Chap. 13) for timber values, Jaimie Kim B. Arias et al. (Chap. 15) for fisheries and Vo Thanh Danh (Chap. 16) for fisheries and agriculture.

Human Capital Approach

This method is frequently used in estimating external damage costs resulting from environmental pollution or other undesirable impacts. Changes in labour productivity can be measured using the human capital approach. Usually, the focus is on adverse health effects, providing estimates of forgone income, costs of medical treatment and costs reflecting psychological discomfort and distress. Reductions in health-related environmental damage costs, resulting from the introduction of environmental protection measures, can be incorporated in a CBA as benefits.

Results obtained using the human capital approach should not be interpreted as the value of human life. Most economists prefer to avoid placing direct values on life, although it is possible to obtain implicit policy values by observing the expenditures undertaken by public authorities to reduce accident rates or serious medical disorders. Another indicator is the willingness to pay by individuals to reduce the *risk* of premature mortality. Rather than the value of life, it is usually the value of statistical life (VSL) that is incorporated in a CBA. The ADB workbook on valuing environmental impacts has a sound discussion of the costs of morbidity and mortality including VSL (ADB 1996).

Defensive Expenditures

Defensive expenditures indicate the minimum amount that people would be willing to pay to prevent an adverse environmental impact. The relevant environmental benefits may exceed the expenditures involved, but if people are observed to be actually undertaking such expenditures (e.g. the construction of levees to prevent flood damage or the introduction of soundproofing in homes), it can be presumed that their valuation of benefits will be at least as great as the costs incurred. Defensive expenditures are assessed for flood control measures in the studies by Jaimie Kim B. Arias et al. (Chap. 15) and Vo Thanh Danh (Chap. 16).

Replacement/Repair Expenditures

These expenditures are typically undertaken after environmental damage has occurred, such as the application of fertilisers to offset a loss of soil nutrients or the clean-up costs incurred by households after flooding. It is spurious to make estimates of such costs and assume ex ante that they can be equated with environmental benefits, as people may not be willing to incur the costs involved. Empirical observations or surveys should be undertaken to demonstrate that the expenditures have been or will be undertaken. It can be assumed that such expenditures represent only a minimum value for the associated benefits. Flood damage estimates are provided in the studies by Jaimie Kim B. Arias et al. (Chap. 15) and Vo Thanh Danh (Chap. 16).

Shadow Projects

Expenditures on shadow projects are a special example of replacement cost. In the environmental context, a shadow project is defined as a man-made substitute for a natural system that may be severely damaged or lost as a result of human activity. The concept is closely related to environmental offset schemes, in which natural ecosystems or ecosystem services displaced by development are replicated elsewhere. When using the costs of shadow projects to value the environment, the same caveats apply as for the defensive expenditure and replacement expenditure approaches.

Surrogate Market Techniques

Property Value Differentials

In a competitive market, property asset prices and rents reflect the value of service from a property, including productive and consumptive environmental services. The property value approach (or so-called *hedonic price* method) can be used to estimate the implicit price of environmental attributes. A common application is the use of house prices to estimate environmental values. The same house with a given set of attributes can be expected to command a much higher price in an attractive or favourable environment, compared with an environment that is degraded.

The extent to which environmental attributes affect property values can be determined using multivariable regression models. Alternatively, and more easily, property value differentials generated by differences in environmental quality may be assessed by property valuation experts. Property owners themselves are often aware of the effects of environmental conditions on the value of their properties and may provide such information by means of interviews or questionnaires. Jaimie Kim B. Arias et al. (Chap. 15) and Saowalak Roongtawanreongsri et al. (Chap. 13) both apply the property value approach when assessing flood damage in their studies.

Travel Cost Method

The travel cost method assumes that the willingness to pay for recreation at a particular site can be inferred from the cost of travel by visitors to the site. The so-called zonal travel cost models are applied by undertaking an on-site survey to ascertain the frequency of visits, distances travelled, the cost of travel (including the implicit value of time), details of each visiting group and other socioeconomic information. Population statistics must be obtained for different zones of trip origin, and visitation rates by zone are calculated. A regression equation is derived showing the relationship between visitation rates and travel costs. This equation is then used to simulate the effect of hypothetical entry charges to derive a demand curve for recreation at the site. Where entry to the site is free, the entire area under the implicit demand curve provides an estimate of the consumers' surplus. Where an entry fee is charged, consumers' surplus comprises the area bounded by the price line and the demand curve.

To estimate changes in benefits for a particular site resulting from a change in environmental quality, such as an improvement in water quality, it is necessary to predict an upward shift in the demand curve, indicating a higher visitation rate and an increase in consumers' surplus for those still visiting the site. The increase in benefits is measured as the difference in total area between the original and new implicit demand curves.

Variations of the travel cost model allow for substitute sites, congestion externalities and individual versus zonal models. The study by Bhadranie Thoradeniya (Chap. 9) features an *individual travel cost model* to estimate recreation values in the Ma Oya River Basin in Sri Lanka.

Wage Differentials

The wage differential method values differences in environmental quality or risk in terms of the wages accepted by workers in different locations. It presumes that workers will accept lower wages in environmentally attractive sites and demand higher wages in degraded sites. Statistical models can be used to estimate the implicit environmental values. However, the method is difficult to apply because wages may be subject to various kinds of labour market rigidities and regulations. In addition, decisions regarding workplace location are typically based on a much wider set of criteria than environmental conditions or wages. None of the studies in the present collection makes use of the wage differential approach.

Stated Preference Models

Contingent Valuation

The contingent valuation method (CVM) establishes a hypothetical market for an environmental good or service and uses a survey questionnaire to elicit people's willingness to pay for some change in the supply or quality of the good or service (Mitchell and Carson 1989; Bateman et al. 2002). CVM is one method of directly measuring existence values and prospective values in an economic evaluation. It can be used to measure use values as well as non-use values.

CVM is subject to a wide range of potential biases; thus careful consideration must be given to the kind of scenario conveyed to respondents, the type of question asked (open-ended questions, payment card method, bidding game techniques, dichotomous choice), the specified payment vehicle and the statistical models used. Application of CVM must be carefully designed and administered to minimise biases. Gunaratne (Chap. 10) uses CVM to determine the WTP by miners for royalties on sand extracted from rivers in Sri Lanka, and Jaimie Kim B. Arias et al. (Chap. 17) apply contingent valuation to determine the WTP for a Technology-Based Flood Early Warning System to be installed along the Sta. Cruz River Watershed in the Philippines.

Choice Modelling

Choice modelling is a stated preference technique in which respondents choose their most preferred resource use option from a number of alternatives. Each alternative exhibits a number of attributes such as land affected, impacts on threatened species or cost to the household. Statistical models are applied to obtain estimates of people's WTP for particular environmental attributes as well as the value of aggregate changes in environmental quality. Choice modelling can thus be used to produce estimates of the value of multiple resource use alternatives.

Choice modelling has traditionally been applied to evaluate choices involving consumer goods, transportation, tourism and the selection of landfill sites. Texts in choice modelling include those by Louviere et al. (2000), Hensher et al. (2005) and Bennett and Blamey (2001). The technique is increasingly being used in developing countries, including applications to value environmental attributes (Bennett and Birol 2010). The study by Cheryl Launio et al. (Chap. 5) applies choice modelling of farmers' preferences regarding the use of rice straw. Gunaratne (Chap. 10) applies a discrete choice experiment (DCE) model to evaluate the preferences of local communities regarding sand mining.

Other Approaches to Valuation

Benefit Transfer Method

The benefit transfer technique borrows values from a so-called *study site* and applies them to a site to be evaluated (the *policy site*). Benefit transfers are used in several of the studies in this volume. In the study of biofuel production by Loan (Chap. 4), pollution damage costs are transferred from other similar studies. Roongtawanreongsri et al. (Chap. 13) use transferred benefits, adjusted for income levels and preferences, to value the ecosystem services provided by a protected forest in Thailand.

Benefit values may be transferred in several ways. The values may be transferred in unadjusted raw form. Examples include the typical value of a recreation visit to a natural area or the value of a rare species. Values may be adjusted as part of the transfer. For example, when the transfer involves two different countries, it is necessary to convert currencies allowing for the exchange rate or purchasing power of the currency. Other adjustments may be required to compensate for the effects of inflation. Differences in living standards may be handled by means of adjustment factors or value weights based on per capita incomes.

A common practice is to transfer data relating to physical factors or relationships at the study site and combine it with data applicable at the policy site. A good example is transferring physical dose–response functions, such as those determining the physical impacts of air pollution on human health or crops, and valuing them in terms of local costs and prices. Such functions may be derived by way of multivariable regression models, in which local data for the explanatory variables can be incorporated. Meta-analyses of comparable study sites (the compilation of large databanks from numerous studies to permit generalised statistical analysis of economic values) are an extension of this approach.

The robustness of the benefit transfer method depends largely on the quality of results for the study sites and the presence of similar conditions at both the study site and the policy site. For reliable use of the benefit transfer technique, the attributes of the study and policy sites should be similar; any environmental change under consideration at the policy site must be similar to that at the study site; and the socioeconomic characteristics and preferences of the population should be similar or at least adjusted as part of the transfer procedure.

Interactive website databases of environmental values have been constructed to facilitate benefit transfers for the purpose of economic analysis (DECCW 2004; Environment Canada 2009). In practical applications, this may be the only easily accessible source of information to fill any gaps in environmental values required to complete a CBA. Various authors have recently explored the use of choice modelling as a basis for transferring environmental values in economic analysis (Morrison et al. 2002; Rolfe and Bennett 2006).

Delphi Technique

This approach uses direct questioning of experts or community representatives to place a value on particular goods or services. It is usually applied in an iterative fashion, in group sessions, to achieve a consensus result.

Valuation Methods Applied in Case Studies

Environmental valuation is a critical step in the procedure for conducting a CBA. The case studies in this volume feature most of the valuation methods previously discussed. In any full CBA, it is customary to begin with market and financial values and modify them in accordance with the principles of welfare economics, including the incorporation of nonmarket values where relevant. The issues, approaches, methodologies and policy recommendations appearing in the studies here have been determined by EEPSEA researchers themselves and accordingly exemplify the kinds of economic assessments that may be carried out in Southeast Asian countries.

It may be noted that in many of the studies, there is an emphasis on productivity changes and market values. Similar observations in a developing country context have been made by other authors (Abelson 1996). This could simply reflect the strong dependence of developing countries on natural resources to support real incomes and livelihoods. Such countries usually have large populations in rural areas, and their production activities typically take the form of cropping, fisheries, forestry, aquaculture and animal production, all of which have close connections with natural resource systems. Other factors may also be responsible, such as low incomes and a limited ability of people to pay for environmental protection in the form of public goods, institutional fragilities that make it difficult to implement schemes that translate nonmarket benefits into actual payment schemes and budgetary constraints that limit the willingness and ability of governments to allocate funding to environmental protection or improvement programmes and projects.

Several of the studies supplement the social economic analysis with financial appraisals to explain the attitudes and behaviour of producers and the effects of policy options on producers' surplus. In other studies, market values are applied in assessing the economic significance of environmental impacts on human health, property and natural assets. The level of economic expertise required to apply market-based valuation methods should not be underestimated, as the application of these methods often requires complex analysis carried out in conjunction with other disciplines.

Where markets do not exist or are subject to significant distortions, it is more appropriate to consider using other valuation methods such as the travel cost method, property value differentials and stated preference modelling. All these methods have been applied in the case studies appearing in this volume. Two of the studies use multinomial logit models to analyse the choice behaviour of individuals, although not aiming specifically to provide environmental values. Some of these techniques, if properly applied, require substantial funding and research input. In developing countries funding support for economic assessments is often limited, thereby necessitating the use of less sophisticated valuation methods. For this reason, techniques such as cost-effectiveness analysis and threshold value analysis are often applied in preference to full cost–benefit analysis.

The benefit transfer technique is applied in some of the studies where nonmarket values are a component of the CBA, such as values for biodiversity and protected areas. In several of the studies, benefit transfers are used to value externalities associated with carbon emissions, although the values have typically been based on abatement costs or the prevailing price of tradable carbon credits in markets for certified emission reductions rather than direct valuation of environmental damage costs.

Significant progress has been made in developing countries, especially in academia, on methods for estimating nonmarket environmental values. There is now a growing awareness of the scope for applying stated preference models such as CVM and choice experiments. As per capita incomes increase over time, it can be expected that nonmarket values for the environment and natural resources will assume even greater importance. However, a high level of skills in experimental design, data collection and econometric methods is required to apply such methods. At present, the level of economic expertise tends to vary between countries. Some countries have reached high standards in their education and training programmes, but others still require further capacity-building.

It must be recognised that where benefit transfers are the only feasible source of nonmarket values, appropriate values may be obtainable only if pre-existing nonmarket valuation studies have been conducted for similar sites or circumstances. This underlines the importance of encouraging nonmarket valuation studies in capacity-building programmes, to help establish databases for economic assessments where full valuation based on primary data cannot easily be carried out.

Stated preference estimates of the WTP for environmental and natural resource attributes or improvements have important potential applications in developing countries. For example, they may guide the design and implementation of actual pricing and charging systems such as payments for environmental services (Pagiola et al. 2002), charges for reticulated water supplies and wastewater treatment schemes and entry fees to national parks. Even where actual pricing is not possible, implicit nonmarket values for natural assets can convey strong messages to policymakers regarding levels of public sector investment that may be warranted to protect such assets and meet community expectations.

The focus of economic analysis should clearly be on facilitating policy decisions, not the derivation of values per se. In practice, valuation methods will tend to be chosen that provide the most effective decision support and can realistically be carried out in the context of available funding, data, technical expertise and the time frame for research.

Strengths and Limitations of BCA

Limitations of Scientific Assessments

In most applications involving natural resources and the environment, the robustness of the economics results depends critically on the integrity and reliability of the scientific modelling and research that has been conducted for the biophysical aspects, especially the interrelationships between the natural environment and human activity. 'Getting the science right' is essential; otherwise economics results will be seriously compromised. Scientific uncertainties continue to hamper economic analyses of issues such as climate change, long-term impacts of persistent pollutants on human health and the behaviour of natural ecosystems in response to human activity.

Multi-objective Trade-Off Analysis

As noted, in reality decision-makers may focus on a broader set of indicators of wellbeing than economic efficiency. This is especially the case where *sustainability* is the professed aim of policy or where outcomes are assessed in terms of triple bottom line accounting: economic, social and environmental. Sustainability is best regarded as a multi-attribute or multi-objective concept, involving considerations of economic efficiency, ecological integrity, prevention of irreversible damage and inter- and intragenerational equity.

Frameworks and procedures for assessing trade-offs among multiple objectives may involve the use of decision support systems, participatory systems analysis models and multi-criteria analysis models (Janssen 1993). Information relating to economic efficiency may be incorporated in such models, alongside other indicators of community wellbeing. One advantage of such models is that the information used may be quantitative or qualitative and monetary or nonmonetary. However, it is essential not to double-count effects within the analysis. For example, if various kinds of physical indicators are adopted to represent the environmental outcomes of options, they should not be double-counted as monetary environmental values in the economic information. Similar caveats apply to social indicators.

Multi-criteria analysis (MCA) is applied in the study by Gunaratne (Chap. 10) using the DEFINITE software program (Janssen et al. 2001) to evaluate options for sand mining in terms of various social, economic, environmental and technical criteria. Multi-criteria trade-off analysis is used also in the study by Wu Jian et al. (-Chap. 14) in exploring options for managing the Qixinghe Wetland in China.

A major difficulty encountered in using multi-objective models is establishing relative importance weights for the criteria adopted in the assessments. The weights may be elicited from technical experts, community representatives or decision-makers themselves. Various mathematical methods and elicitation procedures may be used to derive weights (Janssen 1993). Subjective weights perform a similar function to the monetary values in a benefit–cost analysis, but they clearly relate to a broader set of objectives and decision criteria than economic efficiency.

The main challenge in applying methods such as MCA is to identify which options, which criteria and whose value weights are to be taken into account when setting up the evaluation framework and deriving the results. Success depends as much on the processes of defining the decision problem and ensuring that representative values are incorporated in the analysis as on the mathematical specifications of the calculations undertaken

Ownership of Assessment Process

A final limitation of CBA is that – depending on how the process is coordinated – it may be dominated by technical experts, economists and bureaucrats, with little opportunity for the community or different interest groups to gain a sense of ownership of the assessment process or the results of analysis. Stakeholders may, however, be engaged in assessments by means of surveys, direct consultations, focus group discussions, workshops and other participatory activities. The details of analysis including assumptions adopted, techniques applied, data used and issues of importance should be clearly identified, discussed and reported. Often, local knowledge and experience is a valuable source of information. The study of river basin management by Bhadranie Thoradeniya (Chap. 9) uses an educated trade-off framework (ETF) and participatory decision-making process that includes assessments and results of CBA.

Strengths of CBA

Transparency of Analysis

Perhaps the main strength of CBA is that it forces the analyst to undertake a comprehensive assessment of factors that are likely to affect community wellbeing as a consequence of any kind of planned action. The boundaries of analysis – both spatially and temporally – must be clearly defined. If properly conducted, scientific assessments underlying the economic analysis will be objective and soundly based. Valuations undertaken by economists should also be objective, albeit based on a narrow set of criteria relating to only one indicator of human wellbeing. All results obtained in a CBA should be transparent and available for scrutiny and possible amendment. The ability to explore different data inputs, structural relationships and valuation methods, carry out sensitivity analysis and evaluate the results is a particularly powerful feature of the method.

Promotion of Sustainable Economic Development

Possibly the greatest advantage of CBA is that it converts all information into a single, easily understood indicator using monetary values as the common measuring rod. In the early years of environmental management and environmental impact assessment, limitations of environmental valuation methods meant that the economic aspects of proposed developments were seen mainly in terms of direct financial benefits to private interests, with environmental effects addressed only in descriptive or qualitative form. Thus, the decision context was usually conceptualised as 'the environment versus development' implying that a conflict or trade-off between the two was necessarily involved. Invariably, state treasuries, finance departments and economic development agencies gave greater weight to the direct benefits of development initiatives that could be more easily expressed and measured in monetary terms.

The insights offered by the World Commission on Environment and Development dramatically changed this perception, convincingly demonstrating that environment protection and development should be mutually reinforcing (WCED 1987). CBA has emerged as an important vehicle for integrating the values of natural resources and environment within mainstream development planning and policymaking. As a facilitator of this process in South East Asia, through its capacity-building programmes, EEPSEA has hopefully played a constructive role.

References

Abelson PW (1996) Project appraisal and valuation of the environment: general principles and six case-studies in developing countries. Palgrave Macmillan, London
ADB (1996) Economic evaluation of environmental impacts: a workbook. Asian Development Bank, Manila
ADB (1997) Guidelines for economic analysis of projects. Asian Development Bank, Manila
Aven T (2008) Risk analysis: assessing uncertainties beyond expected values and probabilities. Wiley, Chichester
Bann C (1998) The economic valuation of tropical forest land use options: a manual for researchers. EEPSEA, Singapore
Bateman IJ, Carson RT, Day B, Hanemann M, Hanleys N, Hett T, Jones-Lee M, Loomes G, Mourato S, Ozdemiroglu E, Pearce D, Sugden R, Swanson J (2002) Economic valuation with stated preference techniques: a manual. Edward Elgar, Cheltenham
Bateman IJ, Lovett AA, Brainard JS (2005) Applied environmental economics: a GIS approach to cost-benefit analysis. Cambridge University Press, Cambridge
Bennett J, Birol E (eds) (2010) Choice experiments in developing countries: implementation, challenges and policy implications. Edward Elgar, Cheltenham
Bennett J, Blamey R (2001) Introduction [to the choice modelling approach to environmental valuation]. In: Bennett J, Blamey R (eds) The choice modelling approach to environmental valuation. Edward Elgar Publishing, Cheltenham, pp 1–13
Boardman AE, Greenberg DH, Vining AR, Weimer DL (2010) Cost-benefit analysis: concepts and practice, 4th edn. Prentice Hal, Upper Saddle River
Campbell H, Brown R (2003) Benefit-cost analysis: financial and economic appraisal using spreadsheets. Cambridge University Press, Cambridge
Clark C (2010) Mathematical bioeconomics: the mathematics of conservation, 3rd edn. Wiley, New York
Dasgupta AK, Pearce DW (1972) Cost-benefit analysis: theory and practice. Wiley, New York
DECCW (2004) Envalue: a searchable environmental database – ENVALUE, Department of Environment, Climate Change and Water, NSW. http://www.environment.nsw.gov.au/envalueapp
Dixon JA, Hufschmidt MM (eds) (1986) Economic valuation techniques for the environment: a case study workbook. Johns Hopkins University Press, Baltimore
Dixon JA, James DE, Sherman PB (1989) The economics of dryland management. Johns Hopkins University Press, Baltimore
Environment Canada (2009) Environmental valuation reference inventory – EVRI http://www.evri.ca
Fisher AC (1981) Resource and environmental economics. Cambridge University Press, Cambridge
Freeman AM (1979) The benefits of environmental quality improvement. Johns Hopkins University Press, Baltimore
Freeman AM (1993) The measurement of environmental and resource values: theory and methods. Resources for the Future, Washington, DC
Garnaut R (2008) The garnaut climate change review. Cambridge University Press, Cambridge

Gittinger JP (1982) Economic analysis of agricultural projects, 2nd edn. Johns Hopkins University Press, Baltimore
Glover D (2010) Valuing the environment: economics for a sustainable future. IDRC, Ottawa
Glover D, Jessup T (2006) Indonesia's fires and haze: the cost of catastrophe. Institute of Asian Studies Singapore, and IDRC, Ottawa
Haab TC, McConnell KE (2002) Valuing environmental and natural resources: the econometrics of non-market valuation. Edward Elgar, Cheltenham
Hanley N, Barbier EB (2009) Pricing nature: cost-benefit analysis and environmental policy. Edward Elgar, Cheltenham
Hanley N, Spash C (1995) Cost benefit analysis and the environment. Edward Elgar, Cheltenham
Harberger AC (1978) On the use of distributional weights in social cost-benefit analysis. J Polit Econ 81:S87–S120
Hensher D, Rose J, Greene W (2005) Applied choice analysis: a primer. Cambridge University Press, Cambridge
Hicks JR (1939) The foundations of welfare economics. Econ J 49:696–712
Hicks JR (1943) The four consumers' surpluses. Rev Econ Stud 11(1):31–41
Hufschmidt MM, James DE, Meister AD, Bower BT, Dixon JA (1983) Environment, natural systems and development: an economic valuation guide. Johns Hopkins University Press, Baltimore
James D (1994) The application of economic techniques in environmental impact assessment. Kluwer Academic Publishers, Dordrecht
Janssen R (1993) Multi-objective decision support for environmental management. Kluwer Academic Publishers, Dordrecht
Janssen R, Verwijnen M, Beinat E (2001) DEFINITE for windows: a system to support decisions on a finite set of alternatives. IVM, Vrije Universiteit, Amsterdam
Kaldor N (1939) Welfare comparisons of economics and interpersonal comparisons of utility. Econ J 49:549–552
Krutilla JV, Fisher AC (1985) The economics of natural environments: studies in the valuation of commodity and amenity resources. Johns Hopkins University Press, Baltimore
Little IMD (1957) A critique of welfare economics, 2nd edn. Oxford University Press, Oxford
Little IMD, Mirrlees JA (1974) Project appraisal and planning for developing countries. Basic Books, New York
Louviere J, Hensher D, Swait J (2000) Stated choice methods: analysis and applications. Cambridge University Press, Cambridge
Marglin S, Sen A, Dasgupta P (1972) Guidelines for project evaluation. United Nations, Vienna
Mitchell RC, Carson R (1989) Using surveys to value public goods: the contingent valuation method. Resources for the Future, Washington, DC
Morrison M, Bennett J, Blamey R, Louviere R (2002) Choice modelling and tests of benefits transfer. Am J Agric Econ 84(1):161–170
Pagiola S, Bishop J, Landell-Mills N (eds) (2002) Selling forest environmental services: market-based mechanisms for conservation and development. Earthscan, London
Palisade Corporation (2009) Guide to using @Risk, *Risk Analysis and Simulation Add-in for Microsoft Excel ver 5.5*, New York
Pannell DJ (1997) Sensitivity analysis of normative economic models: theoretical framework and practical strategies. Agric Econ 16:139–152
Pearce DW (1971) Cost-benefit analysis. Macmillan, London
Pearce DW, Nash CA (1981) The social appraisal of projects: a text in cost-benefit analysis. Macmillan, London
Pearce DW, Turner RK (1989) The economics of natural resources and the environment. Harvester-Wheatsheaf, New York
Pearce DW, Whittington D, Georgiou S (1994) Project and policy appraisal: integrating economics and the environment. OECD, Paris

Pearce DW, Pearce C, Palmer C (2002) Valuing the environment in developing countries. Edward Elgar, Cheltenham

Pouliquen LY (1970) Risk analysis in project appraisal. Johns Hopkins University Press, Baltimore

Predo CD, Francisco HA (2008) Improving productivity, profitability and sustainability of degraded grasslands through tree-based land use systems in the Philippines. In: Snelder DJ, Lasco RD (eds) Smallholder tree growing for rural development and environmental services: lessons from Asia, vol 5, Advances in agroforestry series. Springer, Dordrecht

Predo C, James (2006) A user manual for benefit cost analysis using microsoft excel, Manila and Penang, EEPSEA. http://www.eepsea.net

Ray A (1984) Cost benefit analysis: issues and methodologies. World Bank, New York

Richardson JW, Schumann KD, Feldman PA (2006) SIMETAR *simulation and econometrics to analyze risk*. SIMETAR, Inc, College Station

Rolfe J, Bennett J (2006) Choice modelling and the transfer of environmental values. Edward Elgar, Cheltenham

Scarborough H, Bennett J (eds) (2012) Cost-benefit analysis and distributional preferences. Edward Elgar, Cheltenham

Scitovsky T (1941) A note on welfare propositions in economics. Rev Econ Stud 9(1):77–88

Stern N (2007) The economics of climate change: the stern review. Cambridge University Press, Cambridge

Sugden R, Williams A (1979) The principles of practical cost-benefit analysis. Oxford University Press, Oxford

Treasury HM (2003) The green book: appraisal and evaluation in central government. The Stationery Office, London

Treasury Board of Canada Secretariat (1998) Benefit-cost analysis guide. Treasury Board of Canada Secretariat, Ottawa

Treasury UK (1997) The green book: appraisal and evaluation in central government. The Stationery Office, London

US Department of Transportation (2003) Economic analysis primer. U.S. Department of Transportation, Federal Highway Administration, Office of Asset Management, Washington, DC

US EPA (2000) Guidelines for preparing economic analyses. US Environmental Protection Agency, Washington, DC

WCED (1987) Our common future. World Commission on Environment and Development, Oxford University Press, Oxford

Willig R (1976) Consumers' surplus without apology. Am Econ Rev 66(4):589–597

Zerbe RO, Diverly DD (1994) Benefit-cost analysis in theory and practice. Harper Collins, New York

Part II
Agriculture, Forestry and Fisheries

Chapter 3
A Cost-Benefit Analysis of *Jatropha* Biodiesel Production in China

Zanxin Wang

Abstract The seeds of *Jatropha curcas* L. (JCL) can be used to extract oil for direct blending with fossil diesel or be further processed into JCL methyl ester (JME) through transesterification reaction. This study assesses the economic, environmental and energy performance of the production of the two end products using a life cycle analysis method. The results show that at the current level of technology and management, the production of JCL biodiesel is financially and economically unfeasible but that JCL biodiesel has excellent environmental and energy performance. If the seed yield can be improved to above 2.46 tonnes per hectare, it would be economically feasible to produce any of the two end products. Also, the value of carbon emission reduction can justify giving producers a subsidy.

Keywords China • Biodiesel • Life cycle analysis • Seed yield • Economic feasibility

Introduction

Background

The increases in crude oil prices and the concern for environmental protection have spurred the search for renewable alternative sources of oil (Shay 1993; Runge and Senauer 2007; Hazell and Pachauri 2006). As an alternative fuel for diesel engines, biodiesel is attracting greater attention throughout the world. Biodiesel is an environment-friendly alternative to fossil fuel and holds immense potential to assist in meeting the future energy needs of China. As a renewable, biodegradable and nontoxic fuel, biodiesel has low emissions and thus is environmentally beneficial (Krawczyk 1996).

Although biodiesel is a promising fuel, the production of biodiesel is challenged by its cost and the limited availability of fat and oil resources (Ma and Hanna 1999).

Z. Wang (✉)
School of Development Studies, Yunnan University, Kunming, Yunnan Province, People's Republic of China
e-mail: wzxkm@hotmail.com

D. James, H.A. Francisco (eds.), *Cost-Benefit Studies of Natural Resource Management in Southeast Asia*, DOI 10.1007/978-981-287-393-4_3

The present production of biodiesel is only about 4 billion litres per year globally (Rajagopal and Zilberman 2007). Biodiesel has to compete economically with petroleum diesel fuels, and the availability and sustainability of sufficient supplies of less expensive feedstock will be a crucial determinant in delivering a competitive biodiesel to commercial filling stations. One way of reducing biodiesel production costs is to use less expensive feedstock containing fatty acids such as inedible oils, animal fats, waste food oil and by-products of refining vegetable oils (Veljkovic et al. 2006). Fortunately, inedible vegetable oils, mostly produced by seed-bearing trees and shrubs, can provide an alternative.

With no competing food uses, attention has turned to *Jatropha curcas* L. (JCL), which grows in tropical and subtropical climates throughout the developing world (Openshaw 2000). JCL exhibits great advantages for biodiesel production. However, biodiesel production is still an emerging industry. The commercialisation of JCL in China is fairly recent, with commercial seedling production beginning in 2005. JCL has emerged as a high-potential biodiesel feedstock because of its adaptability to diverse growing conditions. Provincial governments in Southwest China have drafted plans to increase the area of JCL planting by over one million hectares in the next decade (Weyerhaeuser et al. 2007). Due to land availability and natural advantages for the growth of JCL, Yunnan Province aims to build the largest biodiesel base in China. According to 'The plan for the development of biodiesel feedstock plantation in Yunnan', the total area of JCL plantation in Yunnan is one million mu (1 ha = 15 mu) at present and will be increased to between 4 and 10 million mu at the end of 2010 and 2015, respectively (Bai 2007).

At present, there are more than 4 million hectares of barren land in Yunnan Province, of which one third is suitable for the growth of JCL. The government strategy for JCL plantation is to focus on these barren lands. Although Yunnan has set an ambitious target of establishing the largest biodiesel base in China and achieving energy independence, there is still a lack of information on the financial and economic performance of biodiesel production.

Biodiesel Production Chain

The biodiesel production chain can be divided into four stages: (1) production of JCL seeds through cultivation, (2) extraction and conversion of biodiesel, (3) distribution and retailing of finished fuels and (4) consumption of biodiesel.

The production of JCL seeds is mainly an agricultural activity in which JCL is grown, harvested and transported to a conversion facility. It involves the establishment and maintenance of JCL plantations and the harvest of JCL seeds and their preliminary treatment. The establishment of JCL plantations includes site preparation, seed treatment, seedling cultivation, nursery management and transplanting. The maintenance of JCL plantations involves irrigation, fertilising, weeding, disease control and pruning. The harvest of JCL seeds includes fruit flickering, drying and transportation.

The processing of JCL seed oil is an industrial activity in which the JCL seed is converted into biodiesel. First, the ripe fruits are plucked from the JCL trees and then are sun-dried and dehusked. To prepare the seeds for mechanical extraction, they should be solar heated for several hours or roasted for 10 min. If chemical extraction is chosen, the shelling of seeds can increase the yield of oil. The oil from JCL seeds is then extracted by mechanical extraction using a screw press or solvent extraction. Since mechanical extraction is more widely used in China, it is assumed that the JCL oil is extracted using mechanical expellers.

The selected expeller specifications for calculation are from a private vegetable oil company. Its processing capacity is 3–5 tonnes of seed per day (in an 8-h period) using power of 7.5 kWh. The equipment for the oil refinery has a processing capacity of 1 tonne of oil per day (in an 8-h period) using power of 7.41 kWh.

The JCL oil can be directly blended with diesel or can be made into biodiesel through transesterification reaction with methanol. Because of its viscosity, JCL oil is not suitable for direct use in engines. The high viscosity of JCL oil may contribute to incomplete fuel combustion and the formation of carbon deposits in engines, resulting in a reduction in the life of an engine and low thermal/energy efficiency (Prasad et al. 2000). However, a significant reduction in viscosity can be achieved by the dilution of vegetable oil with diesel in varying proportions (Pramanik 2003). Among various blends, the blends containing up to 30 % (v/v) JCL oil have viscosity values close to that of diesel fuel, and up to 50 % JCL oil can be substituted for diesel for use in a compression ignition engine without any major operational difficulties. Forson et al. (2004) showed that a 97.4 % diesel/2.6 % JCL fuel blend produces maximum values for brake power and brake-thermal efficiency as well as minimum values for specific fuel consumption and thus can be used as an ignition-accelerator additive for diesel fuel.

In this study, biodiesel refers to the blend of JCL oil and diesel, JCL methyl ester (JME) or its blends with diesel. That is, refined JCL oil can be directly used in engines after it is blended with diesel. Nevertheless, the oil's quality will be better, and there will be fewer long-term problems if it is first converted into biodiesel. This study assessed the costs and benefits for the two end uses.

When the end product is JME biodiesel, the selected production specifications for calculation refer to a biodiesel plant which yearly produces 50,000 tonnes of JME from JCL by transesterification. In the present study, it was assumed that the distance between the oil extraction plant or workshop and the transesterification plant or workshop is negligibly short, and thus no transportation cost is included in calculation.

Although the transesterification process is quite straightforward, the genetic and environmental background of reagents might require the modification of the input ratios of the alcohol reagent and reaction catalyst and alterations to reaction temperature and time, in order to achieve optimal biodiesel production results. Zhou et al. (2006) studied the production of biodiesel using JCL oil and found that the optimal conditions for transesterification reaction were that the molar ratio of JCL oil to methanol is 1:6 and the amount of catalyst is 1.3 % of the weight of the JCL oil, at which the yield of JME was higher than 98 % after a 20-min reaction time at 64 °C. For industrial production, the yields of JME and glycerol were about 96 % and 87 % of JCL oil, respectively (Li et al. 2007). JCL biodiesel has an overall performance close to that of diesel and thus can be used a substitute for diesel (Chen et al. 2006).

The distribution of JCL biodiesel involves the distribution of refined seed oil or JME for blending with fossil fuels. It was assumed the distance between the biofuel plant and the diesel distribution point is 10 km. The biofuel is transported by oil tankers with a carrying capacity of 5 tonnes per trip. The consumption of diesel is 3 l per 10 km when the truck is loaded and 1 l per 10 km when unloaded. Because the shared capital cost and labour cost is negligible, they were not included in the calculation.

The consumption of JCL biodiesel refers to the ultimate end use where the biodiesel enters the fuel tank of a vehicle or other engines. The data of the unit emissions of CO_2, N_2O and CH_4 was taken from the GREET (Greenhouse gases, Regulated Emissions, and Energy use in Transportation) model developed at Argonne National Laboratory (Wang 1999a, b) and other sources, including Agarwal and Agarwal (2007) and Dai et al. (2006). However, since the emissions of N_2O and CH_4 are low, the emission of CO_2 from the combustion of biodiesel was considered neutral because the emitted CO_2 is originally from the atmosphere.

Research Questions

Before massive expansion of *Jatropha* plantation might occur, various economic questions should be clarified. From a private perspective, the main questions are: how much does biodiesel production cost at each stage; and what is the difference in cost when JCL seed yield varies? Is it profitable to produce *Jatropha* biodiesel for producers at different stages? If not, how much should they be subsidised to promote the production of *Jatropha* biodiesel? From a societal standpoint, the main questions are: can the production of *Jatropha* biodiesel be economically justified? What are the main factors affecting its economic feasibility?

The results of the study undertaken here might provide important information to support policymaking for the promotion of JCL biodiesel. Before rapidly scaling up JCL acreage, a systematic study may also avoid unnecessary costs and reduce financial risk for producers.

Methodology

Data Collection

Data was collected from the areas designated for priority development, including Honghe, Chuxiong, Xishuangbanna and Lincang. The following activities were carried out in order to collect data: (1) interviews with producers and entrepreneurs to understand their role in the biodiesel supply chain from extraction, conversion and marketing including price and cost data at each stage; (2) visits to biofuel plants to become familiar with production processes and obtain information on the capital and labour inputs and outputs of products and residuals; and (3) interviews with experts and visits to research organisations involved in the production of JCL biodiesel.

The data covers the inputs and costs of biodiesel production at each stage, as well as the yield data. Field surveys were conducted in Honghe and Chuxiong prefectures of Yunnan Province, China. The data on seed production was collected from Yuanjiang and Shuangbai counties, and the data on oil extraction was from Erkang Science and Technology Co. Ltd. Secondary data includes the emissions of fuels and fertilisers from the GREET model and other technological inputs and outputs from related journal articles.

Price data for fertilisers, fungicides and herbicides was collected through field surveys. The seed price was obtained by interviewing investors in JCL plantations. The data for capital costs were directly sought from producers at different stages, including hoes, spades, shovels, mechanical expellers, refinery equipment and so on. The labour cost per hectare of JCL plantation was calculated by multiplying the local wage rate with the total hours or days involved. The cost of a hectare of land was considered negligible since JCL plantations would be established on barren/marginal land. The energy requirement per kilometre in transportation, obtained from survey of local drivers, was used as basis for estimating energy use and carbon emissions from transportation at different stages of biodiesel production.

Data Analysis

Both the financial and economic feasibilities of *Jatropha* biodiesel production were estimated using the cost-benefit analysis method. Due to lack of shadow price information, market prices were used in both analyses. Moreover, only the external value of carbon emission (reduction) is included in the economic analysis.

It was assumed that: (1) the time horizon for the project is 30 years (the life expectancy of JCL is 30–50 years), the number of working days is 330 days per year and the discount rate is 8 %; (2) the spacing in the JCL plantation is 2 m × 3 m, or the tree density is 1,650 per hectare; (3) the seed yield is 1,485 kg per hectare (500 fruits per tree); (4) 1 l of JCL oil is equivalent to 1 l of diesel because 1 l of diesel has a calorific value equivalent to 0.994 l of JCL oil, calculated according to the average calorific values and specific gravities of diesel and JCL oil; (5) the distance between the JCL plantation and oil refinery or fertiliser plants is 50 km, while the distance between the oil refinery and the oil distribution station is 10 km; the carrying capacity of trucks is 3 tonnes and that of oil tankers is 5 tonnes.

The data analysis consists of five parts. First, an explicit cost accounting model, a spreadsheet budgeting model, was first used to estimate the profitability of an activity for a single price-taking agent, such as an individual farmer or processor; and an analysis was then conducted to estimate the financial feasibility of the whole process of biodiesel production from the perspective of producers. Second, the carbon balance was accounted and valued. Third, an economic analysis was conducted by valuing cost and benefit using shadow prices and incorporating the value of carbon sequestration and the value of fruit husks as a fuel substitute for coal. Fourth, the energy efficiency of JCL biodiesel production was assessed (although that analysis is not presented here). Fifth, sensitivity analysis was carried out to identify significant variables affecting the financial and economic performance of JCL biodiesel production.

Financial Analysis

In each stage of production, the unit cost of production in the *n*th year is given by

$$C_{in} = \sum_{j=1}^{J} Q_{ijn} P_{jn} + A_{in} + D_{in} + E_{in} \tag{3.1}$$

where C_{in} is the unit function of production at the *i*th stage in the nth year; Q_{ij} is the amount of the *j*th input used at the *i*th production stage; P_j is the price of the *j*th input, such as labour and materials; *J* is the total number of input used at the *i*th stage; A_{in} is the abatement cost at the *i*th stage in the *n*th year, if applicable; D_i is the charge for the depreciation of fixed asset at the *i*th production stage in the *n*th year; and E_i is the distributive cost at the *i*th production stage in the *n*th year. The fixed asset was depreciated using the straight-line average service life method, assuming a salvage value to be 5 % of the total.

For producers at each stage, the financial feasibility was assessed using the following equation:

$$\mathrm{FNPV}_i = \sum_{n=0}^{N} \frac{\mathrm{NCF}_{in}}{(1+r)^n} \tag{3.2}$$

$$\text{NCF}_{in} = q_{in}(P_i) + \sum_{m=0}^{M} q_{inm}p_{im} - q_{in}(C_{in}) \tag{3.3}$$

where FNPV_i is the financial net present value of biodiesel production at the *i*th stage; NCF_{in} is the net cash flow at the *i*th stage of production in the *n*th year; *N* is the project horizon; *n* is the *n*th year of the project; *r* is the discount rate; q_i is annual quantity of products, say, seeds at the first stage and extracted oil at the second stage; P_i is the local price of products produced at the *i*th stage; q_{inm} is the quantity of the *m*th by-product produced at the *i*th stage in the nth year; p_{im} is the price of the *m*th by-product; and *M* is the total number of by-products produced at the *i*th stage.

The influences of taxes or subsidies on the net financial return at the *i*th stage can be assessed when tax and subsidy are included in Eq. (3.2).

A financial analysis was conducted to estimate the financial feasibility of whole process of biodiesel production. The financial feasibility was assessed by the net present value of the production, which is given by

$$\text{FNPV} = \sum_{n=0}^{N} \frac{P_nQ_n + \sum_{i=1}^{4}\sum_{m=0}^{M} q_{inm}p_{im} - \sum_{i=1}^{4} C_{in} - \sum_{i=1}^{4} T_{in} + \sum_{i=1}^{4} S_{in}}{(1+r)^n} \tag{3.4}$$

where FNPV is the financial net present value of biodiesel production; P_n is the price of biodiesel in the *n*th year; Q_n is the quantity of biodiesel produced in the *n*th year; and T_{in} and S_{in} are the tax levied and subsidy given at the *i*th stage in the *n*th year, if applicable. Note that *T* and *S* appear in the above equation, but they may occur at different stages of production. This means it is possible that a producer at a certain stage is subsidised, while the producer at another stage is taxed.

Carbon Accounting and Valuation

At each stage of the biodiesel supply chain, there are potential environmental impacts such as habitat destruction, carbon sequestration and emissions of liquid and/or solid hazardous gases. Owing to limited time and financial resources, this study considered only the carbon balance in the production chain for JCL biodiesel. Accounting for the carbon balance was conducted from the cultivation of JCL trees to the combustion of biodiesel.

The carbon balance is the sum of the reduced carbon stock minus the added carbon stock in the whole life cycle of JCL oil. The inventory covered all inputs and processes involving net emissions or sinks of the major GHGs (CO_2, CH_4 and N_2O). All emissions from the cultivation of JCL trees, seed transportation and processing, biodiesel transportation and combustion were accounted for. In the first stage, carbon emissions emanate from fuel consumption and fertiliser application and carbon sequestration by JCL plantations. In the second and third stages, carbon is released as fuel is burned. In the fourth stage, carbon is emitted when biodiesel is combusted. Inventory mass emissions were summed and converted into a final global warming potential measured in CO_2 equivalent considered over a 100-year timescale: $CO_2 = 1$, $CH_4 = 23$ and $N_2O = 296$ (Styles and Jones 2007).

According to *Revised 1996 IPCC Guidelines* (IPCC 1996), CO_2 emissions from biomass combustion are considered to recycle atmospheric CO_2 if biomass is extracted from a sustainable (i.e. replenished) source. These CO_2 emissions are therefore excluded from net emission calculations. The combustion of biodiesel in locomotives is assigned a zero emission factor to account for the carbon sequestered during the cultivation of *Jatropha* trees.

When the end product is JME, glycerine is a coproduct. The carbon emission from synthetic glycerine (9.6 kg CO_2-eq. per kg glycerine) was considered a credit to the carbon balance of JME production.

The CO_{2eq} balance is mathematically expressed as

$$W_{CO_{2e}} = \sum_{i=1}^{4} CR_i - \sum_{i=1}^{4} CE_i \tag{3.5}$$

where W_{CO2eq} is the CO_{2eq} balance of the life cycle of JCL biodiesel, i is the ith stage of production, CR_i is the amount of CO_{2eq} emission reduced at the ith stage and CE_i is the amount of CO_{2eq} emitted at the ith stage.

The monetary value of carbon was estimated using an implicit price for tradable emission permits in the international carbon market and then incorporated in the economic analysis.

Economic Analysis

The economic feasibility of biodiesel production was assessed using the following equation:

$$\text{ENPV} = \sum_{n=0}^{N} \frac{P_n Q_n + \sum_{i=1}^{4} \sum_{m=0}^{M} q_{inm} p_{im} + V_E - \sum_{i=1}^{4} C_{in} - C_E}{(1+r)^n} \tag{3.6}$$

where ENPV is the economic net present value of biodiesel production, V_E is the total environmental benefit and C_E is the total environmental cost.

The production of biodiesel is economically feasible if the ENPV is positive. If the ENPV is greater than the FNPV, provision of subsidy could be justified.

Results

Financial Analysis of JCL Oil Production

The financial analysis is composed of two parts: first, the financial feasibility of the different stages of JCL oil production when the seed producer and oil producer are independent; and second, the financial analysis of JCL oil production when the entire production is handled by a single producer.

Seed Production

The production of JCL seeds starts from seedling cultivation to seeds delivered to oil extraction plants and involves activities that include transplanting, site preparation, tending, seed collection and drying, husk removal and transportation.

Assuming a base yield of 1,485 kg per hectare (500 fruits per tree), the total cost at the seed production stage is 73,609 yuan per hectare. The main costs are associated with fruit drying and husk removal, fruit collection, fertilisation, fungicide spraying and weeding.

To estimate the revenue of seeds, the present price of JCL seed, 2 yuan per kg, was used. Using Eq. (3.2) the financial net present value at seed production stage (FNPV) was calculated to be −8,425.46 yuan per hectare. A breakeven price at the seed production stage (i.e. the price that the producer would need to receive to cover all operating, overhead and establishment costs of production of JCL seed) was calculated as 2.6 yuan per kg.

The results show that the production of JCL seed is not financially feasible. The FNPV tends to increase as the seed yield is improved. When the price of seeds is 2 yuan per kg, the FNPV will be positive only if the seed yield is higher than 3 tonnes per hectare. Efforts to reduce the cost of seed production, particularly labour cost, could enhance seed profitability.

The FNPV is also highly sensitive to any change in seed price. When the seed yield is 1,485 kg per hectare, the breakeven price of seed is 2.6 yuan per kg. At the present seed price of 2 yuan per kg, the FNPV is negative. Obviously, if a target for using biodiesel is established, the gap would have to be bridged by a government subsidy on seed production, as long as there is an economic justification for such a subsidy.

JCL Oil Extraction

When mechanical extraction is used, the oil extraction stage begins with heating seeds into refined JCL oil. When the seed yield is 1,485 kg/ha, the total cost of the processing of JCL seeds is 2,321.91 yuan per tonne. The major cost comes from the purchase of the seeds which accounts for 86.13 % of the total cost, while the sum of all the other costs accounts for only 14.87 %.

The oil percentage ranges from 32.2 to 40.2 % in seeds when oil is extracted using an engine-driven expeller, but this yield comprises only crude oil. According to a survey at an oil extraction plant of Erkang Science and Technology Co. Ltd., the yield of refined oil is about 30.4 %. Based on the specific gravity of JCL oil, the cost of producing 1 l of JCL oil is calculated to be 6.99 yuan per litre when the seed price is 2 yuan per kg.

Subsidies Required for JCL Oil Production

Subsidies would be required for both JCL seed producers and processors if a biodiesel output target is to be achieved.

At the seed production stage, subsidies can be provided based on the weight of seeds or the area of JCL plantation. Assuming that both the seed producers and processors receive a margin of 10 %, the two kinds of subsidies are shown in Table 3.1. Subsidies are required when the seed yield is lower than 3 tonnes per hectare. The required subsidy tends to decrease as the seed yield increases.

At the oil extraction stage, subsidies can be based on the volume of JCL oil. According to the average calorific values and specific gravities of diesel and JCL oil, it was calculated that 1 l of diesel has a calorific value equivalent to 0.994 l of JCL oil, so 1 l of JCL oil is equivalent to 1 l of diesel. As previously calculated, the breakeven price of JCL oil is 6.99 yuan per litre when the seed yield is 1,485 kg per hectare. Based on the present local price of diesel, 5.89 yuan per litre, the subsidy is 1.88 yuan per litre when a margin of 10 % is assumed. An increase in the price of diesel may provide an incentive for investors in JCL oil.

Full-Chain Financial Analysis of JCL Biodiesel Production

When production chains of both JCL oil and JME are operated by single producers, the financial feasibility is shown in Table 3.2. Both JCL oil and JME can be end products. The results reveal that the FNPVs of the production of the two end products are negative. That is, production is not financially feasible for JCL oil or JME.

The major cost in the production of JCL oil or JME is incurred at the seed production stage. As an extension of JCL oil, the transesterification reaction involves additional costs. However, as a coproduct of JME, glycerine shares 8.1 % of the total cost according to credits for allocation which are determined in terms of the market values of JME and glycerine. As a result, the FNPV of JME is slightly higher than that of JCL oil.

When the production chain, beginning with seedling cultivation to end products of JCL oil or JME, is operated by single producers, subsidies can be provided according to the amount of JCL oil the producers produce. The subsidy rate can be

Table 3.1 Subsidies required for the production of JCL seeds

Annual seed yield (kg/ha)	297	891	1,485	079	2,673	267	3,861	4,455
Breakeven price of seed (yuan/kg)	7.4	3.4	2.6	0.3	0.1	2.0	1.9	1.8
Subsidies based on seed weight (yuan/kg)	5.4	1.4	0.6	0.3	0.1	0	0	0
Annual subsidies based on plantation area (yuan/ha)	1,603.8	1,247.4	891	623.7	267.3	0	0	0

Table 3.2 Full-chain financial analysis of JCL biodiesel

Item	Quantity	
	JCL oil	JME
Cost of seed production (%)	88.38	82.83
Cost of oil extraction and refining (%)	11.40	10.89
Transesterification cost (%)		6.10
Cost of biodiesel distribution (%)	0.22	0.19
Total cost (10^4 yuan/ha)	4.11	4.39
Cost of JME		4.03
Cost of glycerine		0.36
JCL oil or JME output (10^4 l/ha)	1.33	1.30
Glycerine output (t/ha)		1.06
Total revenue (10^4 yuan/ha)	2.73	2.89
FNPV (10^4 yuan/ha)	−1.38	−1.37
Breakeven price (yuan/l)	8.87	8.91
Required subsidy rate (yuan/l)	3.97	4.01

determined according to the breakeven price of the end product. The analysis revealed that, assuming a margin of 10 %, the difference between the breakeven prices or required subsidy rates of JCL oil and JME is insignificant.

The production process for JME begins with the pretreatment of JCL oil and continues with the transesterification of the oil and the refinement of JME and glycerol. Based on a production line with an annual capacity of 50,000 tonnes of JME, the inputs were taken from Li et al. (2007). Excluding the cost of seed oil, the cost of production of JME is 674.68 yuan per tonne, among which the major expenditures are methanol, catalyst, electricity, NaOH and coal. Since glycerol is a coproduct of JME, the costs shared by JME and glycerine were found to be 615.31 yuan per tonne of JME and 59.37 yuan per tonne of JME, respectively, by allocating credits between JME and glycerol according to their market values.

As a coproduct of JME, glycerol contributes around 8.1 % of total revenue. Based on the market values of JME and glycerol, the total cost was allocated by assigning 91.9 % to JME and 8.1 % to glycerol. According to the outputs and shared costs, the breakeven prices of JME were calculated to be 8.91 yuan per litre. If they

Table 3.3 Life cycle carbon accounting for JCL biodiesel

Sources of CO_2 reduction	Quantity of CO_2 equivalent	
	JCL oil	JME
Application of fertilisers (for the first 3 years) (kg/ha)	−6719.65	−6719.65
Transportation of seeds and fertilisers (kg/ha)	−972.25	−972.25
JCL plantation (t/ha)	61.93	61.93
Substituting coal with *Jatropha* fruit husks (kg/ha)	321.97	321.97
JCL oil extraction and refining (kg/ha)	−172.27	−172.27
Substituting fertilisers with seed cake (t/ha)	6.16	6.16
Transesterification of JCL oil (t/ha)		−2.37
Substituting synthetic glycerine with bio-glycerine (t/ha)		10.18
Distribution of end products (kg/ha)	−14.3	−13.32
Substituting fossil diesel with JCL biodiesel (t/ha)	37.11	36.21
Life cycle carbon balance per ha plantation (t/ha)	97.64	104.54
Life cycle carbon balance per litre biodiesel (kg/l)	7.34	8.04

purchase JCL seeds to produce JME, producers should be provided with a subsidy rate of 4.01 yuan per litre, assuming a margin of 10 % of the breakeven price.

Carbon Balance for JCL Biodiesel

When different technology options are chosen, the carbon balance of the end product will differ. Analysis of the carbon balances for end products of JCL oil and JME follows below.

When JCL oil is directly used by blending it with fossil diesel, its life cycle carbon balance is as shown in Table 3.3. The production and use of JCL oil has a positive carbon balance. JCL plantations and JCL oil are major contributors to the carbon balance. As coproducts of JCL oil, the combustion of fruit husks can also reduce carbon emission s when they are used as a substitute for coal.

The major GHG emitters are the application of fertilisers and the transportation of seeds and fertilisers. Although a lot of direct energy is used in the oil extraction stage, the GHG emissions are relatively small.

The production process for JME is an extension of that for JCL oil. The carbon sources of the former differ from that of the latter starting with the transesterification reaction. The life cycle carbon balance of JME is positive, similar to that of JCL oil.

Intuitively, more GHG will be emitted for the production of JME because more inputs are used. However, as a coproduct of JME, glycerine adds a credit to the life cycle carbon balance of JME because it can be considered a substitute for synthetic glycerine. Consequently, the life cycle carbon balance of JME is higher than that of JCL oil.

Table 3.4 Economic feasibility of JCL biodiesel production

Item	Quantity	
	JCL oil	JME
Total cost (10^4 yuan/ha)	4.11	4.39
Value of carbon emission reduction (10^4 yuan/ha)	0.50	0.52
Value of fruit husks (10^4 yuan/ha)	0.28	0.28
Value of JCL biodiesel (10^4 yuan/ha)	2.73	2.66
Value of glycerine (10^4 yuan/ha)		0.24
Total revenue (10^4 yuan/ha)	3.51	3.70
ENPV (10^4 yuan/ha)	−0.60	−0.69

Economic Feasibility of the Production and Use of JCL Biodiesel

The economic feasibility study included the carbon value and the value of fruit husks and seed shells, where fruit husks and seed shells are considered to be substitutes for coal. According to Xing and Wang (2009), the price of CER in China's CDM market was around €10/t or 91.8 yuan per tonne. This was the most recent price since the global financial crisis although it was previously much higher. The fruit husks were first converted to the coal equivalent based on their calorific value, and then the coal price, 450 yuan per tonne, was used as the proxy price of fruit husks since fruit husks are not traded in the market.

Table 3.4 shows that, when no other external values are included, the ENPVs of JCL oil and JME are both negative, with the former slightly higher than the latter. The total value of carbon emission reduction and fruit husks is 7,800 yuan per hectare, which is an important external benefit in the production of JCL biodiesel. However, the ENPVs are slightly lower than zero, and there is good potential for providing an economic justification for the production of JCL biodiesel if the production technology is improved.

Sensitivity Analysis

Sensitivity analysis assesses risks by identifying the variables that most influence a project's economic feasibility and quantifying the extent of their influence. The analysis was conducted by varying the value of certain variables. These included biodiesel price, yields of JCL seeds, discount rate and total cost.

Sensitivity of Financial Feasibility

As a determinant of biodiesel output, seed yield has a significant effect on the FNPV. The FNPVs tend to increase as the seed yield is enhanced, as shown in Table 3.5. Because the marginal cost is higher than the marginal revenue, the FNPVs are still negative even when the seed yield is as high as 3,861 kg/ha. Since the unit cost of JME is higher than that of JCL oil, changes in seed yield have a bigger effect on the FNPV of JCL oil than that of JME. The gaps between the FNPVs of two end products increase as the seed yield improves. When the seed yield is lower than 1.66 t/ha, the NPV of JME is higher than that of JCL oil. The contribution of glycerine to the NPV is higher when the seed yield is lower and tends to decrease as the seed yield increases.

The FNPV is also affected by the discount rate, representing the opportunity cost of investment. The FNPVs tend to increase as the seed yield is improved. However, the NPV of JME is more sensitive to a change in the discount rate than that of JCL oil. When the discount rate is 6.8 %, the production of JME and JCL oil will yield the same NPV. The NPV of JME is lower than that of JCL oil when the discount rate is lower than 6.8 %, and the relationship reverses when the discount rate is higher than 6.8 %.

The FNPVs of JCL oil and JME are also very sensitive to changes in the production cost. That is, a slight change in production cost will result in a sharp decrease in the FNPVs of both end products. Thus, more effort should be made to reduce the total cost to enhance the financial feasibility of the production of JCL biodiesel. In particular, major efforts should be focused on the seed production stage because more than 80 % of the total cost is incurred at the seed production stage.

Sensitivity of Economic Feasibility

Despite many other external benefits that the production of JCL biodiesel may bring, only the values of carbon emission reduction and fuel coproducts were considered. The value of carbon emission reduction was based on the price of CERs (Certified Emission Reductions) in the CDM (Clean Development Mechanism) market. Fuel coproducts refer to the fruit husks, which were valued according to the price of coal based on the amount of calorific value. Because the carbon balance and coproduct output is affected by the seed yield, the economic value is thus affected by seed yield.

The ENPV increases as the seed yield increases; it becomes positive if the seed yield is higher than 2.27 t/ha for the production of JCL oil and 2.46 t/ha for the production of JME. Since they tend to increase as the seed yield is improved, the values of carbon emission reduction and fruit husks play an important role in making the ENPV positive. Positive ENPV may provide an economic justification for providing subsidies to producers.

Table 3.5 Financial feasibility of producing JCL biodiesel

Annual seed yield (kg/ha^{-1})	297	91	485	079	673	3,267	3,861	
JCL oil	NPV (10^4 yuan/ha^{-1})	−1.62	−1.50	−1.38	−1.26	−1.14	−1.03	−0.91
	Breakeven price (yuan/l^{-1})	3.32	1.28	.87	.83	7.26	.90	.64
JME	NPV (10^4 yuan/ha^{-1})	1.50	1.44	1.37	1.30	−1.23	1.16	1.09
	Breakeven price (yuan/l^{-1})	2.53	1.18	.91	.94	7.40	.06	.82

The gap in ENPVs between the end products of JCL oil and JME is small. However, the gap between the ENPV of the two end products tends to become larger as the seed yield increases. Obviously, from an economic perspective, it is more desirable to produce JCL oil as a final product when the seed yield is high. However, the long-term effects of JCL oil on engines are unknown.

Other factors may also affect the ENPV, but their effects are the same as their effects on the FNPV and carbon balance. Thus, only the effect of changes in seed yield was analysed in the study.

Conclusions and Policy Implications

JCL seeds can be used to extract oil for direct blending with fossil diesel or be further processed into JME through transesterification reaction. This study assessed the financial and economic feasibilities of the production of these two end products using a cost-benefit analysis method.

The results show that, at the current level of technology and management, the production of JCL biodiesel is financially and economically not viable, noting that external values other than that of carbon emission reductions were not included. However, if seed yield can be improved to above 2.46 t/ha, it would be economically feasible to produce either of the two end products.

To promote the development of the JCL biodiesel industry, government support is indispensable. When JCL seed and oil is produced by different producers, the required annual subsidies are about 881.66 yuan per hectare for seed producers and 2.68 yuan per litre for oil producers when the seed yield is 1,485 kg/t and the diesel price is 5.22 yuan per litre, assuming a margin of 10 % for producers. When JME is the end product and producers purchase JCL seeds to produce JME, about 4.01 yuan per litre should be made available to producers to make enough revenue and break even on costs. If the seed yield is higher than 1,485 kg/t, the compensation rate for producers at different stages of the production chain tends to decrease. In particular, no compensation is required if the seed yield is higher than 3,267 kg/t and the seed price (i.e. 2 yuan per kg) prevails. If the production chain is run by a single producer, then compensation rates for the end products of JCL and JME are 3.97 yuan per litre and 4.01 yuan per litre, respectively.

For both end products of JCL oil and JME, the major costs are incurred at the seed production stage, which accounts for more than 80 % of total costs. In detail, these costs are from the purchase of fertiliser, fruit collection and drying, husk removal and weeding.

Based on the above-mentioned results, the life cycle of the financial and economic performance of JCL biodiesel can be improved in the following areas: (1) at the current level of oil extraction technology, since much oil is still in the seed cake after pressing, oil yield could be improved by optimising the production process; (2) seed yield can be improved through the selection of high-yield seed varieties and genetic improvement; (3) the creation of machines for dehusking and fruit collection would reduce the cost of seed production; (4) the labour cost of fruit collection can be reduced by cultivating high-yield and dwarf varieties; (5) financial and economic feasibility can also be improved by developing high-value-added products from coproducts – for example, the seed cake can be used to produce foodstuff and pesticides, and glycerine is an important raw material for some high-value pharmaceuticals, such as antivirus medicines. In particular, other positive external values could be generated when biodiesel production is coupled with ecological restoration. The results of the study have the following policy implications:

China's *Renewable Energy Law*, which was passed by the Congress on 28 February 2005 and took effect on 1 January 2006, laid a solid foundation for the production of JCL biodiesel. This law shaped an integrated renewable energy policy framework by providing a set of directives encouraging renewable energies, including national renewable energy targets, a special fiscal fund, tax relief or exemption, and public research and development support as well as education and training. Although it provides a framework, China's *Renewable Energy Law* requires relevant governmental authorities to formulate specific measures for the production and use of biofuels, especially JCL biodiesel, which is targeted for promotion by many provinces.

Despite its lack of financial feasibility, the production and use of JCL biodiesel as a labour-intensive economic activity could generate many job opportunities and increase farmers' income. The positive carbon balance and significant energy efficiency of JCL provide a way for governments to respond to climate change and energy security. These outcomes are consistent with governments' expectations and thus have important policy implications for policymakers.

In 2006, the State Administration of Taxation announced that biodiesel would be exempt from consumption tax. In 2008, the Ministry of Finance, State Administration of Taxation and National Development and Reform Committee jointly announced that 90 % of the income tax on biodiesel would be reduced. In the same year, the Ministry of Finance and the State Administration of Taxation enacted a policy that if more than 70 % of the raw material of biodiesel is from plant oil and waste animal fat, the value-added tax would be refunded to producers. Within this policy context, no tax was assumed for the production and consumption of JCL biodiesel. The results show that to promote the industrialisation of JCL biodiesel, tax exemption is not enough.

Although China has set up a directive to subsidise woody feedstock producers of 3,000 yuan per hectare (*China Daily* 2007), it is clearly insufficient to make up the loss in JCL seed production. Moreover, the government has set up the subsidy rate for bioethanol plants, but no policy is yet available for the subsidisation of biodiesel plants. The experience of other countries shows that government support plays a critical role in the promotion of biofuels. For example, to improve the competitiveness of biofuels, many countries, such as Germany and France (Manuel and Peters 2007), levy a consumption tax on gasoline and diesel and reduce or exempt tax on biofuels. Mandatory blending obligations can also be adopted to promote biofuel production.

Considering that most of the provinces that are putting a great deal of effort into promoting JCL biodiesel are poverty stricken, there is an opportunity to integrate rural development with the production of JCL biodiesel. In particular, as the cost accounting shows, the production of JCL biodiesel is labour intensive and can be integrated with government strategies for rural development, such as job creation and poverty alleviation.

The production and use of JCL biodiesel can greatly reduce GHG emissions. This positive environment performance has significant implications for China's commitment to carbon emission reduction, that is, for the carbon emission per unit of GDP in 2020 to be 40–45 % lower than that of 2005 (set before the Copenhagen Climate Change Summit). To substitute fossil fuel with biofuel could be an effective measure in the implementation of China's energy saving and emission reduction programme.

The effect of reducing GHG emissions is a positive externality which can provide partial justification for government to subsidise JCL biodiesel producers. The economic performance of JCL biodiesel can be greatly improved if it is possible to integrate the development of the biodiesel industry with CDM projects. However, such integration requires new mechanisms and institutional arrangements.

References

Agarwal D, Agarwal AK (2007) Performance and emissions characteristics of JCL oil (preheated and blends) in a direct injection compression ignition engine. Appl Therm Eng 27:2314–2323

Bai C (2007) The development of jatropha industry in Yunnan. Yunnan For 28(2):12–18

Chen Y, Mao Z, Wu Z, Zhu H, Tang Z (2006) Comprehensive exploitation and utilization of jcl oil plants. China Oils Fats 31(3):36–56

China Daily (2007) Biofuel makers to get subsidy, December 8, vol.27, No. 8620

Dai D, Hu Z, Pu G, Li H, Wang C (2006) Energy efficiency and potentials of cassava fuel ethanol in Guangxi region of China. Energy Convers Manag 47:1686–1699

Forson FK, Oduro EK, Hammond-Donkoh E (2004) Performance of JCL oil blends in a diesel engine. Renew Energy 29(7):1135–1145

Hazell P, Pachauri RK (eds) (2006) Bioenergy and agriculture: promises and challenges, 2020 Focus no. 14. International Food Policy Research Institute, Washington, DC

IPCC (Intergovernmental Panel on Climate Change) (1996) *Revised 1996* IPCC guidelines for national greenhouse gas inventories. In: Houghton JT, Meira Filho LG, Lim B, Treanton K, Mamaty I, Bonduki Y, Griggs DJ, Callender BA (eds) IPCC/OECD/IEA. UK Meteorological Office, Bracknell. http://www.ipcc-nggip.iges.or.jp/public/gl/invs1.htm

Krawczyk T (1996) Biodiesel-alternative fuel makes in roads but hurdles remain. Inform 7:80–829

Li C, Jiang L, Cheng S (2007) Biodiesel – green energy. Chemical Industry Press, Beijing

Ma F, Hanna M (1999) Biodiesel production: a review. Bioresour Technol 70:1–15

Manuel F, Peters J (2007) Biodiesel: a new oildorado? Energy Policy 35:1675–1684

Openshaw K (2000) A review of *Jatropha curcas*: an oil plant of unfulfilled promise. Biomass Bioenergy 19:1–15

Pramanik K (2003) Properties and use of *Jatropha curcas* oil and diesel fuel blends in compression ignition engine. Renew Energy 28:239–248

Prasad CMV, Krishna MVSM, Reddy CP, Mohan KR (2000) Performance evaluation of non-edible vegetable oils as substitute fuels in low heat rejection diesel engines. Proc Inst Mech Eng Part D J Automob Eng 214:181–187

Runge C, Senauer B (2007) How biofuels could starve the poor. Foreign Aff (86):41–53

Rajagopal D, Zilberman D (2007) Review of environmental, economic and policy aspects of biofuels, Policy research working paper 4341. The World Bank, Washington, DC

Shay EG (1993) Diesel fuel from vegetable oils: status and opportunities. Biomass Bioenergy 4:227–242

Styles D, Jones M (2007) Energy crops in Ireland: quantifying the potential life -cycle greenhouse gas reductions of energy-crop electricity. Biomass Bioenergy 31:759–772

Veljkovic VB, Lakicevic SH, Stamenkovic OS, Todorovic ZB, Lazic KL (2006) Biodiesel production from tobacco (*Nicotiana tabacum* L.) seed oil with a high content of free fatty acids. Fuel 85:2671–2675

Wang MQ (1999a) *GREET* 1.5-transportation fuel-cycle model (Vol. 1): methodology, development, use, and results. Center for Transportation Research, Energy Systems Division, Argonne National Laboratory, Argonne. http://greet.anl.gov/publications.html

Wang MQ (1999b) *GREET 1.5*-transportation fuel-cycle model (Vol. 2): appendices of data and results. Centre for Transportation Research, Energy Systems Division, Argonne National Laboratory, Argonne

Weyerhaeuser H, Tennigkeit T, Su Y, Kahrl F (2007) Biofuels in China: an analysis of the opportunities and challenges of *Jatropha Curcas* in Southwest China, ICRAF working paper, no 53. ICRAF China, Beijing

Xing B, Wang P (2009) Declining price frustrating China's carbon transaction temporally, Xinhua net, April 23. http://news.xinhuanet.com/fortune/2009-04/23/content_11240310.htm

Zhou H, Lu H, Tang S, Liang B (2006) The transesterification conditions for the production of biodiesel with *Jatropha curcas* L. Oil. Appl Chem Eng 35(4):284–287

Chapter 4
Comparing the Social Costs of Biofuels and Fossil Fuels: A Case Study of Vietnam

Loan Le Thanh

Abstract Biofuel production has been promoted to save fossil fuels and reduce greenhouse gas (GHG) emissions. This paper investigates biofuel's energy efficiency, GHG emission performance, and cost-effectiveness energy for transportation. Energy and GHG balances are calculated for a functional unit of 1 km using life-cycle assessment and considering effects of land use change (LUC) and managed soils in feedstock production. In comparison with biodiesel, the production and utilisation of ethanol performs better in terms of cost-effectiveness and energy efficiency. It also shows acceptable GHG emission performance. The results confirm that, with their current fuel consumption, it is possible to substitute gasoline and diesel with E5 and E10 ethanol and B5 and B10 biodiesel, respectively, to achieve energy efficiency and GHG emission saving.

Keywords Vietnam • Biofuel • Cost-effectiveness • Energy efficiency • Social cost

Introduction

Background

The global transportation sector is relying heavily on fossil fuels, which contributed 96.3 % of the sector's energy consumption in 2009 (Davis et al. 2010). Fossil fuel-related CO_2 emissions from the global transportation sector accounted for 23 % of total CO_2 emissions from fuel combustion in 2009 (IEA 2011a). Interest in biofuels as substitutes for fossil fuels has increased worldwide for three reasons. Firstly, biofuels potentially substitute for fossil fuels in the context of an increase in energy price due to an increase in energy demand and insecurity of supply (Greene 2007; Rajagopal and Zilberman 2007; Edwards et al. 2008; IEA 2011a, b). Secondly, biofuels are suggested as a solution for climate change mitigation (IEA 2011a, b; Edwards et al. 2008; Schipper et al. 2009; Leather 2009). Thirdly, biofuel

L. Le Thanh (✉)
Faculty of Economics, Nong Lam University, Ho Chi Minh City, Vietnam
e-mail: ltloan@hcmuaf.edu.vn

D. James, H.A. Francisco (eds.), *Cost-Benefit Studies of Natural Resource Management in Southeast Asia*, DOI 10.1007/978-981-287-393-4_4

production has the potential to foster rural economic development (Greene 2007; Rajagopal and Zilberman 2007).

Biofuel substitution has been recommended in the literature and promoted in many countries. However, there are concerns about its economic viability (Schipper et al. 2009; Leather 2009; Kovacevic and Wesseler 2010; Jung et al. 2010; Ajanovic and Haas 2010; Jaeger and Egelkraut 2011; Hu et al. 2008). To make biofuels competitive with fossil fuels, subsidies have been implemented in many countries (Jung et al. 2010; Ajanovic and Haas 2010; Jaeger and Egelkraut 2011). Nevertheless, a comparison of cost-effectiveness between biofuels and fossil fuels has not yet been conducted properly in many studies (Jaeger and Egelkraut 2011; Hu et al. 2008; Duer and Christensen 2010; Wang et al. 2011).

In previous studies a functional unit (FU) in terms of MJ or L has been used. This would be appropriate if biofuels were utilised in the form of heating energy or pure fuels (Gnansounou et al. 2009), but not in the form of blends for transportation, because the fuel efficiency should be considered. The use of substitution ratios between fossil fuels and biofuels based on the fuel efficiency of fossil fuels and blends (not pure biofuels) is also not appropriate (Gnansounou et al. 2009; Lechon et al. 2009; Nguyen et al. 2007). In addition, the external costs and benefits of biofuel production and utilisation have often not been considered in previous studies, with the exception of Kovacevic and Wesseler (2010). The GHG emissions associated with the effects of land use change and managed soils in biofuel feedstock plantation are either considered in terms of physical units or overlooked in comparison with fossil fuels (Reijnders 2011). In Le et al. (2013) the energy and greenhouse gas balances of ethanol were reported.

Objectives of Study

In this paper the aim has been to compare the social costs (i.e. the sum of private and external costs) of biofuels and fossil fuels for an FU, defined as 1 km of vehicle transportation. This FU embodies the fuel efficiency, and it is proper for the comparison of biofuels and fossil fuels in transportation.

The study contributes to the existing literature on the cost comparison of biofuels and fossil fuels by considering both private and non-private costs. Empirical results are based on a case study in Vietnam, where cassava-based ethanol and jatropha-based biodiesel are most promising (Nguyen 2010; Le 2010; Government of Vietnam 2007, 2010; MARD 2008). The study compares two biofuels and their alternative fossil fuels, ethanol and gasoline, and biodiesel and diesel, with a focus on the blends of E5 and E10 for ethanol and B5 and B10 for biodiesel. The blend of E5 is a 5 % ethanol (E100) blended with 95 % gasoline in volume; and B5 is a 5 % biodiesel (B100) blended with 95 % diesel. E10 and B10 are 10 % biofuels blended with 90 % fossil fuels in volume.

Biofuel Systems

Properties of Biofuel Systems

Figure 4.1 shows the life-cycle systems of production and utilisation of biofuels. A life-cycle assessment was used in this study to estimate the GHG and non-GHG emissions from the production and utilisation of biofuels, which are then expressed in monetary term as an external cost.

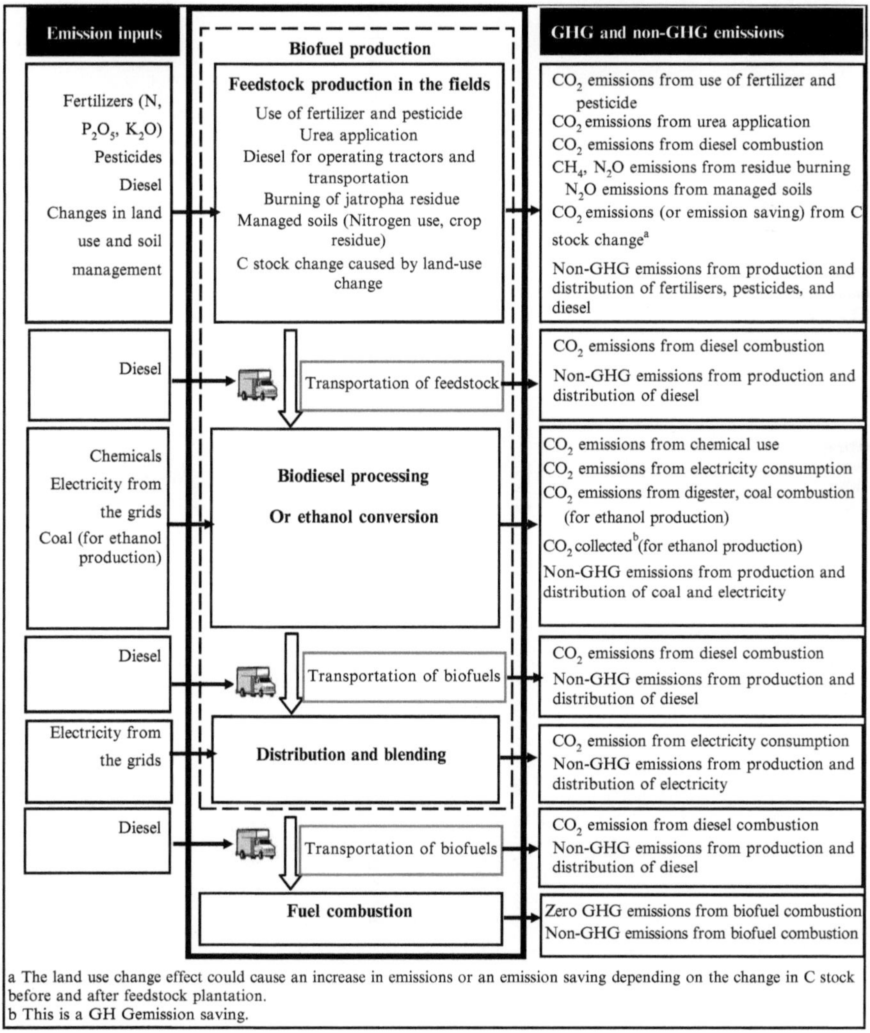

Fig. 4.1 Life-cycle system of biofuel production and utilisation

Table 4.1 Properties of fuels and blends (Gnansounou et al. 2009)

Properties	Unit	Gasoline	Ethanol	E5[a]	E10[a]	Diesel	Biodiesel	B5[a]	B10[a]
Density	$g\ L^{-1}$	743.0	790.0	745.4	747.7	832.0	879.0	834.4	836.7
LHV	$MJ\ L^{-1}$	32.2	21.1	31.6	31.1	35.9	32.6	35.7	35.6

[a]The properties of blends are calculated from those of gasoline and ethanol for E5 and E10 and from diesel and biodiesel for B5 and B10 according to the volume shares

Table 4.2 Percentage changes in fuel consumption of ethanol blends (E5 and E10) w.r.t. gasoline and of biodiesel blends (B5 and B10) w.r.t. diesel

Blends	E5		E10	
Fuel consumption indicators	L km^{-1}	g kW h^{-1}	L km^{-1}	g kW h^{-1}
Vehicle				
Ford Laser Ghia 1.8		−5.2[a]		−4.2
Honda Super Dream 100 cc		−6.4		−5.4
1.4i SI engine	5.2[b]	2.8–0.2[c]	5.5[b]	3.6–1.5[c]
Ford Focus	−1.2[d]			
Renault Megane	−0.6			
Various car models			−5.6[d]	
Toyota 1.6 L/2000			1.1	
XU7JP/L3 engine			5.1	
Vehicle	B5		B10	
6-cylinder MAN diesel engine		2.5		
Mitsubishi-6D14		0.4		
Perkins D3.152		2.7		
Various trucks	0			
Ford Focus 1.8 Tddi 90 VC	0.3		0.6	
Renault Laguna 1.9 dCi passenger car			1.0	
Mitsubishi-6D14				0.7

Sources available from author

[a]A minus sign means the lower fuel consumption of ethanol blends w.r.t. gasoline

[b]Ref. Ozsezen and Canakci (2011)

[c]Ref. Eyidogan et al. (2010). These two values are measured at the vehicle speeds of 80 km h^{-1} and 100 km h^{-1}, respectively

[d]These are averaged from the figures in Ref. Gnansounou et al. (2009)

Table 4.1 shows the properties of fuels, to convert from fuel consumption (L km^{-1}) to fuel efficiency (MJ km^{-1}). Table 4.2 presents the fuel consumption of blends with respect to gasoline and diesel. It is argued that the lower heating values (LHVs) of ethanol blends cause higher fuel consumption, while their higher octane values and compression ratios improve the thermodynamic properties and may reduce fuel consumption (Gnansounou et al. 2009; Nguyen et al. 2007; Le et al. 2009; Ozsezen and Canakci 2011; Eyidogan et al. 2010; Delgado 2003; Reading et al. 2002; Roayaei and Taheri 2009). The higher fuel consumption of biodiesel blends is explained by their lower LHVs and higher viscosity causing lower atomisation and combustion properties (Jindal et al. 2010; Agarwal and Agarwal 2007; Buyukkaya 2010; Lin et al. 2008; Aliyu et al. 2011). In reality, the fuel efficiency is affected not only by fuel properties but also by other factors such as vehicle speed and gear, vehicle models, and road conditions.

Functional Unit

Following the suggestion by Gnansounou et al. (2009), the study applied the FU of travelling 1 km using biofuels or fossil fuels as energy for road vehicles. The efficiencies in terms of MJ/km of biofuel components in blends are distinguished from the efficiencies of the fossil fuel components and those of the blends. It was assumed that the efficiencies of gasoline and diesel components in the blends are the same as their own standard efficiencies and that the efficiencies of ethanol and biodiesel are explained by their contributions to the blends after deducting those of the gasoline and diesel components, respectively (Gnansounou et al. 2009).

Sensitivity Analysis

A sensitivity analysis was conducted to evaluate the effects of different blends of biofuels and their fuel consumption. On the basis of the testing results, the percentage change in fuel consumption of ethanol blends with respect to gasoline was considered at three levels, formulating six scenarios: S1, S2, and S3 are the cases of E5 with 5 % higher, the same, and 5 % lower levels of fuel consumption per kilometre, respectively; S4, S5, and S6 are the cases of E10 with 5 % higher, the same, and 5 % lower levels of fuel consumption per km, respectively.

The testing results show that the percentage changes in fuel consumption of the blends of B5 and B10 with respect to diesel range between 0 and 5 %. Four scenarios for biodiesel were formulated: S7 and S8 are the cases of B5 with the same and 5 % higher levels of fuel consumption compared to diesel, respectively; S9 and S10 are the cases of B10 with the same and 5 % higher levels of fuel consumption, respectively. The efficiencies of biofuel components in blends are distinguished in Table 4.3. The social costs of the fuels were accordingly compared in terms of US dollars for a functional unit of 1 km ($\$\ km^{-1}$).

Methodology

Cost-Effectiveness Analysis

Cost-effectiveness analysis was undertaken in the study to compare alternative fuels (ethanol with gasoline and biodiesel with diesel) in terms of their social costs of production and utilisation for an FU. To calculate the social cost for an FU, the social cost of 1 GJ of fuel ($\$\ GJ^{-1}$) was first calculated and then multiplied by the amount of GJ needed for an FU ($GJ\ km^{-1}$) for each scenario in Table 4.3.

Table 4.3 Fuel efficiencies of ethanol and biodiesel components in blends

3a – Fuel efficiency of ethanol component in blends								
	Blend	Gasoline component		Gasoline component		Ethanol component		
	Fuel consumption (L km^{-1})	Fuel efficiency (MJ km^{-1})	Fuel efficiency (km MJ^{-1})	Percentage of gasoline energy in blend (%)	Fuel efficiency (km MJ^{-1})	Percentage of ethanol energy in blend (%)	Fuel efficiency (km MJ^{-1})	Fuel efficiency (MJ km^{-1})
Indicator	(1)	(2)	(3) = 1:(2)	(4)	(5) = (3) for gasoline	(6)	(7) = ((3) × 100 – (4) × (5)):(6)	(8) = 1:(7)
Gasoline	0.080[b]	2.56[a]	0.39	100	0.39			
Ethanol								
S1 (E5, +5 %)	0.084[c]	2.65	0.38	96.66	0.39	3.34	0.03	35.32
S2 (E5, 0 %)	0.080[d]	2.52	0.40	96.66	0.39	3.34	0.59	1.68
S3 (E5, −5 %)	0.076[e]	2.39	0.42	96.66	0.39	3.34	1.22	0.82
S4 (E10, +5 %)	0.084[c]	2.60	0.38	93.21	0.39	6.79	0.31	3.21
S5 (E10, 0 %)	0.080[d]	2.48	0.40	93.21	0.39	6.79	0.59	1.68
S6 (E10, −5 %)	0.076[e]	2.35	0.43	93.21	0.39	6.79	0.91	1.10
Diesel	0.054[f]	1.94[g]	0.52	100	0.52			
Biodiesel								
S7 (B5, 0 %)	0.054[h]	1.93	0.52	95.43	0.52	4.57	0.57	1.76
S8 (B5, +5 %)	0.057[i]	2.02	0.49	95.43	0.52	4.57	0.03[j]	37.01

(continued)

Table 4.3 (continued)

3a – Fuel efficiency of ethanol component in blends

	Blend	Gasoline component		Gasoline component		Ethanol component		
	Fuel consumption	Fuel efficiency		Percentage of gasoline energy in blend (%)	Fuel efficiency (km MJ^{-1})	Percentage of ethanol energy in blend (%)	Fuel efficiency	
	(L km^{-1})	(MJ km^{-1})	(km MJ^{-1})				(km MJ^{-1})	(MJ km^{-1})
Indicator	(1)	(2)	(3) = 1:(2)	(4)	(5) = (3) for gasoline	(6)	(7) = ((3) × 100 − (4) × (5)):(6)	(8) = 1:(7)
S9 (B10, 0 %)	0.054[h]	1.92	0.52	90.82	0.52	9.18	0.57	1.76
S10 (B10, +5 %)	0.057[i]	2.02	0.50	90.82	0.52	9.18	0.30	3.36

[a]Ref. Gnansounou et al. (2009)
[b]This figure is calculated from 2.56 MJ km^{-1} and the LHV of gasoline
[c]These figures equal the consumption of gasoline multiplied by 1.05
[d]These figures equal the consumption of gasoline multiplied by 1
[e]These figures equal the consumption of gasoline multiplied by 0.95
[f]Ref. Lechon et al. (2009)
[g]This figure is calculated from 0.054 L km^{-1} and the LHV of diesel
[h]These figures equal the consumption of diesel multiplied by 1
[i]These figures equal the consumption of diesel multiplied by 1.05
[J]S8 appears a very low contribution of biodiesel to the blend B5

Break-Even Price Calculation

The social costs of fuels were calculated as the break-even price, identified by setting the net present values of fuel projects equal to zero at a given discount rate. These break-even prices comprise the average costs for every GJ of fuels produced and utilised. The study follows Kovacevic and Wesseler (2010) by considering both private and non-private costs and benefits in the social cost calculation. The net present value (NPV) can be calculated as follows:

$$\mathrm{NPV} = \sum_{t=0}^{T} \frac{p\mathrm{F}(t)}{q^t} - \left(\sum_{t=0}^{T} \frac{C(t) - B(t)}{q^t} \right),$$

or, because p is constant:

$$\mathrm{NPV} = p \sum_{t=0}^{T} \frac{F(t)}{q^t} - \left(\sum_{t=0}^{T} \frac{C(t) - B(t)}{q^t} \right)$$

Setting NPV equal to zero yields the following:

$$p = \frac{\left(\sum_{t=0}^{T} \frac{C(t)-B(t)}{q^t} \right)}{\sum_{t=0}^{T} \frac{F(t)}{q^t}},$$

where p is the break-even price or the average cost for 1 GJ of fuels produced and utilised; $C(t)$ is the annual cost of biofuel production at year t; $B(t)$ is the annual benefit of by-products; $F(t)$ is the annual fuel production in terms of GJ; q^{-t} is the discount factor with $q = 1 + i$, and i is the discount rate; and T is time frame of the project.

Private Production Costs and Benefits of Fuels

The gas station prices in 2010 exclusive of tariffs, taxes, and fees represent the private costs of gasoline and diesel. The private costs are calculated as the sum of the import cost (including cost, insurance, and freight) and the transportation cost from the dock-warehouse to gas stations using the average national transport distance of 50 km. For biofuels, private production costs are incurred in the three phases of feedstock production, biofuel processing, distribution, and blending (Fig. 4.1).

The feedstock production phase incorporates the costs of land rental, seedlings, fertilisers, pesticides, diesel for tractors and water pumping machine, maintenance, labour, and seed transportation. The inputs in this phase were collected per hectare and then converted to inputs per GJ of biofuel output using the projected average yields (t ha^{-1}), the processing ratios (kg L^{-1}), and LHVs of biofuels (GJ L^{-1}).

For the biofuel processing phase, cost items include capital, electricity, coal, labour, water, and chemicals. The revenues from glycerine and compost are considered as private benefits of biodiesel production. The private benefits of ethanol production include the revenues of by-products of cassava stillage, CO_2 from fermentation sold for further utilisation and fixation as an alternative for long-term storage of CO_2, and the certified emission reductions (CERs) which the ethanol plants have earned from the Clean Development Mechanism project. For the processing, distribution, and blending phases, the inputs are calculated for 1 l on the basis of the production capacity of 10^8 L $year^{-1}$ and converted to inputs per GJ of output using the LHVs of biofuels. For the distribution and blending phase, the cost of electricity equals its average price in 2010 multiplied by the quantity on the basis of the pumping capacity of 10,884 L kW h^{-1}, collected from the survey (see section on Data Collection for details of the survey).

The transportation costs are calculated from transportation distances, truck capacity, and diesel consumption. The three stages of transportation include transporting (1) feedstock to biofuel plants, (2) biofuels from processing plants to blending stations, and (3) blends from blending stations to gas stations. To calculate the diesel consumption for transportation, required information includes national transportation distances for feedstock, biofuels, and blends, truck capacities, and diesel consumption. Each national distance is the average of three regional distances with the weights of corresponding capacities.

External Costs and Benefits of Fuels

Three externalities are considered in the calculation of external costs and benefits: (1) GHG emissions from fuel production, distribution, and combustion; (2) non-GHG emissions from fuel production, distribution, and combustion; and (3) security of supply of fossil fuels.

GHG Emissions

The GHGs consisting of carbon dioxide (CO_2), methane (CH_4), and nitrous oxide (N_2O) were aggregated into the CO_2 equivalent (CO_2e) using global warming potential (GWP) factors of 1 for CO_2, 25 for CH_4, and 298 for N_2O. The external cost of GHG emissions is calculated by multiplying the CO_2e emissions by the global external cost of CO_2e emissions of 33.2 \$ t^{-1}, which represents the avoidance costs of climate change. It was estimated for the year 2010 on the basis of cost-effectiveness analysis to determine the least-cost option to achieve the emission reduction target set under the Kyoto protocol (Mailbach et al. 2008).

For gasoline and diesel, the GHG emissions from the production and combustion in terms of CO_2e are 83.8 g MJ^{-1} [41, 44]. For biofuels, the GHG emission from biofuel combustion is zero according to the Renewable Energy Directive (RED) European Parliament (2009). The GHG emissions from biofuel production are calculated using the guidelines from the IPCC, RED, and the Biograce project (see Fig. 4.1) (Biograce 2011; IPCC 2006; Mailbach et al. 2008; European Parliament 2009; European Commission 2010). The GHG emissions from plant construction, production of equipment, and vehicles for transportation are not taken into account under the cut-off criteria suggested in the foregoing references.

Non-GHG Emissions

The four non-GHG emissions of HC, NO_x, PM, and SO_2 are considered in the study. The external costs equal the amounts of non-GHG emissions from fuel production, distribution, and combustion multiplied by the unit damage costs of non-GHG emissions (Fig. 4.1).

Security of Supply

The security of supply of fossil fuels is a motivation for biofuel production in most countries (IEA 2011a, b; Greene 2007; Rajagopal and Zilberman 2007; Mailbach et al. 2008). It is defined as the supply reliability at affordable prices (Greene 2007). The external costs of security of supply were formulated in various aspects and estimated ranging from 0.03 to 0.19 $ L^{-1} at the 2010 price (Table 4.5).

Biofuels in Vietnam

Data Collection

The primary data for the case study were collected through two surveys in the harvesting seasons of cassava and jatropha in Vietnam from January to June in 2011. For the ethanol survey, four of the top ten cassava-producing provinces were selected: Binh Phuoc, Tay Ninh, Dong Nai, and Dak Nong featuring four of the total eight ethanol plants. For the biodiesel survey, three jatropha-producing provinces were selected: Binh Thuan, Ninh Thuan, and Dong Nai with two jatropha-based biodiesel plants. Interviews were conducted with farmers, managers in biofuel plants, stakeholders including agricultural input suppliers, labourers, transporters, and key informants. Data were obtained relating to (1) farm inputs, (2) on-site conversion ratios in processing, (3) biofuel processing inputs, (4) LUC estimation, and other information.

Biofuel Industry in Vietnam

Biofuel production has rapidly developed under government policies. Up to 2010, eight ethanol plants had started operations with a total annual capacity of 680 million litres. Seven of the eight plants are located in the Central Highlands, South Central Coastal, and Southeast regions, which contributed 72 % of total cassava output in the period from 2006 to 2010 General Statistics Office of Vietnam (2011). Jatropha has been planted mostly in unused barren land in the North and Coastal regions, in compliance with the direction of the Ministry of Agriculture and Rural Development (MARD 2010, 2011).

Private Production Costs and External Costs for GHG Emissions

The private production costs and the external costs were calculated in relation to the production structure for ethanol and biodiesel production in Vietnam. A detailed description of the calculations for energy and GHG emissions for ethanol is given in Le et al. (2013).

External Costs of Non-GHG Emissions

Firstly, for the production and distribution phases of fossil fuels, the amounts of non-GHG emissions per GJ were collected from the GREET database (Argonne National Laboratory 2011). For the production and distribution phases of biofuels, the amounts of non-GHG emissions were calculated by multiplying the emission factors from GREET database and the amounts of production inputs including fertilisers, pesticide, electricity, coal, and diesel. Chemical inputs were ignored due to their expected insignificant amounts and the lack of data on their emission factors. Secondly, for the fuel combustion phase, the amounts of non-GHG emissions were derived from the official emission standards in Vietnam that are based on European emission standards for vehicles (European Parliament 1996, 1998, 2003, 2007).

The external costs equal the non-GHG emissions from fuel combustion multiplied by the unit damage costs of non-GHG emissions. The non-GHG emissions from fossil fuel combustion follow the "Euro 2 and Euro 3 standards" (European Parliament 1996, 1998; Government of Vietnam 2011) for motorbikes for the periods of 2010–2016 and 2017–2025, respectively, and the "Euro 2, Euro 4, and Euro 5 standards" for cars for the periods from 2010 to 2016, from 2017 to 2021, and from 2022 to 2025, respectively European Parliament 1996, 2003, 2007;

Government of Vietnam 2011). While diesel is used by cars, gasoline is used by both cars and motorbikes in Vietnam.

Therefore, the shares of gasoline consumption are estimated at 80 % and 20 % for the period of 2017–2021 and at 65 % and 35 % for the period of 2022–2025, for motorbikes and cars, respectively, using the numbers of motorbikes and cars, the ratios of gasoline and diesel engines, and the ratios of diesel and gasoline consumption for transportation (Davis et al. 2010; Schippers et al. 2008). The non-GHG emissions from biofuel combustion are calculated using the non-GHG emissions from fossil fuel combustion and the relative change of non-GHG emissions from combustion of biofuels compared to fossil fuels (−15 % HC, −10 % NO_x, −20 % PM, −80 % SO_2 for ethanol compared to gasoline and −67 % HC, +10 % NO_x, −47 % PM, −100 % SO_2 for biodiesel compared to diesel (USEPA 2011, 2002)).

Due to the lack of data on the external damage costs of non-GHG emissions in Vietnam, proxies were used for these costs. External damage costs for EU countries reported by the European Commission were adjusted for Vietnam. The scaled unit damage costs at the 2010 price are obtained for the year period 2010–2025 and adjusted to reflect the differences between Vietnam and the EU countries concerning the willingness-to-pay and the physical damage scale per ton of pollutants (Leiby 2007). Assuming that the willingness-to-pay and the physical damage scale per ton of pollutants are proportional to the gross domestic product (GDP) per capita and the population density, respectively, these two adjusting factors were measured by the ratios of GDPs (purchasing power parity) per capita and population densities of the EU and Vietnam (Nguyen 2008) (Table 4.4).

External Costs of Security of Supply

Due to the lack of data on the external costs related to the security of supply of fossil fuels for Vietnam, the study considered the external cost of security of supply from the literature and applied the lowest estimation of 0.03 $ L^{-1} in the calculation (Table 4.5).

It is worth noting that in the absence of data on the external costs of non-GHG emissions and fossil fuel security of supply for Vietnam, the empirical analysis is based on the adjustment results from estimation for EU countries for the former and on the lowest estimation found in literature for the latter. The reader should be aware of the uncertainty concerning these two estimates.

Table 4.4 Converting external damage costs of non-GHG emissions for the EU to scale external damage costs for Vietnam (University of Stuttgart 2008; World Atlas 2011; Global Finance 2011)

Pollutants	Damage costs for the EU (€ t^{-1})		Adjusting factors[a]		Damage costs for Vietnam (\$ t^{-1})[b]	
	2010	2025	GDP per capita ratio	Population density ratio	2010	2025
HC	1148	564	0.09	2.24	314	154
NO_x	7793	7350	0.09	2.24	2,135	2,014
PM	29,006	23,454	0.09	2.24	7,948	6,427
SO_2	7501	6718	0.09	2.24	2,055	1,841

[a]Population densities are 116.0 and 260.3 persons per square kilometre, and GDPs (PPP) per capita are 33,729 \$ and 3104 \$, respectively, for the EU and Vietnam in 2010
[b]The costs in Euro currency are multiplied by the average exchange rate in 2010 from the European state bank (1.33 \$ $€^{-1}$)

Table 4.5 External costs of fossil fuel security of supply

External cost formulation	External costs in 2010 (\$ L^{-1})
Incremental benefits of the import reduction to society	0.03
Costs of keeping stocks for a period needed to start up biofuel programme without subsidy	0.16–0.19
Direct economic costs of the transfer of wealth from the USA to oil-producing countries, GDP losses due to the oil price higher than its competitive level, and macroeconomic adjustment	0.19

Sources available from author

Results and Discussion

Overview of Results

In this section results are presented for the cost-effectiveness of the biofuels in comparison with the fossil fuels. The private, external, and social costs of production and utilisation are first shown in terms of \$ GJ^{-1}, followed by the costs per functional unit, i.e. the costs in US dollars per kilometre (\$ km^{-1}). Three discount rates of 4 %, 8 %, and 10 % are used to investigate the effect of discount rates on the results.

Cost-Effectiveness of Cassava-Based Ethanol and Gasoline

At a discount rate of 4 %, the private cost of ethanol is 18.6 \$ GJ^{-1} (Table 4.6), of which the cassava production cost amounts to 57.9 %, the conversion cost accounts

Table 4.6 Costs of production and utilisation of ethanol and gasoline ($ GJ^{-1})

	Ethanol			Gasoline		
	Discount rate			Discount rate		
Cost items	4 %	8 %	10 %	4 %	8 %	10 %
Private cost	18.57	20.13	20.93	15.78	15.78	15.79
Cassava production	10.76	11.24	11.48			
Ethanol conversion	7.33	8.37	8.91			
Distribution and blending	0.48	0.52	0.53			
External cost	1.63	1.70	1.74	4.11	4.12	4.13
GHG emissions	1.22	1.29	1.33	2.76	2.76	2.76
Non-GHG emissions	0.40	0.41	0.41	0.25	0.26	0.26
Security of supply	NA[a]	NA[a]	NA[a]	1.10	1.10	1.10
Social cost	20.20	21.83	22.67	19.89	19.90	19.91

[a]Not applied

for 39.5 %, and the cost of distribution and blending contributes 2.6 %. The private cost of ethanol is 17.7 % higher than that of gasoline, but its external cost is 60.3 % lower than that of gasoline. The social cost of ethanol is 20.2 $ GJ^{-1}, which is 1.6 % higher than that of gasoline at a discount rate of 4 %.

Table 4.6 shows that the social cost per GJ of ethanol is higher than that of gasoline due to its higher private cost component. The higher external costs related to emissions and the security of supply lead to the higher external cost for gasoline; however, the social cost of ethanol is eventually higher than that of gasoline. These findings hold for the three discount rates and the differences are even larger at the higher discount rate. For instance, the social cost of ethanol is 13.9 % higher than that of gasoline at a discount rate of 10 %.

Regarding the external costs of ethanol production, the non-GHG emissions are mostly caused by the ethanol conversion, particularly coal combustion and electricity use Le et al. (2013). This finding encourages the ethanol plants to make the best use of by-product biogas or use the environmentally friendly substitutes for coal and electricity together with the agricultural practices suggested in Le et al. (2013) so as to reduce the external cost. The contributions of the external costs of non-GHG emissions to the social costs are less than 3.0 % for all fuels. These small contributions result in a relatively insignificant effect on the overall results.

If the fuel efficiency in transportation is taken into account, the costs can be obtained per functional unit, i.e. in terms of $ km^{-1}. The social costs in terms of $ GJ^{-1} are multiplied by the fuel efficiency (GJ km^{-1}) to obtain the cost-effectiveness ($ km^{-1}). In terms of a functional unit, the ethanol substitution for gasoline is cost-effective for scenarios S2, S3, S5, and S6, but not for scenarios S1 and S4 (Table 4.7). At a discount rate of 4 %, the ethanol substitution for gasoline in the form of E5 would save 0.02 $ km^{-1} or 33.4 % of the social cost per functional unit compared to gasoline (see columns 5 and 6 in Table 4.7) for scenario S2. The saving performance also holds for scenarios S3, S5, and S6. For scenarios S1 and S4, the

Table 4.7 Cost-effectiveness of ethanol and gasoline

Scenarios	Social cost ($ GJ^{-1})	Fuel efficiency (GJ km^{-1})	Social cost ($ km^{-1}) Ethanol	Social cost ($ km^{-1}) Gasoline	Cost difference ($ km^{-1})	Cost difference (%)
	(1)	(2)[a]	(3) = (1) × (2) for ethanol	(4) = (1) × (2) for gasoline	(5) = (3) – (4)	(6) = (5) × 100: (4)
At a discount rate of 4 %						
Gasoline	19.89	0.0026		0.05		
Ethanol						
S1 (E5, 5 %)	20.20	0.0353	0.71	0.05	0.66	1298.53
S2 (E5, 0 %)	20.20	0.0017	0.03	0.05	–0.02[b]	–33.40
S3 (E5, –5 %)	20.20	0.0008	0.02	0.05	–0.03	–67.56
S4 (E10, 5 %)	20.20	0.0032	0.06	0.05	0.01	27.14
S5 (E10, 0 %)	20.20	0.0017	0.03	0.05	–0.02	–33.40
S6 (E10, –5 %)	20.20	0.0011	0.02	0.05	–0.03	–56.37
At a discount rate of 8 %						
Gasoline	19.90	0.0026		0.05		
Ethanol						
S1 (E5, 5 %)	21.83	0.0353	0.77	0.05	0.72	1410.55
S2 (E5, 0 %)	21.83	0.0017	0.04	0.05	–0.01	–28.07
S3 (E5, –5 %)	21.83	0.0008	0.02	0.05	–0.03	–64.96
S4 (E10, 5 %)	21.83	0.0032	0.07	0.05	0.02	37.32
S5 (E10, 0 %)	21.83	0.0017	0.04	0.05	–0.01	–28.07
S6 (E10, –5 %)	21.83	0.0011	0.02	0.05	–0.03	–52.87
At a discount rate of 10 %						
Gasoline	19.91	0.0026		0.05		
Ethanol						
S1 (E5, 5 %)	22.67	0.0353	0.80	0.05	0.75	1468.30
S2 (E5, 0 %)	22.67	0.0017	0.04	0.05	–0.01	–25.32
S3 (E5, –5 %)	22.67	0.0008	0.02	0.05	–0.03	–63.62

(continued)

Table 4.7 (continued)

	Social cost	Fuel efficiency	Social cost ($ km^{-1})		Cost difference	
	($ GJ^{-1})	(GJ km^{-1})	Ethanol	Gasoline	($ km^{-1})	(%)
Scenarios	(1)	(2)[a]	(3) = (1) × (2) for ethanol	(4) = (1) × (2) for gasoline	(5) = (3) – (4)	(6) = (5) × 100: (4)
S4 (E10, 5 %)	22.67	0.0032	0.07	0.05	0.02	42.57
S5 (E10, 0 %)	22.67	0.0017	0.04	0.05	−0.01	−25.32
S6 (E10, −5 %)	22.67	0.0011	0.02	0.05	−0.03	−51.07

[a]These are the figures in column 8 in Table 4.3 divided by 1,000
[b]A minus sign means cost-effectiveness

cost differences are positive, meaning that the ethanol substitution for gasoline is not cost-effective. Seeking for break-even points, the zero cost difference is found at 1.7 % and 3.5 % higher fuel consumption of E5 and E10 compared to gasoline, respectively, at a discount rate of 4 %, provided that other factors are constant. This means that the ethanol substitution for gasoline in the form of E5 or E10 would be cost-effective if the fuel consumption of E5 and E10 in terms of L km^{-1} does not exceed the consumption of gasoline by more than 1.7 % and 3.5 %, respectively, compared to gasoline at a discount rate of 4 %.

Similar results are found at discount rates of 8 % and 10 %. The ethanol substitution for gasoline is also cost-effective for scenarios S2, S3, S5, and S6, but not for S1 and S4. In view of the break-even points, the higher discount rate requires the lower fuel consumption of ethanol blends in terms of L km^{-1} to achieve the cost-effectiveness of the ethanol substitution for gasoline. For instance, the fuel consumption of E5 and E10 compared to gasoline is allowed to increase up to 1.7 % and 3.5 %, respectively, at a discount rate of 4 %; these figures are, respectively, 1.3 % and 2.6 % at a discount rate of 10 %.

Cost-Effectiveness of Jatropha-Based Biodiesel and Diesel

At a discount rate of 4 %, the private cost of biodiesel is 29.2 $ GJ^{-1} (Table 4.8), of which the jatropha production cost amounts to 97.3 %, the processing cost accounts for 1.7 %, and the cost of distribution and blending contributes to 1.0 %. The private cost of biodiesel is 95.8 % higher than that of diesel. Biodiesel could produce an external benefit of 2.19 $ GJ^{-1}, while diesel incurs an external cost of 4.2 $ GJ^{-1}. The social cost of biodiesel is 27.0 $ GJ^{-1}, which is 41.3 % higher than that of diesel at a discount rate of 4 %.

Table 4.8 Costs of production and utilisation of biodiesel and diesel (\$ GJ^{-1})

	Biodiesel			Diesel		
	Discount rate			Discount rate		
Cost items	4 %	8 %	10 %	4 %	8 %	10 %
Private cost	29.20	30.64	31.40	14.91	14.91	14.91
Jatropha production	28.41	28.86	29.11			
Biodiesel processing	0.50	1.47	1.96			
Distribution and blending	0.29	0.31	0.32			
External cost	−2.19	−2.24	−2.26	4.21	4.24	4.26
GHG emissions	−2.58	−2.65	−2.69	2.76	2.76	2.76
Non-GHG emissions	0.40	0.42	0.43	0.46	0.49	0.50
Security of supply	NA[a]	NA[a]	NA[a]	0.99	0.99	0.99
Social cost	27.02	28.41	29.14	19.12	19.15	19.17

[a]Not applied

Table 4.8 shows that the social cost per GJ of biodiesel is much higher than that of diesel due to its higher private cost component. The higher external costs related to emissions and the security of supply lead to the higher external cost for diesel, while the positive effect of land use change due to jatropha plantation achieves an emission saving or an external benefit for biodiesel. Even with this external benefit, the social cost of biodiesel is much higher than that of diesel. These findings hold for the three discount rates, and the differences are larger at the higher discount rates. For instance, the social cost of biodiesel per GJ is 52.0 % higher than that of diesel at a discount rate of 10 %.

Regarding the external costs of biodiesel production, the non-GHG emissions are mostly caused by the use of chemical fertilisers and diesel for seed transportation. For the external cost reduction, this finding encourages farmers to minimise their use of chemical fertilisers and the biodiesel plants to better locate surrounding the feedstock areas so as to shorten the transport distance and thus the amount of diesel use for transporting dried seed.

In terms of a functional unit, the biodiesel substitution for diesel is not cost-effective for all scenarios (Table 4.9). At a discount rate of 4 %, the biodiesel substitution for diesel in the form of B5 would increase social cost of 0.01 \$ km^{-1} or 28.6 % of the social cost per functional unit compared to diesel (see columns 5 and 6 in Table 4.9) for scenario S7. Searching for break-even points, the zero cost difference is found at 1.4 % and 2.8 % lower fuel consumption of B5 and B10 compared to diesel, respectively, at a discount rate of 4 %, provided that other factors are constant. This means that the biodiesel substitution for diesel in the form of B5 or B10 would be cost-effective if the fuel consumption of B5 and B10, in terms of L km^{-1} compared to diesel, would decrease by more than 1.4 % and 2.8 % for B5 and B10, respectively, at a discount rate of 4 %.

Similar results are found at discount rates of 8 % and 10 %. The biodiesel substitution for diesel is not cost-effective for all scenarios. In view of the break-even points, the higher discount rate requires lower fuel consumption of biodiesel

Table 4.9 Cost-effectiveness of biodiesel and diesel

Scenarios	Social cost ($ GJ^{-1})	Fuel efficiency ($GJ\ km^{-1}$)	Social cost ($ km^{-1})		Cost difference	
			Biodiesel	Diesel	($ km^{-1})	(%)
	(1)	(2)[a]	(3) = (1) × (2)	(4) = (1) × (2)	(5) = (3) – (4)	(6) = (5) × 100:(4)
At a discount rate of 4 %						
Diesel	19.12	0.0019		0.04		
Biodiesel						
S7 (B5, 0 %)	27.02	0.0018	0.05	0.04	0.01[b]	28.58
S8 (B5, +5 %)	27.02	0.0370	1.00	0.04	0.96[c]	2600.22
S9 (B10, 0 %)	27.02	0.0018	0.05	0.04	0.01	28.58
S10 (B10, +5 %)	27.02	0.0034	0.09	0.04	0.05	145.47
At a discount rate of 8 %						
Diesel	19.15	0.0019		0.04		
Biodiesel						
S7 (B5, 0 %)	28.41	0.0018	0.05	0.04	0.01	34.98
S8 (B5, +5 %)	28.41	0.0370	1.05	0.04	1.01	2734.64
S9 (B10, 0 %)	28.41	0.0018	0.05	0.04	0.01	34.98
S10 (B10, +5 %)	28.41	0.0034	0.10	0.04	0.06	157.69
At a discount rate of 10 %						
Diesel	19.17	0.0019		0.04		
Biodiesel						
S7 (B5, 0 %)	29.14	0.0018	0.05	0.04	0.01	38.31
S8 (B5, +5 %)	29.14	0.0370	1.08	0.04	1.04	2804.43
S9 (B10, 0 %)	29.14	0.0018	0.05	0.04	0.01	38.31
S10 (B10, +5 %)	29.14	0.0034	0.10	0.04	0.06	164.04

[a]These are the figures in column 8 in Table 4.3 divided by 1,000
[b]A plus sign means cost-ineffectiveness
[c]The high cost-ineffectiveness in S8 due to low contribution of biodiesel to the blend B5

blends in terms of L km^{-1} to achieve the cost-effectiveness of the biodiesel substitution for diesel. For instance, the fuel consumption of B5 and B10 compared to diesel would decrease more than 1.4 % and 2.8 % compared to diesel, respectively, at a discount rate of 4 %; these figures are, respectively, 1.9 % and 3.7 % at a discount rate of 10 %. However, these fuel consumption levels of biodiesel blends have not been achieved in reality.

Conclusions

In this study, the social costs of biofuels and fossil fuels have been comprehensively compared: ethanol and gasoline, and biodiesel and diesel for an FU of 1 km of vehicle transportation with a focus on the blends of E5 and E10 for ethanol and B5 and B10 for biodiesel in Vietnam.

In terms of per MJ, the social costs of biofuels are higher than those of their alternative fossil fuels due to higher private cost components. With the consideration of fuel efficiency in transportation, different results are obtained. The ethanol substitution for gasoline in the form of E5 and E10 saves 0.02 \$ km^{-1} or 33.4 % of social cost per km of vehicle transportation compared to gasoline if the fuel consumption of E5 and E10 in terms of L km^{-1} is equal to the fuel consumption of gasoline at a discount rate of 4 %. The lower the fuel consumption of E5 and E10 compared to gasoline, the higher the achievement of this saving. The biodiesel substitution for diesel in the form of B5 or B10 remains not cost-effective if the fuel consumption of B5 and B10 remains the same or 5 % higher compared to diesel.

Examining the cost-effectiveness of biofuels under efficiency levels of blends, the fuel consumption of blends required to make biofuel cost-effective compared to fossil fuels has been identified. For ethanol to be cost-effective, the fuel consumption of E5 and E10 must not exceed, in terms of L km^{-1}, by more than 1.7 % and 3.5 %, respectively, in comparison with gasoline, assuming a discount rate of 4 %. For the cost-effectiveness of biodiesel, the fuel consumption of B5 and B10 compared to diesel would need to decrease by more than 1.4 % and 2.8 % for B5 and B10, respectively, at a discount rate of 4 %.

It can be concluded that the cost-effectiveness of using biofuel in comparison to fossil fuel depends on the efficiency of biofuel production and blended fuel combustion. To develop a sustainable biofuel market in Vietnam, further investments will be needed for improving the efficiency of biofuel production and blended fuel combustion.

References

Agarwal D, Agarwal AK (2007) Performance and emissions characteristics of Jatropha oil (preheated and blends) in a direct injection compression ignition engine. Appl Therm Eng 27 (13):2314–2323

Ajanovic A, Haas R (2010) Economic challenges for the future relevance of biofuels in transport in EU countries. Energy 35(8):3340–3348

Aliyu B, Shitanda D, Walker S, Agnew B, Masheiti S, Atan R (2011) Performance and exhaust emissions of a diesel engine fuelled with Croton megalocarpus (musine) methyl ester. Appl Therm Eng 31(1):36–41

Argonne National Laboratory (2011) The greenhouse gases, regulated emissions, and energy use in transportation (GREET) computer model 1.8d., Argonne National Laboratory. http://greet.es.anl.gov Downloaded on 5 Dec 2011

Biograce (2011) Biograce project: harmonised calculations of biofuel greenhouse gas emissions in Europe Biograce, Utrecht. www.biograce.net. Downloaded on 5 Dec 2011

Buyukkaya E (2010) Effects of biodiesel on a di diesel engine performance, emission and combustion characteristics. Fuel 89(10):3099–3105

Davis SC, Diegel SW, Boundy RG (2010) Transportation energy data book: Edition 29, Report no ORNL-6985, Contract No DE-AC05-00OR22725, Oak Ridge National Laboratory, Tennessee

Delgado R (2003) Comparison of vehicle emissions at European Union annual average temperatures from E10 an E5 petrol, Report No LM030411. Idiada Automotive Technology, Tarragona

Duer H, Christensen PO (2010) Socio-economic aspects of different biofuel development pathways. Biomass Bioenergy 34(2):237–243

Edwards R, Szekeres S, Neuwahl F, Mahieu V (2008) Biofuels in the European context: facts and uncertainties, Report no.: JRC44464. Joint Research Centre, European Commission, Petten

European Parliament (1996) Directive 96/69/EC of the European Parliament and of the Council, On the amending Directives 70/220/EEC on the approximation of the laws of the Member States relating to measures to be taken against air pollution by emissions from motor vehicles. Off J Eur Commun No L 282/64:64–67

European Parliament (1998) Directive 98/69/EC of the European Parliament and of the Council, On the relating to measures to be taken against air pollution by emissions from motor vehicles and amending Council Directive 70/220/EEC. Off J Eur Commun, No L 350/1:1–57

European Parliament (2003) Directive 2003/76/EC of the European Parliament and of the Council, On the amending Council Directive 70/220/EEC relating to measures to be taken against air pollution by emissions from motor vehicles. Off J Eur Union, No L 206/29:29–30

European Parliament (2007) Regulation (EC) No 715/2007 of the European Parliament and of the Council, On type approval of motor vehicles with respect to emissions from light passenger and commercial vehicles (Euro 5 and Euro 6) and on access to vehicle repair and maintenance information. Off J Eur Union, No L 171/1:1–16

European Commission (2010) Decision 2010/335/EU, On guidelines for the calculation of land carbon stocks for the purpose of Annex V to Directive 2009/28/EC. 2010.06.10. Off J Eur Union L 151:19–41

European Parliament (2009) Directive 2009/28/EC of the European Parliament and of the Council, On the promotion of the use of energy from renewable sources and amending and subsequently repealing Directives 2001/77/EC and 2003/30/EC. Off J Eur Union L 140:16–62

Eyidogan M, Ozsezen AN, Canakci M, Turkcan A (2010) Impact of alcohol-gasoline fuel blends on the performance and combustion characteristics of an SI engine. Fuel 89(10):2713–2720

General Statistics Office of Vietnam (2011) Agriculture, forestry, and aquiculture, General Statistics Office of Vietnam, Hanoi. http://www.gso.gov.vn/default.aspx?tabid=427&idmid=3. Downloaded on 5 Dec 2011

Global Finance (2011) European Union Economic Report: GDP data and GDP forecasts; economic, financial and trade information. http://www.gfmag.com/gdp-data-country-reports/631-the-european-union-gdp-economic-report.html#axzz1UDSH9sxx

Gnansounou E, Dauriat A, Villegas J, Panichelli L (2009) Life cycle assessment of biofuels: energy and greenhouse gas balances. Bioresour Technol 100(21):4919–4930

Government of Vietnam (2007) Decision No. 177/2007/QD-TTg of Government of Vietnam: approval of the scheme on development of biofuel up to 2015 and a vision to 2025, Hanoi Vietnam

Government of Vietnam (2010) Document No. 4377/VPCP-KGVX of Government of Vietnam: implementation of the scheme on development of biofuel up to 2015 with a vision to 2025, Hanoi Vietnam

Government of Vietnam (2011) Decision No. 49/2011/QD-TTg of Government of Vietnam, roadmap for application of emission standards to cars, motorcycles manufactured assembled and newly imported, Hanoi Vietnam

Greene DL (ed) (2007) Modelling the oil transition. In: Proceedings of the DOE/EPA workshop on the economic and environmental implications of global energy transitions, February 2007, Oak Ridge National Laboratory, Tennessee. http://www.afdc.energy.gov/pdfs/epa_ethanol.pdf. Downloaded 5 Dec 2011

Hu Z, Tan P, Yan X, Lou D (2008) Life cycle energy, environment and economic assessment of soybean-based biodiesel as an alternative automotive fuel in China. Energy 33(11):1654–1658

IEA (2011a) Statistics CO_2 emissions from fuel combustion, International Energy Agency, Paris. http://www.iea.org/co2highlights/. Downloaded on 5 Dec 2011

IEA (2011b) World energy outlook 2011. International Energy Agency, Paris

IPCC (2006) IPCC guidelines for national greenhouse gas inventories. In: Eggleston HS, Buendia L, Miwa K, Ngara T, Tanabe K (eds) Agriculture, forestry and other land use, vol 4. Inventories Programme Institute for Global Environmental Strategies, Hayama

Jaeger WK, Egelkraut TM (2011) Biofuel economics in a setting of multiple objectives and unintended consequences. Renew Sustain Energy Rev 15(9):4320–4333

Jindal S, Nandwana BP, Rathore NS (2010) Comparative evaluation of combustion, performance, and emissions of Jatropha methyl ester and Karanj methyl ester in a direct injection diesel engine. Energy Fuel 24(3):1565–1572

Jung A, Dörrenberg P, Rauch A, Thöne M (2010) Biofuels – at what cost? Government support for ethanol and biodiesel in the European Union – 2010 Update. Report No 978-1-894784-40-5, International Institute for Sustainable Development, Geneva

Kovacevic V, Wesseler J (2010) Cost-effectiveness analysis of algae energy production in the EU. Energy Policy 38(10):5749–5757

Le MD (2010) Cassava feedstock demand for biofuel production in Vietnam. In: Proceedings of the National workshop on Policy for biofuel development in Vietnam, 2010 Dec. 9, Hanoi, Institute of Policy and Strategy for Agriculture and Rural Development of Vietnam, Hanoi

Le AT, Tuan PM, Vinh TQ, Luong NT (2009) Impacts of using gasohol E5 and E10 on performance and exhaust emissions of in-used motorcycle and car in Vietnam. In: Proceedings of the world alternative energy sciences, 2009 March 5–8, Bangkok

Le TL, van Ierland EC, Zhu X, Wesseler J (2013) Energy and greenhouse gas balances of cassava-based ethanol. Biomass Bioenergy 51:125–135

Leather J (2009) Clean Air Initiative for Asian Cities Center team. Rethinking transport and climate change, ADB sustainable development working paper series no. 10. Asian Development Bank, Manila

Lechón Y, Cabal H, de la Rúa C, Caldés N, Santamaría M, Saez R (2009) Energy and greenhouse gas emission savings of biofuels in Spain's transport fuel: the adoption of the EU policy on biofuels. Biomass Bioenergy 33(6–7):920–932

Leiby PN (2007) Estimating the energy security benefits of reduced US Oil Imports. Tennessee (USA), Report No ORNL/TM-2007/028, Oak Ridge National Laboratory, Tennessee

Lin YC, Tsai CH, Yang CR, Wu CHJ, Wu TY, Chien GPC (2008) Effects on aerosol size distribution of polycyclic aromatic hydrocarbons from the heavy-duty diesel generator fueled with feedstock palm-biodiesel blends. Atmos Environ 42(27):6679–6688

Mailbach M, Schreyer C, Sutter D, van Essen HP, Boon BH, Smokers R et al (2008) Handbook on estimation of external cost in the transport sector: internalisation measures and policies for all external cost of transport, Report No.: 07.4288.52, CE Delft, Delft

MARD (2008) Decision No.1842/2008/QD-BNN-LN of Ministry of agriculture and rural development of Vietnam: approval of the scheme: research, development and usage products of Jatropha Curcas L. in Vietnam in the period from 2008 to 2015 and a vision to 2025, Hanoi Vietnam

MARD (2010) Official Dispatch No.3435/BNN-TCLN of Ministry of agriculture and rural development of Vietnam, Assessment on the status of jatropha curcas L. development in Vietnam, Hanoi Vietnam

MARD (2011) Official Dispatch No. 706/TTr-BNN-KH of Ministry of agriculture and rural development of Vietnam, Planning for development of agriculture, forestry and fishery in Vietnam to 2020 and a vision to 2030

Nguyen KQ (2008) Internalizing externalities into capacity expansion planning: the case of electricity in Vietnam. Energy 33(5):740–746

Nguyen AP (2010) Biofuel development from cassava and food security in Vietnam. In: Proceedings of the National workshop on Policy for biofuel development in Vietnam, 2010 Dec 9, Hanoi, Institute of Policy and Strategy for Agriculture and Rural Development of Vietnam, Hanoi

Nguyen TTL, Gheewala SH, Garivait S (2007) Energy balance and GHG-abatement cost of cassava utilization for fuel ethanol in Thailand. Energy Policy 35(9):4585–4596

Ozsezen AN, Canakci M (2011) Performance and combustion characteristics of alcohol-gasoline blends at wide-open throttle. Energy 36(5):2747–2752

Rajagopal D, Zilberman D (2007) Review of environmental, economic and policy aspects of biofuels, Policy research working paper 4341. World Bank, Washington, DC

Reading AH, Norris JOW, Feest EA, Payne EL (2002) Ethanol emissions testing, Report No E&E/DDSE/02/021. AEA Technology Plc, Harwell, Oxfordshire, UK

Reijnders L (2011) The life cycle emission of greenhouse gases associated with plants oils used as biofuel. Renew Energy 36(2):879–880

Roayaei E, Taheri K (2009) Test run evaluation of a blend of fuel-grade ethanol and regular commercial gasoline: its effect on engine efficiency and exhaust gas composition. Clean Technol Environ Policy 11(4):385–389

Schipper L, Fabian H, Leather J (2009) Transport and carbon dioxide emissions: forecasts, options analysis, and evaluation, ADB sustainable development working paper series no. 9. Asian Development Bank, Manila

Schippers L, Cordeiro M, Liska R, Le AT, Orn H, Wei-shiuen NG (2008) Measuring the invisible: quantifying emissions reductions from road transport solutions – Hanoi case study. World Resources Institute, Washington DC

University of Stuttgart (2008) Cost assessment for sustainable energy systems (CASE), WP2 report on methodology for estimating external costs to human health crops and materials, Institute of Energy Economics and the Rational Use of Energy, University of Stuttgart. www.feem-project.net/cases/documents/2ExternalCosts_per_unit_emission_080821.xls

USEPA (2002) A comprehensive analysis of biodiesel impacts on exhaust emissions, Report No EPA420-P-02-001. Environmental Protection Agency, Washington, DC

USEPA (2011) Clean alternative fuels: ethanol. (USA): Environmental Protection Agency, Washington, DC

Wang Z, Calderon MM, Lu Y (2011) Lifecycle assessment of the economic, environmental and energy performance of Jatropha curcas L. biodiesel in China. Biomass Bioenergy 35 (7):2893–2902

World Atlas (2011) Countries of the world, Texas. http://www.worldatlas.com/aatlas/populations/ctypopls.htm

Chapter 5
Economic Analysis of Rice Straw Management Alternatives and Understanding Farmers' Choices

Cheryll C. Launio, Constancio A. Asis, Rowena G. Manalili, and Evelyn F. Javier

Abstract The negative effects of open-field rice straw burning on the environment and human health are well documented in local and international literature. This research project assesses the environmental consequences of rice straw burning and other straw management practices in terms of greenhouse gas (GHG) emissions, particularly that of methane (CH_4) and nitrous oxide (N_20). It also evaluates the cost-effectiveness of adopting selected rice straw management alternatives. On a per hectare basis, considering a time horizon of 5 years with associated assumptions on cost savings and secondary benefits, incorporating stubble more than 30 days before crop establishment and incorporating composted rice straw in the field yielded the lowest cumulative CH_4 and N_2O emissions. The study found that the most cost-effective option for farmers is to incorporate stubble and straw in the soil more than 30 days before crop establishment. This option is followed by rapid straw composting and incorporation of rice straw into the field. The study recommends looking for alternative uses of rice straw and finding ways to reduce the cost of collection and transportation of rice straw, coupled by the strict enforcement of laws banning rice straw burning.

Keywords Philippines • Rice straw • Cost-effectiveness analysis • Multinomial logit • GHG emission

C.C. Launio (✉) • R.G. Manalili
Socioeconomics Division, Philippine Rice Research Institute (PhilRice), Muñoz, Nueva Ecija, Philippines
e-mail: ccasiwan@yahoo.com; rg.manalili@philrice.gov.ph

C.A. Asis • E.F. Javier
Soils and Plant Physiology Division, Philippine Rice Research Institute (PhilRice), Muñoz, Nueva Ecija, Philippines
e-mail: ca.asis@philrice.gov.ph; ef.javier@philrice.gov.ph

D. James, H.A. Francisco (eds.), *Cost-Benefit Studies of Natural Resource Management in Southeast Asia*, DOI 10.1007/978-981-287-393-4_5

Introduction

Background

The practice of open-field burning of rice straw (including both 'stubbles' – the lower part of the rice plant left in the field after harvesting – and 'straw', the loose straw output after threshing) is a major source of air pollutants (Gupta et al. 2004; Wassman and Dobbermann 2006; Tipayarom and Oanh 2007) including greenhouse gases (GHG), particulate matter and other elements such as dioxins and furans that impact human health (Torigoe et al. 2000; Gadde et al. 2009). The Philippine Department of Science and Technology (DOST) study even reports that burning of rice straw and other agricultural waste contributes more dioxins and furans to air and land than vehicle emissions (DOST 2006). Other studies also show that rice straw burning causes loss of major nutrients in the soil: almost complete nitrogen (N) loss, phosphorous (P) losses of about 25 %, potassium (K) losses of 20 % and sulphur (S) losses of 5–60 % (Dobbermann and Fairhurst 2002).

Given these negative effects of open-field rice straw burning on the environment and human health as documented in local and international literature, farmers have been encouraged to refrain from burning rice straw and adopt more environment- and human-friendly rice straw management practices. In fact, both the existing solid waste management law in the Philippines (RA 9003) and the Philippine Clean Air Act of 1999 prohibit in principle open-field burning which includes burning of rice straw. Ordinances specifically against rice straw burning have also been passed in some provinces and towns.

Scientists and researchers, on the other hand, have studied the potential impacts of various rice straw management technologies as alternatives to rice straw open-field burning. Crop residues properly returned to the soil can maintain or enhance soil quality and productivity through favourable effects on soil properties and life-support processes (Lal 1995). Continuous rice stubble and straw incorporation into the soil returns most of the nutrients and helps to conserve soil nutrient reserves in the long term (Dobbermann and Fairhurst 2002). Eagle et al. (2000) concluded that the retention of straw in rice fields can result in increased soil N supply, suggesting that nitrogen fertiliser application rates can be reduced. Javier (2009) also reported that several studies have proven that continuous rice straw incorporation can build up potassium reserves which are essential for plant vigour.

However, literature also recognises that the incorporation of straw and stubble into wet soil (during ploughing) also results in temporary immobilisation of N and a significant increase in methane (CH_4) emission from rice paddies, a practice that contributes again to GHG emissions (Dobbermann and Fairhurst 2002). The quantity of methane emission varies widely depending on the rice straw-based nutrient management practice used (Dobbermann and Fairhurst 2002; Wassman and Vlek 2004; Vibol and Towprayoon 2010). Hence, rice straw management can significantly impact GHG emissions from rice fields but also can provide a mitigation option. Using the proper rice straw-based nutrient management options may result in soil rejuvenation and environmental conservation (Javier 2009).

In practice, many farmers still burn their rice straw or, if not, incorporate it into submerged soil during ploughing. Local governments are also unable to implement the law especially with the large volume of rice straw produced each year in rice-producing regions. Rice straw burning thus continues to contribute to air pollution, compounding the already high methane emission levels inherent in irrigated rice-based farming systems. This raises the question of why farmers choose to burn their rice straw and which straw management alternatives are cost-effective to farmers and to society as a whole.

Aims of the Study

The study aimed to assess the environmental consequences of rice straw burning and evaluate the adoption and cost-effectiveness of selected rice straw management alternatives. Specifically, it:

- Estimated potential environmental impacts in CO_2-equivalent emissions of existing and potential rice straw management alternatives to open-field burning
- Compared selected rice straw management alternatives to open-field burning using cost-effectiveness analysis
- Determined factors and constraints driving adoption of rice straw management practices

Methodology

Data and Information Sources

Both secondary information and primary data were used in the study. Five-year (2006–2010) average data on area harvested, yield and production were obtained from the Bureau of Agricultural Statistics. Other assumptions used in the estimation were obtained from existing literature and previous studies in the Philippines and from results of key informant surveys and technical expert consultations. Household surveys were conducted in four major rice-producing provinces, namely: (1) Nueva Ecija, top producer; (2) Leyte, to represent the Visayas and a province with an existing provincial ordinance banning rice straw burning; (3) North Cotabato, largest rice producer in Mindanao; and (4) Ilocos Norte, to represent a province where there is an existing rice straw market. The questionnaire elicited information on rice stubble and straw management practices, perceptions of rice-based farming households on the environmental impacts of rice straw burning and other rice straw management practices. Two focus group discussions (FGDs) were also conducted in each of the four study provinces, covering farmers and farmer-leaders in irrigated and rainfed rice ecosystems. The purpose was to have thorough discussions on the perceived environmental impacts of rice straw burning and the existing and potential policy and institutional options to reduce the negative impacts and promote the positive impacts.

A multistage stratified area probability sample design was used in the household survey. The primary sampling unit (PSU) was the barangay (village) with the ecosystem (irrigated/rainfed) as a stratification variable. Ten barangays were randomly sampled from the entire population of rice-producing barangays in each province. The number of irrigated and rainfed barangays was prorated depending on secondary data of rice area harvested in the province. Ten farming households were randomly sampled in each sample barangay for a total of 100 household samples per province.

Data Analyses

GHG Emission Analysis

To estimate the GHG emissions from rice straw open burning and other straw management practices, methodologies and guidelines set by the Intergovernmental Panel on Climate Change (IPCC) Guidelines (2006) were used. Thus, while CO_2 emission is affected by straw management practice, CO_2 emissions from biomass burning were not included since the carbon released during the combustion process was assumed to be reabsorbed by the vegetation during the next growing season (IPCC 2006).

The following equation was used to determine the annual quantity of rice straw or rice stubble per hectare:

$$Q = \sum\sum P_{ij}\mathrm{SGR}_{ijj} \quad (5.1)$$

where Q is the annual quantity of rice straw or rice stubble (tons/ha/year), P_{ij} is the rough rice production (tons/ha) in season i and rice ecosystem j and SGR_{ij} is the straw to grain ratio or stubble to grain ratio in season i and rice ecosystem j.

Since in the Philippines the management of rice stubble usually differs from rice straw, the straw-grain ratio was differentiated from the stubble-grain ratio. Based on total biomass and grain yield data from an average of four WS and five DS experimental data in the Philippines by Corton et al. (2000) and stubble to straw ratio derived from special sampling at PhilRice by one of the co-authors, straw-grain ratios of 0.74 and 0.36 were derived and used and stubble-grain ratios of 1.21 and 0.59 for WS and DS, respectively. Equation 5.2 below was then used to calculate GHG emissions from rice straw burning:

$$E_a = (Q)(f_{Co})(EF_a) \quad (5.2)$$

where a = type of GHG, E_a = emission of a in kg/ha/year, f_{Co} = combustion factor and EF_a = emission factor of a in g/kg of dry straw.

Table 5.1 GHG emission factors for different rice straw management practices

Rice straw/stubble management practice	Name of pollutant	EF	Unit	Reference
Open burning	CH_4	1.2	g/kg$_{dryfuel}$	Based on EFs compiled by Gadde et al. (2009)
	N_20	0.07	g/kg$_{dryfuel}$	
	Combustion factor	0.8		IPCC Guidelines (2006)
Scattering and incorporation of rice stubble and straw in the soil (wet condition)	CH_4 (WS)	129.77	kg/ton yield	Corton et al. (2000) (Incremental CH_4 due to straw incorporation)
	CH_4 (DS)	36.99	kg/ton yield	
	Baseline EF for continuously flooded fields without organic amendments	1.3		IPCC Guidelines (2006)
	Conversion factor for rice straw amendment	1.0 for straw incorporated <30 days before cultivation; 0.29 for straw incorporated >30 days before cultivation		IPCC Guidelines (2006)
	Scaling factor to account for differences in water regime during cultivation period	0.78 for irrigated		IPCC Guidelines (2006)
		0.27 for rainfed		
	Scaling factor to account for differences in water regime in the preseason before cultivation period	1 for irrigated (<180 days); 1.22 for rainfed		IPCC Guidelines (2006)
Composting and incorporation	CH_4 (WS)	13.37	kg/ton yield	Corton et al. (2000) (Incremental CH_4 due to incorporated compost)
	CH_4 (DS)	2.1	kg/ton yield	
	Conversion factor for rice straw compost amendment	0.05		IPCC Guidelines (2006)
Rice straw used as animal feed	CH_4	10,000–20,000 gCH_4/ton dw		Singhal et al. (2005) as cited by Truc (2011)

For incorporation in the soil of rice straw and rice straw compost, the IPCC Guidelines were used for rice cultivation and emission from managed agricultural soils, adjusting the conversion factors for organic amendments. The estimates were compared with results using GHG emission changes from experiments conducted in the Philippines by Corton et al (2000) and considered the more conservative estimate. Table 5.1 shows the summary of emission factors and other parameters used and corresponding references.

In the analysis of GHG emissions, the study focused on estimating the incremental change in the GHG emissions from paddy areas as a result of rice straw management alternatives, not the GHG emissions from paddy rice systems per se.

Rice Management Options

Rice straw management options available to households engaged in rice production are shown in Table 5.2. The various alternatives were analysed on a per hectare basis considering the potential effects on GHG emission over a 5-year period. The base case scenario is the current major practice where rice stubble is incorporated into the soil during the first land preparation activity in a flooded condition, and rice straw is burned at the threshing site. Based on the household survey, the average duration from the start of land preparation to crop establishment is 21 days. The baseline case accordingly assumes that stubble is incorporated less than 30 days before cultivation.

While power generation and biofuel production are potential uses of rice straw removed from the field, a full-scale analysis was beyond the scope of the study. The analysis focuses only on the potential impact on GHG emissions in the paddy sector if stubble and straw are removed from the field.

Cost-Effectiveness Analysis (CEA)

CEA was employed to compare the selected rice straw management alternatives shown in Table 5.2. The analysis entailed the following steps:

1. Based on incremental GHG emission analysis, the net reduction in total CO_2-equivalent (CO_2-eq.) emission was determined if farmers switch from the baseline scenario to each management option, fulfilling the CO_2-eq. abatement capacity for each alternative. The total global warming potential (GWP) was estimated by multiplying the estimated emissions of each kind of GHG *(i)* from

Table 5.2 Rice straw management options evaluated for emission and economic analysis

	Rice straw management options	Stubble	Straw	Timing of incorporation
M1	Late stubble incorporation and straw burning (baseline)	Incorporated	Burned	<30 days before cultivation
M2	Late stubble and straw incorporation	Incorporated	Scattered and incorporated	<30 days before cultivation
M3	Late stubble incorporation and straw removal for animal feed	Incorporated	Removed from field for animal feed	<30 days before cultivation
M4	Early stubble and straw incorporation	Incorporated	Scattered and incorporated	>30 days before cultivation
M5	Late stubble incorporation and straw compost incorporation	Incorporated	Rapidly composted and incorporated as compost	<30 days before cultivation
M6	Early stubble incorporation and straw compost incorporation	Incorporated	Rapidly composted and incorporated as compost	>30 days before cultivation
M7	Stubble and straw composting and straw compost incorporation	Removed from field, composted with the straw and incorporated	Rapidly composted and incorporated as compost	>30 days before cultivation

paddy rice systems under each rice straw management practice (j) by its corresponding GWP, as in Equation (5.3) below:

$$\text{Total GWP}_j = \sum \left(\text{GHG}_{ij} \times \text{GWP}_{ij}\right) \tag{5.3}$$

where i is the type of GHG such as CH_4 or N_20 and GWP is an internationally accepted scale of equivalence for other GHGs in units of tons of CO_2-eq. (Lv et al. 2010). The GWP values used were $CO_2 = 1$, $CH_4 = 23$ and $N_20 = 296$ (IPCC 2006) measured in metric ton CO_2-eq. per year per hectare.

2. The cost of each rice straw management option expected to result in changes in GHG emission was estimated. It included the incremental operating and maintenance cost for the farmer implementing the management option as well as the secondary benefits due to changes in straw management practice. A 5-year period was considered in the cost analysis of the options. The shadow wage rate used was 60 % of the market wage rate as suggested by the Philippine NEDA (2004).
3. The cost-effectiveness or abatement cost per ton of CO_2-eq. was calculated and compared for each rice straw management alternative. The following generic

formula for abatement cost calculation similar to that cited in Davidson and van Essen (2009) was used:

$$\text{Cost-effectiveness} = \frac{I^{an} + \Delta_{O\&M} - \text{secondary benefits}}{\text{Annual reduction in } CO_{2}\text{-eq emissions}} \quad (5.4)$$

where I^{an} is the annuity of the investment cost (if applicable) and $\Delta_{O\&M}$ is the additional annual operating and maintenance cost. The formula also includes monetised secondary benefits (e.g. value of the composted rice straw) to be subtracted from the costs. The most cost-effective option is that with the lowest cost per reduction in tons CO_2-eq. (Php/ton CO_2-eq.). Costs in the study were calculated for a 5-year scenario to account for effects expected at around 5 years. A 7.5 % real discount rate was used, based on the discount rate suggested by the Philippine National Economic Development Authority (NEDA 2004) adjusted for inflation using the average from 2000 to 2010. The CEA was conducted from the farmers' perspective, without any assumed external social costs or benefits.

Results

GHG Emission Abatement Capacity of Options

Table 5.3 shows the estimated potential GHG emissions and global warming potential (tons CO_2-eq.) by rice straw management option for farmers, calculated per hectare for a 5-year period based on IPCC 2006 Guidelines, compiled factors from literature and various assumptions. For the baseline scenario, the total estimated emissions considering only CH_4 and N_2O emissions totalled around 51 tons CO_2-eq. per hectare over a 5-year period.

Results indicate that shifting from rice straw burning to rice straw incorporation will not necessarily reduce global warming potential if straw is incorporated less than 30 days before cultivation especially in a flooded condition after incorporation (M1 to M2). This is due to the higher CH_4 emission calculated. In the longer term, however, the increase in CH_4 emission due to straw incorporation can also be offset by the reduction in N_2O emission from potential reduction in the use of chemical fertilisers (Bird et al. 2002). At the same time, while there is less CH_4 emission in straw burning, the loss of nutrients from the straw might eventually lead to increased use of chemical fertilisers that can significantly increase N_2O emission. Although not considered in the GHG accounting here, Wang et al. (2011) also reported significant interactive effects of straw with nitrogen on N_2O emissions, i.e. straw amendment significantly reduced, on average, N_2O emissions from rice fields when they received nitrogen fertiliser.

Table 5.3 Estimated incremental GHG emissions (ton CO_2-eq. ha^{-1}) from rice fields for different rice straw management options

		Greenhouse gas emission (tons ha^{-1} $year^{-1}$)							
		Years 1–4			Year 5				
No.	Rice straw management option	CH4	N_2O	Total GWP	CH4	N_2O	Total GWP	5-year total (tons CO_2-eq.)	% change from baseline
M1	Late stubble incorporation and straw burning (baseline)	7.60	0.16	7.77	7.60	12.32	19.92	50.99	
M2	Late stubble and straw incorporation	12.55	0.08	12.63	12.55	−12.07	0.48	51.01	−0.05
M3	Late stubble incorporation and straw removal for animal feed	9.02	0.07	9.10	9.02	12.23	21.25	57.63	−13.03
M4	Early stubble and straw incorporation	4.45	0.08	4.53	4.45	−12.07	−7.62	10.51	79.38
M5	Late stubble incorporation and straw compost incorporation	7.61	0.48	8.09	7.61	−11.67	−4.06	28.28	44.53
M6	Early stubble incorporation and straw compost incorporation	2.83	0.48	3.31	2.83	−11.67	−8.84	4.39	91.39
M7	Stubble and straw composting and compost incorporation	0.34	1.10	1.44	0.34	−11.01	−10.67	−4.91	109.64

Note: Details of calculation can be obtained from the authors

Merely shifting from burning rice straw (M1) or changing the timing of the stubble and straw incorporation from less to more than 30 days before cultivation (M2 to M4) can reduce the incremental GHG emissions due to straw management by around 80 % largely because of its effect on CH_4 emission, assuming other factors are constant. A significant reduction of more than 90 % in total incremental GHG emissions is also expected if the rice stubble is incorporated more than 30 days after cultivation and the rice straw is removed from the soil and returned as rice straw compost. This is even more significant when the rice stubble is completely removed and returned as rice straw compost. Corton et al. (2000) reported from their experiments in the Philippines that addition of rice straw compost increased CH_4 by only 23–30 % as compared with the 162–250 % increase in emissions with the use of fresh rice straw.

When both stubble and straw are completely removed from the field and used for nonfarm uses such as power generation or bioethanol production, GHG emissions from paddy rice systems can also be potentially reduced. While the study could not afford to conduct a life cycle assessment of GHG emissions from using rice straw for power generation and bioethanol production, literature is available on this issue. Delivand et al. (2011) in the case of Thailand indicate that the life cycle GHG emission of the straw combustion process chain is 30 kg CO_2-eq. per ton of dry straw or 0.043 kg CO_2-eq./kWh, i.e. 0.613 kWh net electricity per kg straw. This considers both the logistics and grate boiler combustion. Using this as a factor and considering a scenario where the stubble is incorporated into the soil more than 30 days before cultivation, the approximate net GHG emission was estimated at 1,263 tons per year per hectare.

They also found that substituting the natural gas or coal fuels with rice straw fuels for power generation would result in a considerable fossil fuel savings and a lower GHG emission. For example, for the case of substituting natural gas with rice straw fuel, it is estimated that 0.368 ton CO_2-eq. will be avoided, while for imported coal, 0.683 ton CO_2-eq. can be avoided (Delivand et al. 2011). Using rice straw as raw material for power generation will thus avoid emissions of GHG and other air pollutant particulates when compared with open straw burning and methane emissions when compared with incorporation.

Regarding the use of rice straw for bioethanol production, a review by Cheng and Timilsina (2011) reported that all advanced biofuel technologies have the advantage of producing fuels with almost zero or very little net carbon dioxide emissions to the atmosphere. The same authors stated that the advantage of lignocellulosic materials such as agricultural residues is that these are abundant in most land areas and their generation does not have to compete for arable land against food and feed production. Using 0.28 L kg^{-1} as ethanol yield (Kim and Dale 2004), the approximate ethanol production per hectare is calculated at around 1,250 l per season. Assuming a 0.5 kg CO_2-eq. reduction per litre ethanol, the estimated net emission reduction is 625 kg per hectare.

Cost Estimates for Rice Stubble and Straw Management Options

Cost estimates were based on the incremental differences between the baseline practice of incorporating stubble during land preparation and burning straw and other stubble-straw management options. Some of the relevant cost savings and secondary benefits were expected to be incurred or realised after around 5 years of continuous practice. Based on existing literature, the total cost per hectare was calculated for a 5-year period as shown in Table 5.4. For most of the options, the rice stubble and straw will be incorporated into the soil but will vary in form and timing; the cost analysis largely revolves around changes in labour costs and some material costs in the case of composting. The expected changes in labour use associated with the management option and prevailing wage rate at Php200 per man-day were based on results of the FGDs and household survey conducted for the study. In the economic analysis, 60 % of the market wage rate was used as the shadow wage rate.

Removal of stubble and straw from the field requires significant labour cost so that at current prices and based on existing practices, composting of both the straw and stubble entails the highest cost followed by the options where the straw will be composted and incorporated into the soil as compost (see Table 5.4). Based on the household survey, manual cutting of rice stubble takes 20–21 man-days per hectare depending on the season. Collecting and gathering of the cut rice stubble was estimated to take another 7–10 man-days. For collecting and gathering piled rice

Table 5.4 Estimated costs for each rice straw management option (Php ha^{-1} year^{-1})

	Rice straw management option	Years 1–4			Year 5		
		WS	DS	Total	WS	DS	Total
M1	Late stubble incorporation and straw burning (baseline)	156	156	312	1,615	1,615	3,230
M2	Late stubble and straw incorporation	552	552	1,104	552	552	1,104
M3	Late stubble incorporation and straw removal for use as animal feed	876	756	1,632	2,335	2,215	4,550
M4	Early stubble and straw incorporation	552	552	1,104	552	552	1,104
M5	Late stubble incorporation and straw compost incorporation	3,134	1,729	4,863	3,134	1,729	4,863
M6	Early stubble incorporation and straw compost incorporation	3,134	1,729	4,863	3,134	1,729	4,863
M7	Stubble and straw composting and straw compost incorporation	10,654	7,188	17,842	10,654	7,188	17,842

Note: Details of calculation can be obtained from the authors

straw and hauling it onto the roadside ('roadsiding'), farmers estimated this to take about 12 man-days per hectare during DS and 14 man-days during WS. Rapid composting was considered, where rice straw is composted in a designated area on farmers' fields using inoculants such as manure, effective organism activated solution (EMAS) or *Trichoderma* spp. An alternative, common to developed countries and some Southeast Asian countries, is the use of combine harvester and baling machines, but these are capital intensive and their use is still relatively unpopular.

Cost-Effectiveness Analysis of Rice Straw Management Options

Based on the estimates of GHG emission and potential abatement of GHG emission on farmers' fields from Table 5.3 and the estimated cost for each management option in Table 5.4, Table 5.5 presents abatement costs per ton CO_2-eq. From among the options that result in net reduction in incremental CO_2-eq., shifting from the current farmers' practice of late stubble incorporation or incorporating less than 30 days before cultivation and straw burning (M1) to early incorporation of both stubble and straw incorporation in soil more than 30 days before cultivation (M4) appears to be the most cost-effective with a negative abatement cost (net benefit) at Php21 or around US$0.50 per ton of CO_2-eq. reduction. Results also suggest that incorporating the stubble more than 1 month before cultivation and rapidly composting the straw and applying it back to the field (M6) is also a cost-effective option for farmers at around Php300/ton CO_2-eq. It significantly mitigates GHG emissions while improving the soil condition; thus, even with the additional labour cost of piling and composting, it is the next most cost-effective abatement option.

Incorporating both the stubble and straw less than 30 days before cultivation (M2), on the other hand, results in a slight net increase in emission. The increase in cost for the option is lower than the secondary benefit of savings from fertiliser given the time horizon. The amount of potential savings if a ton of CO_2-eq. is not emitted is very significant. Incorporating stubble and removing straw for use as animal feed (M3) will also result in a net increase in emissions, but the straw value as animal feed is very significant assuming rice straw can be sold at Php5/bundle of 5 kg. Stubble and straw removal and application as compost significantly reduce emissions, but the abatement cost is more than Php1000 per ton of CO_2-eq. which may not be affordable to farmers.

It was beyond the scope of the study to conduct a full-scale economic feasibility analysis of using rice straw for power generation and biofuel production which are potential uses of rice straw. However, based on reviewed literature, second-generation lignocellulosic technologies have considerably more potential for avoiding many of the GHG emission and other environmental shortfalls and

Table 5.5 Estimated abatement costs (Php/ton CO_2-eq.) by rice straw management option

No	Rice straw management option	Total emission (tons CO_{2e})	Abatement relative to baseline (tons CO_{2e}-eq/ha)	Incremental abatement cost (Php/ha)	Secondary benefit (Php/ha)	Abatement cost[a] (Php/ton CO_2-eq.)	Abatement cost[b] (US$/ton CO_{2e}-eq.)
M1	Late stubble incorporation and straw burning (baseline)	50.99					
M2	Late stubble and straw incorporation	51.01	−0.02	1,172	2,033		
M3	Late stubble incorporation and straw removal for use as animal feed	57.63	−6.64	5,341	18,122		
M4	Early stubble and straw incorporation	10.51	40.48	1,172	2,033	−21	−0.50
M5	Late stubble incorporation and straw compost incorporation	28.28	22.71	16,380	2,033	632	14.87
M6	Early stubble incorporation and straw compost incorporation	4.39	46.60	16,380	2,033	308	7.25
M7	Stubble and straw composting and straw compost incorporation	−4.91	55.90	68,892	2,033	1,196	28.14

Note: Details of calculation can be obtained from the authors
[a]Values for M2 and M3 were not included since there is no net abatement
[b]1US$ = 42.50Php – average for 2011 (Source: www.nscb.gov.ph)

perform better in terms of energy efficiency (CGIAR Science Council 2008). The same report, however, stated that cost-effective second-generation biofuels are at least 10 years down the road and bio-refineries and bio-based economies are even further afield. Gadde et al. (2009) studied the overall potential and limitations of energy contribution and greenhouse gas mitigation for rice straw as a renewable energy source in India, Thailand and the Philippines. They found that the energy potential from rice straw in theory was different from the realisable energy due to the fact that residue collection efficiency or capacity and its transportation were challenging. Delivand et al.'s (2011) study on the feasibility of using rice straw residues for power generation in Thailand concluded that straw-based combustion facilities are financially feasible and profitable if the capacity of the power plant is 8 MWe or greater. In order to ensure secure fuel supply, they report that smaller-scale power plants, i.e. 8 and 10 MWe, are more practicable.

On the part of Filipino rice farmers, if the demand for rice straw is created through its use as raw material for power or bioethanol production plants, they may be able to supply the straw with a direct cost of around Php 5,200/ha for two seasons or Php 1.20/kg of straw considering the labour cost of gathering and moving to a storage area. Removing even the stubble, on the other hand, will cost an additional Php 8,000/ha so that the rice straw cost will more than double at approximately Php 3/kg loose straw.

Factors Explaining Farmers' Choices of Rice Straw Management Options

The previous analysis attempted to explore the potential of selected rice stubble and straw management practices that can be implemented by rice farmers in order to reduce the incremental non-CO_2 GHG emissions and estimated the costs associated with them. This section uses farm survey data in four major rice-producing provinces to ascertain factors explaining actual farmers' choices of straw management options. Since there are some options with no adopters, the choices are generalised to straw burning, incorporation and removal options.

Table 5.6 shows the results obtained from the estimated multinomial logit model, reported as relative risk ratios. The model is statistically significant based on the likelihood ratio chi-square criteria and overall correctly predicts 76 % of respondents in terms of their choice of rice straw management practice.

While season did not appear to be a significant factor affecting choice of straw management in the surveyed provinces, farm type was a significant factor with a 0.51 relative risk ratio. This means that if a rainfed farmer becomes an irrigated farmer, the ratio of the probability for him to choose to incorporate straw in soil relative to the probability to practise burning is expected to decrease by a factor of 0.51 given that the variables in the model are held constant. Irrigated farmers are also less likely to remove rice straw from the field as shown by the highly

Table 5.6 Results of multinomial logit model (baseline category: straw burning)

Variable	Straw incorporation			Straw removal		
	RRR	SE (RRR)	RRR	SE (RRR)		
Season dummy (1 – dry season)	1.35	0.33		1.02	0.32	
Farm type dummy (1– irrigated)	0.51	0.18	*	0.23	0.10	***
North Cotabato dummy	2.20	0.94	*	0.03	0.02	***
Leyte dummy	16.51	7.50	***	0.46	0.33	
Ilocos Norte dummy	5.11	2.29	***	5.44	2.84	***
Attendance in training (1 – attended training)	1.60	0.42	*	0.90	0.32	
Age	1.01	0.01		1.00	0.01	
Educational attainment	0.95	0.04		1.04	0.06	
No. of household members with age >13	1.25	0.10	***	1.38	0.14	***
Income in non-rice farming	1.00	0.00		1.00	0.00	***
Total area cultivated	0.97	0.07		0.67	0.10	***
Cow ownership (1 – owner)	2.50	0.91	***	4.41	1.78	***
Tenure status (1 – owner)	1.51	0.40		1.77	0.61	*
Distance from farm to house	0.87	0.07	*	0.79	0.11	*
Yield	1.00	0.07		1.07	0.09	
Perceptions						
Negative impacts of open-field burning	1.48	0.24	**	0.70	0.13	**
Positive effects of rice straw incorporation	1.78	0.41	***	1.08	0.31	
Negative effects of rice straw incorporation	0.72	0.11	**	0.86	0.17	
Awareness of environmental regulations	2.23	0.51	***	2.33	0.64	***
Attitudes towards incentives/market	0.36	0.05	***	0.37	0.07	***
Burning ordinance in province	1.17	0.32		0.47	0.18	**
Solid waste management programme in province	1.23	0.34		3.23	1.58	**

No. of observations = 687; *** $p < 0.01$; **$p < 0.05$; *$p < 0.10$
Log-likelihood: −390.81
Pseudo-R^2: 0.41

significant negative coefficient. A possible explanation is that intensive rice-rice cropping is largely practised in irrigated farms in the country so that the turnaround or interval period between rice crops is shorter and farmers have to dispose rice straw and stubbles more quickly. Flinn and Mariano (1984) in studying straw management practices in irrigated and rainfed environments concluded that stacking of rice straw for livestock consumption is higher in rainfed compared with irrigated areas. In relation to this, the model also shows that cow ownership is a significant variable related to choosing straw incorporation and straw removal over burning.

Relative to the largest rice-producing province of Nueva Ecija, farmers in North Cotabato, Leyte and Ilocos Norte are more likely to incorporate straw than burn when all other factors are held constant. For Leyte, for example, the risk for farmers incorporating relative to burning is higher by a factor of 17 given that other factors in the model are constant. For Ilocos Norte, the relative risk of farmers removing straw from the field relative to burning is higher by a factor of 5.

Considering economic factors, an increase in the number of adult household members who can be proxy for available family labour is positively related to choice of straw incorporation or straw removal relative to straw burning. This is consistent with the descriptive results that straw incorporation and removal are much more labour intensive than straw burning. Related to this, the total area cultivated is also significantly related with the choice of straw removal over straw burning possibly because the larger the total area cultivated, the more labour required in removing straw from the field. This result is consistent with the result of a binary analysis conducted using data from a separate national survey. Income from non-rice farming which can also be a proxy for household engagement in vegetable or other crop farming is a highly significant variable related to rice straw removal relative to burning. Rice straw has value as mulch to households engaged in non-rice farming especially onion and garlic production.

The increased distance from house to farm was found to be negatively related with the log odds of choosing straw incorporation and straw removal relative to burning. A possible explanation is the higher cost of collection and transportation when the farm is located farther from the house where they usually stack the rice straw for livestock feed or other purposes.

In terms of perception-related variables, the more farmers agreed with the statements on the negative effects of open-field burning on global warming and air pollution, the higher the relative risk ratio of choosing to incorporate straw relative to burning straw. An inconsistent result is that agreement with the statements on the adverse effects of open-field burning is negatively related to the choice of straw removal over burning. A possible explanation is that even while they agree with the negative impacts on air quality and climate, the additional cost of straw removal is much higher than their willingness to mitigate these perceived negative impacts. Also, a higher awareness of environmental regulations and policies is significantly and positively related to the choice of straw incorporation or straw removal relative to burning.

On the other hand, for the attitudinal statements on possible behaviour given incentives or increased demand of rice straw, farmers who require incentives are less inclined to practise straw incorporation or straw removal relative to burning unless such incentives are given. Finally, the presence or absence of a provincial ordinance against rice straw burning appears to be insignificantly related to the choice of farmers to practise straw incorporation; respondents in provinces with an ordinance banning rice straw burning are less likely to remove their rice straw. This result is probably because in Leyte province where there is a specific ordinance against rice straw burning, the more common practice is straw incorporation. The implementation of a provincial solid waste management programme is significantly and positively related to the practice of straw removal relative to straw burning.

Conclusions and Recommendations

This research project was generally aimed at assessing the environmental consequences of rice straw burning and other straw management practices and evaluating the cost-effectiveness and adoption of selected rice straw management alternatives. Based on the emission inventory considering the incremental GHG emissions from rice straw management practices, the estimated contribution of rice residues is around 16 M tons of CO_2-eq. considering only CH_4 and N_2O emissions. This is approximately 8 % of the total projected CO2-eq. emissions for 2008 based on the Philippines' first initial national communication on climate change. The current practice of incorporating rice straw/stubble into soil during land preparation in wet or flooded conditions less than a month before transplanting is the largest contributory rice residue management practice on a per year basis not considering potential long-term impacts.

On a per hectare basis calculated for a 5-year period based on IPCC 2006 Guidelines, compiled factors from literature and various assumptions, the total estimated emission level, considering only CH_4 and N_2O gases, is around 51 tons CO_2-eq. for the baseline scenario. This is not very different if the practice changes from straw burning to straw incorporation less than 30 days before crop establishment. Aside from GHG, rice straw burning, however, emits particulates that are known to be harmful to human health. Changing the timing of straw incorporation from less than 30 days to more than 30 days before crop establishment will significantly reduce the incremental GHG emissions in paddy fields due to straw management by around 80 %.

Economic analysis from the point of view of rice farmers indicated that merely shifting from incorporating stubble before transplanting and straw burning to incorporating both stubble and straw in the soil over 30 days before crop establishment appears to be the most cost-effective option with a negative abatement cost (i.e. benefit) of Php 21 (US$ 0.50) per ton CO_2-eq. reduction. Incorporating both rice stubble and straw less than a month before cultivation, on the other hand, appears to result in a slight net increase in emissions. Rapid composting and incorporating the rice straw compost in the field entail significantly higher additional costs, but significantly mitigate GHG emissions; hence, it is the next most cost-effective option. While a full-scale cost-effectiveness analysis of using rice straw for power generation and bioethanol production was not included in this study due to data limitations, creating higher rice straw demand given the same supply will result in higher economic value for rice straw which is favourable for rice-based farmers. Further study on the technical and economic feasibility of using rice straw for power generation and bioethanol production and the consequent GHG emissions from these options in the case of the Philippines is recommended.

Also, there is evident uncertainty about the rate of growth of emissions, their general effects and their local effects (GEF 1992). The results of this study, particularly on GHG emissions, are only indicative, and the cost-effectiveness analysis must be interpreted in relative terms. For rice-based farmers, there are other mitigation options such as water management and tillage options. Since affordability of technologies is a prime issue in the mitigation of GHG emissions, further study of the economics of these other mitigation options relative and complementary to rice straw management practices would provide broader information useful for farmers, policymakers and other rice stakeholders.

On the determinants of rice straw management practice or why farmers choose to burn, incorporate or remove rice straw, a mix of socio-economic, farm and awareness and attitude variables are prominent. Continually providing training on rice production to farmers is a particular policy direction that significantly relates to choice of straw incorporation over straw burning, as shown by the significantly positive coefficient of the training variable and perceptions of the benefits of straw incorporation. Increasing the demand for rice straw for other uses such as mulch is also another significant factor, as shown by the significance of non-rice farm income used as proxy for households engaged in vegetable and other non-rice farming.

While awareness of the environmental impacts of rice straw burning appears to be a significant positive factor for choosing straw incorporation over straw burning, farmers' choice of straw removal relative to burning may not be affected in the same manner owing to the significantly larger cost of straw removal. Options for reducing the cost of collection and transportation of rice straw such as mechanisation of harvesting and baling activities may reduce costs in the long run and increase the probability of the adoption of straw removal options including composting. Awareness of environmental laws and regulations appears to be a consistent positive factor that can increase the relative risk of choosing straw incorporation and removal over burning, so that increasing information campaigns and drives regarding these laws and strict implementation of sanctions are also recommended.

References

Bird J, Eagle A, Horwath W, Hair M, Zilbert E, Van Kessel C (2002) Long-term studies find benefits, challenges in alternative rice straw management. Calif Agric 56:69–75

CGIAR Science Council Secretariat (2008) Biofuels research in the CGIAR: a perspective from the science council. CGIAR Council Policy Statement on Biofuels Production, Rome, pp 1–34

Cheng J, Timilsina G (2011) Status and barriers of advanced biofuel technologies: a review. Renew Energy 36:3541–3549

Corton TMJB, Bajita FS, Gorospe RR, Pamplona CA, Asis R, Wassman RSL, Buendia LV (2000) Methane emission from irrigated and intensively managed rice fields in Central Luzon (Philippines). Nutr Cycl Agroecosyst 58:37–53

Davidson M, van Essen H (2009) Methodological issues related to assessing cost effectiveness of climate change abatement options. Paper produced as part of contract ENV.C.3/SER/2008/0053 between European Commission Directorate-General Environment and AEA Technology. www.eutransportghg2050.eu. Accessed 13 July 2010

Delivand M, Barz M, Garivait S (2011) Overall analyses of using rice straw residues for power generation in Thailand: project feasibility and environmental GHG impacts assessment. J Sustain Energy Environ, Special Issue:39–46

Department of Science and Technology (DOST) (2006) First national inventory of dioxins and furans study. Available at http://www.philstar.com/agriculture/350114/burning-rice-straw-agri-waste-threatens-environment. Accessed 13 July 2010

Dobbermann A, Fairhurst TH (2002) Rice straw management. Better Crops Int 16(Special Supplement):7–11

Eagle AJ, Bird JA, Horwath WR, Linquist BA, Brouder SM, Hill JE, van Kessel C (2000) Rice yield and nitrogen utilization efficiency under alternative straw management practices. Agron J 92:1096–1103

GEF (Global Environment Facility) (1992) Economic costs of carbon dioxide reduction strategies, Working paper series number 3. Accessed from http://www.scribd.com/doc/55417518/Economic-Costs-of-Carbon-Dioxide-Reduction-Strategies on 3 Feb 2012

Flinn JC, Mariano VP (1984) Organic matter and rice. International Rice Research Institute, Los Baños

Gadde B, Bonnet S, Menke C, Garivait S (2009) Air pollutant emissions from rice straw open field burning in India, Thailand and the Philippines. J Environ Pollut 157:1554–1558

Gupta P, Sahai S, Singh N, Dixit C, Singh D, Sharma C, Tiwari M, Gupta R, Garg S (2004) Residue burning in rice-wheat cropping system: causes and implications. Curr Sci 87 (12):1713–1717

Intergovernmental Panel on Climate Change (IPCC) (2006) IPCC guidelines for national greenhouse gas inventories. http://www.ipcc-nggip.iges.or.jp/public/2006gl/index.html. Accessed 21 Sept 2011

Javier E (2009) Rice straw-based nutrient management for irrigated lowland rice, PhilRice Technology Bulletin, No. 60, ISSN 0117–9799:20p

Kim S, Dale B (2004) Global potential bioethanol production from wasted crops and crop residues. Biomass Bioenergy 26:361–375

Lal R (1995) The role of residues management in sustainable agricultural systems. J Sustain Agric 5(4):51–78

Lv Y, Gu S, Guo D (2010) Valuing environmental externalities from rice wheat farming in the lower reaches of the Yangtze River. Ecol Econ 69:1436–1442

National Economic Development Authority (NEDA) (2004) ICC Project Evaluation Procedures and Guidelines. Accessed from http://www.neda.gov.ph/progs_prj/ICC/Latest_ICC_FormsJan_2004/ICC%20project%20evaluation%20procedures%20and20guidelines.pdf on 13 Feb 2012

Tipayarom D, Oanh N (2007) Effects from open rice straw burning emission on air quality in the Bangkok Metropolitan Region. ScienceAsia 33:339–345

Torigoe K, Hasegawa S, Numata O, Yazaki S, Matsunaga M, Boku N, Hiura M, Ino H (2000) Influence of emission from rice straw burning on bronchial asthma in children. Pediatr Int 42:143–150

Truc N (2011) Comparative assessment of using rice straw for rapid composting and straw mushroom production in mitigating greenhouse gas emissions in Mekong Delta, Vietnam and Central Luzon, Philippines. Unpublished PhD thesis, University of the Philippines at Los Baños, Philippines

Vibol S, Towprayoon S (2010) Estimation of methane and nitrous oxide emissions from rice fields with rice straw management in Cambodia. Environ Monit Assess 161:301–313

Wang J, Jia JX, Xiong ZQ, Khalil MAK, Xing GX (2011) Water regime nitrogen fertilizer-straw incorporation interaction: Field study on nitrous oxide emissions from a rice agroecosystem in Nanjing, China. Agricu Ecosyst Environ 141:437–446

Wassman R, Dobbermann A (2006) Greenhouse gas emissions from rice fields: what do we know and where should we head for? In: Proceedings of the 2nd joint international conference on sustainable energy and environment (SEE), 21–23 November 2006, Bangkok, Thailand. Kyoto University, Japan and the Joint Graduate School of Energy and Environment, Thailand

Wassman R, Vlek PLG (2004) Mitigating greenhouse gas emissions from tropical agriculture: scope and research priorities. Environ Dev Sustain 6(1–2):1–9

Chapter 6
Optimal Forest Management for Carbon Sequestration: A Case Study of *Eucalyptus urophylla* and *Acacia mangium* in Yen Bai Province, Vietnam

Nghiem Thi Hong Nhung

Abstract The study's main objective is to identify optimal management strategies for planted forests when carbon benefits are considered. The Faustmann model is extended to include multi-stands and spatial arrangement of forest stands of *Eucalyptus urophylla* and *Acacia mangium* in Yen Bai province in northern Vietnam. Farmers' current practice is to cut trees at age five, whereas the optimal rotation age of *E. urophylla* (when only timber has a market value) is 10 and 9 years for households and forest enterprises, respectively, at a 5 % discount rate. For *A. mangium*, the optimal rotation age is 13 years for both households and forest enterprises. The NPV is VND 16 million per hectare for households of both *E. urophylla* and *A. mangium*. For forest enterprises, the NPV is VND 57 million and VND 62 million per hectare for *E. urophylla* and *A. mangium*, respectively. Adding carbon values makes the optimal rotation age slightly shorter and the NPV higher. Similarly, taking into account the spatial arrangement of forest stands shortens the rotation age and improves the NPV. To encourage forest owners to lengthen the rotation age, the government can use a lump sum payment of carbon benefits at the beginning of the rotation. A planting cost subsidy will also help to narrow the actual and the optimal rotation age.

Keywords Vietnam • Faustmann model • *Acacia mangium* • Sensitivity analysis • Net present value (NPV)

Introduction

Background

Productive planted forests (as defined by FAO 2006) constitute approximately 13.1 % of the total forest area in Vietnam and have increased by 11.9 % per

N.T.H. Nhung (✉)
Department of Public Health, University of Otago, Wellington, New Zealand
e-mail: nhung.nghiem@otago.ac.nz

D. James, H.A. Francisco (eds.), *Cost-Benefit Studies of Natural Resource Management in Southeast Asia*, DOI 10.1007/978-981-287-393-4_6

annum during the period 2002–2006 (Forest Protection Dept 2007). In time of rapid climate change, managing planted forests for multiple-use purposes, i.e. timber production and carbon sequestration, is increasingly important. The global cost of a climate policy with forestry as an abatement option is $3.0 trillion cheaper than a policy without forestry (Tavoni et al. 2007). In addition, Vietnam is claimed to gain a lot from engaging in the international carbon market; however, the benefits of carbon sequestration have not been adequately studied in the management of productive planted forests (Bui and Hong 2006).

During the period 2007–2015, the area of productive planted forests will continue to expand as the result of the Five Million Hectare Afforestation Program (Vietnamese Government 2007a) and because of a large demand for wood domestically (Department of Forestry 2006). To increase the value of carbon sequestration while ensuring timber production, there is an urgent need to adopt an optimal management strategy for productive planted forests.

The optimal rotation age has been identified for some dominant tree species in Vietnam that maximizes the net present value from timber production. Nguyen and Doan (2005) showed that the rotation age of *hybrid Acacia* is 7–8 years for small-sized timber and 14–15 years for large-sized timber. For *Acacia mangium*, Nguyen et al. (2006) suggested that the optimal age should be longer than the rotation age that is applied by forest farmers. However, the optimal management strategy for planted forests when carbon sequestration is considered remains unknown in Vietnam.

In Yen Bai province, most of the fast-growing tree species in productive planted forests are cut at the age of 5 years (Nguyen et al. 2006). This cutting age is less than the optimal harvesting age that takes into account both timber production and carbon sequestration (Diaz-Balteiro and Rodriguez 2006; van Kooten et al. 1995). Moreover, the majority of forest farmers apply clear cutting harvesting practice (Bui and Hong 2006) that destroys habitat and causes serious loss of biodiversity (Pawson et al. 2006). These consequences are the result of forest farmers' daily basic needs, lack of investment capital (Nguyen et al. 2006) and no payment for environment services (Bui and Hong 2006) among other factors.

The optimal rotation age, for both single and multi-stand forests, when carbon value is included has been widely known in the international literature. The literature shows that carbon rotation length may differ from Faustmann rotation. With the growing concern over climate change, Englin and Callaway (1993) included carbon value into Faustmann rotation and found that the optimal rotation differs from the standard Faustmann rotation. Interestingly, they showed that the optimal rotation increases as the discount rate increases. They explained that carbon emission after harvest lengthens the optimal rotation, as the discount rate rises, rather than shortens it. Moreover, van Kooten et al. (1995) studied the effect of carbon taxes and subsidies on the optimal forest rotation age and concluded that the carbon optimal rotation is a little longer than the Faustmann rotation age. Diaz-Balteiro and Rodriguez (2006) employed a dynamic programming approach and showed that the optimal rotation age is sensitive to changes in the discount rate and carbon price. Gutrich and Howarth (2007) confirmed that when carbon storage brings benefits to society, optimal rotation ages are extended depending on the forest types.

Aim of Study

This study has applied the theoretical model of optimal forest management developed by Faustmann (1849), i.e. a stand-level model, and extended to include a multiple-stand forest or a forest-level model, in the Vietnamese context. This model also included spatial arrangement among forest stands and uncertainty analyses via simulation.

The study considered policy recommendations to further develop multiple-use forests in Vietnam. In particular, it is expected that the study could identify policy for the Ministry of Agriculture and Rural Development to achieve desired management strategies for different tree species and ownerships, to cope with a change in market conditions and to encourage carbon storage in Vietnam. The optimal forest management strategy can also be used to evaluate Clean Development Mechanism (CDM) projects in terms of their net present value and the amount of carbon sequestered.

Research Questions

The main research questions addressed in the study are the following:

1. Which pattern of planting and harvesting trees should a forest owner follow in order to maximize the net present value from timber harvesting and carbon sequestration?
2. What are the differences in the net present value (NPV) and the rotation age (T) between two optimal management strategies that consider or do not consider the value of carbon sequestration?
3. What are the differences in the net present value and the rotation age of the optimal management strategies for two dominant tree species: for households and for forest enterprises?
4. How does the optimal management strategy vary with a change in forest area, timber price, carbon price or discount rate, ceteris paribus?
5. What are the gaps in NPV and T between the current forest management strategy and the optimal strategy? What policy instruments are relevant to fill the gaps?

Methodology

The Model

Assume a forest consists of n stands with n equal to 1 for the Faustmann model and is greater than 1 for the forest-level model. The objective of the model is to maximize the net present value from harvesting timber and sequestering carbon.

The dependent variables (i.e. the optimal management strategy) include the net present value and its associated rotation length. For further details of the model, see Nghiem (2011, 2013, 2014).

Let

$v(.)$: the discounted sum of timber and carbon values
a_s : the area of stands in hectares
x_{st} : age of stands in period t

The model objective is to maximize the discounted revenue from timber and carbon:

$$v(a_1, ..a_n; x_{10}, .., x_{n0}) = \max \sum_{t=0}^{50} (1 + r)^{-t}[U_t + A_t] \tag{6.1}$$

subject to

$d_{st} = 0 \ or \ 1, s = 1, \ldots, n,$	binary decision variables
$x_{s,t+1} = (x_{st} + 1).(1 - d_{st}), s = 1, \ldots, n,$	age of stand s at period $t + 1$
$x_{st} \geq 0, s = 1, \ldots, n,$	nonnegative constraint of ages

where

r	discount rate
$U_t \sum_{s=1}^{n} [q(x_{st}). a_{s.} d_{st}.P(x_{st})] - G(h_t)$	timber value at the period t
$q(x_{st})$	timber volume per ha of stand s in period t
d_{st}	the binary decision variable, $d_{st} = 1$: stand s is clear-cut in period t $d_{st} = 0$: stand s is kept in period t
$P(x_{st})$	price of timber per unit volume, which varies with timber's age
$G(h_t)$	planting cost of timber at period t, which varies with planting size
$G(h_t) = \beta h_t^{\lambda}$	β and λ are estimated from survey data
$h_t = \sum_{s=1}^{n} d_{st}.a_s$:	planting size in hectare in period t
$A_t = \left[\sum_{s=1}^{n} Qc(x_{st}).a_s.d_{st.} \right]. Pc$:	amount of carbon sequestered at period t
Qc	amount of carbon sequestered, metric tonne
P_c	per ha carbon price per metric tonne

The model assumes no carbon tax is imposed on the forest owner at harvest. The carbon emission and transaction costs are also neglected. This formula implies carbon sequestration is calculated at the end of rotation for simplicity. Sensitivity to different payment schemes is provided in the section on sensitivity analysis.

To estimate the cost functions for two species, the planting costs obtained from the survey were adjusted for appropriate inflation rates. The model assumes a fixed carbon price of 3 USD (equivalent to 51,000 VND) per tonne CO_2.

Timber growth functions of *Eucalyptus urophylla* and *Acacia mangium* for site index II are estimated from the data obtained from the Department of Science and Technology-MARD (2003).

Eucalyptus urophylla:

$$q(x_{st}) = -1.38x_{st}^2 + 40.33x_{st} - 94.07, \quad R^2 = 0.99 \tag{6.2}$$

Acacia mangium:

$$q(x_{st}) = -0.3x_{st}^2 + 28.06x_{st} - 63.33, \quad R^2 = 0.99 \tag{6.3}$$

where x_{st} denotes timber age and $q(x_{st})$ represents timber volume at age x_{st}.

The survey conducted for the study showed that the household's timber productivity is approximately one third of the productivity provided in the Yield table (Department of Science and Technology 2003). The study used a timber volume of household equal to $q(x_{st})/3$. This number is also reported by the Vietnam Ministry of Natural Resources and Environment (MONRE 2005).

To identify the amount of carbon uptake, carbon sequestration functions from Vo et al. (2009) are used. Since the carbon sequestration functions from Vo et al. were not suitable to apply for planted forests aged beyond 7 years, some adjustments were carried out on the carbon sequestration functions as follows. First, using tree density data from Tables of Site Index (Department of Science and Technology 2003) and tree density at planting from the survey, the relationship between stand age and tree density for households and enterprises was calculated. Tree density of both species at planting was 2,189 and 1,660 trees/ha for households and enterprises, respectively.

Second, the study ran the carbon sequestration functions from Vo et al. with respect to stand age and tree density for site index I for *Eucalyptus urophylla* and site index II for *Acacia mangium*. The carbon sequestration from Vo et al. is $LnQc = 22.036 + 0.313lnX - 1.584lnN$, where Qc is the amount of carbon uptake (kg/ha), X is the age of the tree and N is the tree density (per ha). Tree density at age 12/13 for *Eucalyptus urophylla/Acacia mangium* was used for the rest of rotation so that the amount of carbon sequestered was not too high.

Finally, based on these figures, the study re-estimated the carbon sequestration functions with respect to stand age only, and this was used in the analysis. The final functions are as follows. For households with *Eucalyptus urophylla*:

$$Qc(x_{st}) = -0.071x_{st}^2 + 6.0155x_{st} + 11.567, R^2 = 0.9633 \tag{6.4}$$

For enterprises with *Eucalyptus urophylla*:

$$Qc(x_{st}) = -0.1101x_{st}^2 + 9.3232x_{st} + 17.928, R^2 = 0.9633 \quad (6.5)$$

For households with *Acacia mangium*:

$$Qc(x_{st}) = -0.0305x_{st}^2 + 4.9714x_{st} + 66.114, R^2 = 0.9921 \quad (6.6)$$

For enterprises with *Acacia mangium*:

$$Qc(x_{st}) = -0.0271x_{st}^2 + 4.42x_{st} + 58.781, R^2 = 0.9921 \quad (6.7)$$

where x_{st} denotes timber age and $Qc(x_{st})$ represents the carbon amount (tonne/ha) sequestered up to age x_{st}.

The study adjusted the carbon sequestration functions from Vo et al. (2009) so that these functions fit to the study. The adjusted carbon sequestration functions from Vo et al. for *Eucalyptus urophylla* are given below:

$$Qc(x_{st}) = 15.987e^{0.1516x_{st}} \quad (6.8)$$

$$Qc(x_{st}) = e^{22.036+0.313\ln(x_{st})-1.584\ln(N)} \quad (6.9)$$

$$Qct(x_{st}) = \left(e^{(4.1343-4.6438/x_{st})}\right)/1,000 \quad (6.10)$$

where $Qct(x_{st})$ is the carbon amount uptake by individual trees (kg/tree) and N denotes tree density (trees/ha). Equation (6.8) is the carbon sequestration function for households, adjusted using Eq. (6.9). In Eq. (6.10), it was assumed that tree density is 1,000 trees/ha.

To find the optimal strategy, all scenarios with different rotation ages were compared, based on the objective variable. The scenario that had the highest net present value was chosen as the optimal strategy. The model was coded in GAMS and was used to analyse the optimal forest management strategy for case studies in Yen Bai province, Vietnam.

Field Survey

The final questionnaires implemented for the study contain five parts. Part I obtained information about age, gender, ethnic, education, size of household, labour and financial status. Part II included questions about sizes of forest lands and other lands, cutting practice, initial stage of forest and understorey plants. Part III requested information about timber production costs, capital and subsidy from the government. Part IV is related to the income from forest and other sources in 2007. Part V investigated the management strategy regarding rotation length, government subsidy and method of payment for carbon sequestration and carbon pooling.

The study sites comprised Tran Yen, Van Yen and Yen Binh districts. Two communes in each of the three districts were selected based on consultations with local government officials. They comprised Viet Cuong and Cuong Thinh in Tran Yen district, Ngoi A and Yen Thai in Van Yen district and Vu Linh and Vinh Kien in Yen Binh district. The sample population consisted of households who grew *Eucalyptus urophylla* or *Acacia mangium* and owned a productive forest of at least 0.5 ha. Households were selected randomly. Surveys were implemented in September 2008. A total of 271 usable questionnaires out of 291 distributed were collected. The study also collected data from four state forest enterprises, a private forest enterprise, a special management board, the Yen Bai Forestry Department and other forestry departments at district level and commune level.

Results and Discussion

Study Area and Forest Management in Vietnam

Yen Bai is located 180 km northwest of Hanoi, with a total area of 688,630 ha of which more than 70 % is covered by mountains and highlands. Yen Bai is a major timber supply area in Northern Vietnam. The area of productive planted forest in Yen Bai is 116,472 ha, accounting for 59.1 % of the total area and 6.94 % of the total productive planted forest area in Vietnam (Forest Protection Department 2007). Nine state enterprises and 41,000 households are involved in forest plantation (Yen Bai Forestry Department 2008; Nguyen et al. 2006). In Yen Bai, 50 % of the productive forest area is managed by a commune committee, 42.1 % by households and 7.6 % by forest enterprises, while the remaining area belongs to other institutions. Planted forests are dominated by fast-growing trees such as *Eucalyptus* spp., *Acacia* spp., *Styrax tonkinensis*, *Manglietia conifera*, *Cinnamomum cassia blume* and other native species (Yen Bai Forestry Department 2008).

All forest land in Vietnam is under state ownership. According to Government Decree No. 163/1999/ND-CP (Vietnamese Government 1999), forest lands are allocated and leased to individuals and organizations for long-term forestry purposes. Forest land allocated to households shall not exceed 30 ha and a 50-year use right. After 50 years, the right to use the land can be extended at the request of the forest owner. Government Decree No. 135/2005/ND-CP (Vietnamese Government 2005a) allows state forest enterprises to contract their land to households for a maximum of 30 years. Decision No. 40/2005/QD-BNN (Vietnamese Government 2005b) allows forest owners to make their own decisions about the harvest age of timber and clear-cut size. Decision No. 100/2007/QD-TTg (Vietnamese Government 2007b) states that the government encourages forest owners to lengthen the rotation ages.

Planting Cost and Timber Revenue

For households, the planting cost of *Eucalyptus urophylla* is 6.85 million VND per ha, 0.07 million VND higher than that of *Acacia mangium*. For forest enterprises, the average planting cost (including planting cost, tending cost up to year 3 and other costs) is 17.02 million VND per ha for *Eucalyptus urophylla* and 14.09 for *Acacia mangium*. These costs are higher than those of households since enterprises use more fertilizers and bear other associated costs such as management, design and interest payments. These cost data were used to calculate the Faustmann rotation for households and for enterprises.

The timber revenue per ha per rotation for *Eucalyptus urophylla* is nearly seven million VND higher than that for *Acacia mangium* (Table 6.1). The study used the average timber price of 0.37 and of 0.33 million VND per m^3 for *Eucalyptus urophylla* and *Acacia mangium* to calculate Faustmann rotations for households and enterprises.

From household survey data, the study estimated the following function of economies to planting scale for *Eucalyptus urophylla*:

$$G(h_t) = 5.16 h_t^{0.98}, \quad R^2 = 0.66 \tag{6.11}$$

where h_t denotes planted area (ha) and $G(h_t)$ represents planting cost (million VND per ha). In this case, the level of economies to scale λ equals *0.98*.

Similarly, the cost function for *Acacia mangium* is as follows:

$$G(h_t) = 4.56 h_t^{1.07}, \quad R^2 = 0.66 \tag{6.12}$$

In the case of *Acacia mangium*, the cost function exhibited increasing planting cost to scale. Intuitively, this can be explained by households with a large area having more money to invest in their plantations; hence, an increasing cost to scale exists. However, it was hard to identify economies of scale since households employed different techniques and amount of capital, labour and fertilizer in their forests. Also, more than 70 % of Yen Bai province comprises mountainous and highland terrain, making it difficult to apply modern technology. Hence, the study used economies to scale for planting cost of nearly 1 for both tree species. The study

Table 6.1 Timber price and revenue in 2007

Species	Selling place	Average timber revenue per ha (million VND)	Average timber price per m^3 (million VND)
Eucalyptus urophylla	Stumpage price	19.38	**0.37**
	Landing 1	22.06	0.44
Acacia mangium	Stumpage price	12.89	**0.33**
	Landing 1	10.39	0.35

still used the function of economies to planting scale of *Eucalyptus urophylla* to calculate forest-level rotation. Sensitivity of the optimal rotation age to the value of λ is presented in the section on sensitivity analysis.

Faustmann Rotation

The optimal ages for both species are higher than the actual harvesting ages (i.e. 5 years) based on the survey. At positive discount rates, the optimal rotation length ranges from 8 to 10 years, and the NPV is between 3 and 189 million VND per ha for *Eucalyptus urophylla* and from 9 to 17 years and between 3 and 209 million VND/ha for *Acacia mangium*. The Faustmann rotation for households is a little longer than that for enterprises, but the NPV is much smaller. This result is interesting since most planted forest areas of forest enterprises are contracted to households. The terms of conditions vary among forest enterprises. The NPV for forest enterprises is higher than that for households because of timber growth. Households whose land is allocated from the government tend to plant trees with higher density, use less fertilizers and may apply inappropriate growing techniques.

When carbon has value, the household's rotation and the enterprise's rotation are nearly identical. For both households and enterprises, the carbon optimal rotation is slightly shorter compared to the Faustmann rotation, but the carbon NPV is higher than the Faustmann's (see Tables 6.2 and 6.3). This finding (i.e. carbon sequestration making the rotation age shorter) is in contrast to the result of van Kooten et al. (1995). The difference could be due to the different tree species in the two studies. In this study, the trees are fast-growing trees; hence, carbon uptake and

Table 6.2 Faustmann rotation age with and without carbon value for *Eucalyptus urophylla* in forest households and enterprises

	Households				Enterprises			
	Timber only		With carbon		Timber only		With carbon	
Discount rate (%)	*T* (years)	NPV (m. VND/ ha)	*T*	NPV	*T*	NPV	*T*	NPV
1	10	51.97	10	64.44	10	170.52	10	189.85
2	10	38.17	10	47.82	10	126.86	10	141.81
3	10	28.41	10	36.03	10	95.87	10	107.68
4	10	21.36	10	27.49	10	73.42	10	82.92
5	10	16.15	**9**	21.24	9	56.86	9	65.17
6	10	12.24	**9**	16.68	9	44.90	9	51.84
7	9	9.36	9	13.15	9	35.61	9	41.48
8	9	7.10	9	10.35	9	28.25	9	33.29
9	9	5.29	8	8.15	9	22.35	**8**	27.04
10	9	3.81	8	6.42	8	17.78	8	21.95

Note: *Carbon price is 51,000 VND per metric tonne (equivalent to 3 USD per metric tonne)

Table 6.3 Faustmann rotation age with and without carbon value for *Acacia mangium* in forest households and enterprises

	Households				Enterprises			
	Timber only		With carbon		Timber only		With carbon	
Discount rate (%)	*T* (years)	NPV (m. VND/ha)	*T*	NPV	*T*	NPV	*T*	NPV
1	17	59.95	17	75.65	17	195.76	17	209.71
2	17	42.60	13	55.22	17	141.63	13	153.24
3	17	30.51	13	41.19	13	105.76	13	115.57
4	13	22.34	13	31.06	13	80.58	13	88.33
5	13	16.60	10	24.11	13	62.00	10	68.52
6	13	12.30	10	18.87	11	48.23	10	54.53
7	13	9.01	10	14.83	10	38.27	10	43.73
8	11	6.55	9	11.71	10	30.59	10	35.26
9	11	4.67	9	9.29	10	24.46	9	28.51
10	10	3.15	9	7.32	10	19.49	9	23.27

timber growth are higher in early years. The native tree species in van Kooten et al. has a totally different pattern of growth throughout its life. Therefore, it is optimal to cut and replant trees sooner to maximize profit from selling timber and sequestering carbon.

Forest-Level Rotation

To estimate forest-level rotation, the study used Microsoft Excel to randomly choose 5 case studies for *Eucalyptus urophylla* from the survey data (Table 6.4). The number of stands ranges between 1 and 5 in the database, so the study randomly chose one case that has 2, 4 and 5 stands and two cases that have 3 stands. For the initial ages of forest stands, the study analysed two cases: (1) zero initial ages and (2) actual initial ages. Forest stand areas are the actual households' stand areas.

For *Eucalyptus urophylla*, in all ten cases, the optimal rotation age with or without carbon is shorter by up to 2 years than the Faustmann rotation, and the NPV is greater than that of the Faustmann. In particular, when the initial age of the forest is greater than zero (i.e. cases 1, 3, 5, 7 and 9), at a 5 % discount rate, carbon sequestration makes the rotation age shorter. When the initial age equals zero, at 8 % discount rate, adding carbon value shortens the rotation age. The optimal rotation decreases with an increasing discount rate as in the Faustmann model (Table 6.5).

Table 6.4 Case studies of forest-level rotation for *Eucalyptus urophylla*

Cases	Codes*	No. of stands	Area of stands (ha)					Age of stands (years)				
			S1	S2	S3	S4	S5	S1	S2	S3	S4	S5
1	A086	5	3.1	2.9	1	0.8	0.8	3	3	3	3	3
2	A086	5	3.1	2.9	1	0.8	0.8	0	0	0	0	0
3	A075	4	1	0.3	0	0.3	...	4	2	2	2	...
4	A075	4	1	0.3	0	0.3	...	0	0	0	0	...
5	A038	3	3	1	1	...	...	1	1	1	...	...
6	A038	3	3	1	1	...	...	0	0	0	...	...
7	A017	3	2	0.5	1	...	...	2	1	1	...	...
8	A017	3	2	0.5	1	...	...	0	0	0	...	...
9	A007	2	1	1	...	...	...	4	2	...	...	...
10	A007	2	1	1	...	...	...	0	0	...	...	...

Note: *Coding number used in the database

Table 6.5 Forest-level rotation with and without carbon value for *Eucalyptus urophylla*

Discount rate (%)	5				8			
	Timber		Carbon		Timber		Carbon	
Case	*T* (years)	NPV (m.VND/ha)	*T*	NPV	*T*	NPV	*T*	NPV
1	10	21.42	8	26.81	8	12.16	8	15.93
2	9	17.82	9	22.48	9	8.64	8	11.47
3	10	21.27	8	26.64	8	12.08	8	15.93
4	9	17.42	9	22.08	9	8.34	8	11.14
5	10	18.51	9	23.32	8	9.52	8	12.75
6	9	17.70	9	22.36	9	8.55	8	11.37
7	10	19.12	8	24.01	8	10.07	8	13.44
8	9	17.61	9	22.27	9	8.48	8	11.30
9	10	20.97	8	26.27	8	11.79	8	15.57
10	9	17.48	9	22.14	9	8.38	8	11.19

Sensitivity Analysis

Carbon Price

The study performed a sensitivity analysis of the Faustmann model with a carbon price from 3 USD (0.051 million VND) to 70 USD (1.19 million VND) per metric tonne and with the carbon price assumed constant over the planning horizon. In the case of *Eucalyptus urophylla* (Table 6.6), the carbon rotation age decreases significantly with an increasing carbon price (up to 75 %) and moderately with the discount rate (up to 33 %). The level of decrease is slightly different between households and enterprises. In general, the rotation age ranges from 2 to 10 years. For *Acacia mangium* (Table 6.7), the relationship between the carbon rotation age

Table 6.6 Sensitivity analysis of Faustmann rotation age with constant carbon price for *Eucalyptus urophylla* in forest households and enterprises

		Households		Enterprises	
Discount rate (%)	Carbon price (USD/tonne)	*T* (year)	NPV (m. VND/ha)	*T*	NPV
1	3 (0.051 m.VND)	10	64.44	10	189.85
	10 (0.17)	10	93.54	10	234.95
	20 (0.34)	9	135.70	10	299.37
	70 (1.19)	4	385.18	6	645.93
5	3 (0.051)	9	21.24	9	65.17
	10 (0.17)	9	33.75	9	84.56
	20 (0.34)	7	52.75	8	112.56
	70 (1.19)	3	174.62	5	277.39
8	3 (0.051)	9	10.35	9	33.29
	10 (0.17)	8	18.43	8	45.92
	20 (0.34)	6	31.22	7	64.72
	70 (1.19)	3	115.12	5	174.52
10	3 (0.051)	8	6.42	8	21.95
	10 (0.17)	7	12.91	7	31.69
	20 (0.34)	6	23.20	7	46.92
	70 (1.19)	2	92.84	4	135.57

Table 6.7 Sensitivity analysis of Faustmann rotation age with constant carbon price for *Acacia mangium* in forest households and enterprises

		Households		Enterprises	
Discount rate (%)	Carbon price (USD/tonne)	*T* (year)	NPV (m. VND/ha)	*T*	NPV
1	3 (0.051 m.VND)	17	75.65	17	209.71
	10 (0.17)	6	123.87	13	245.22
	20 (0.34)	1	533.47	10	304.19
	70 (1.19)	1	2,924.46	1	1,953.58
5	3 (0.051)	10	24.11	10	68.52
	10 (0.17)	4	51.29	10	86.71
	20 (0.34)	1	258.31	7	117.21
	70 (1.19)	1	1,416.04	1	945.93
8	3 (0.051)	9	11.71	10	35.26
	10 (0.17)	3	31.84	8	47.31
	20 (0.34)	1	178.04	6	70.15
	70 (1.19)	1	976.01	1	651.99
10	3 (0.051)	9	7.32	9	23.27
	10 (0.17)	3	24.81	8	33.29
	20 (0.34)	1	146.97	5	52.77
	70 (1.19)	1	805.67	1	538.20

and the carbon price or the discount rate is similar to that of *Eucalyptus urophylla*. However, the households' rotation age is very sensitive to carbon price. It decreases to 1 year at the carbon price of 20 USD/tonne at all discount rates. The spread of the optimal rotation age is larger than that of *Eucalyptus urophylla*, ranging from 1 to 17 years.

The relationship between the rotation length and the carbon price contrasts to the findings of van Kooten et al. (1995) but is in agreement with the results of Englin and Callaway (1993) including amenity value. Diaz-Balteiro and Rodriguez (2006) find that carbon price has little effect with no clear pattern regarding the optimal rotation. The relationship between the rotation age and discount rate agrees with the literature, with an exception for Englin and Callaway, in which carbon rotation (without timber value) increases in relation to the discount rate. For *Eucalyptus urophylla*, at a 5 % discount rate and carbon price of 20 USD per metric tonne, the difference between the net present value with and without a carbon scheme is approximately 21 million VND per ha, about equal to the compensation level required by households to lengthen the rotation age.

The study also analysed the optimal management strategies under stochastic carbon prices. Carbon prices were generated randomly in GAMS over a 50-year planning horizon and then were used to run the optimization model. This procedure was carried out ten times. There were three magnitudes of carbon price: from 0 to 1.2 million VND per metric tonne (equal to 0–70 USD/tonne), from 0 to 0.5 m. VND/tonne (equal to 0–29.5 USD/tonne) and from 0 to 0.3 m. VND/tonne (equal to 0–17.7 USD/tonne). The relationship between the optimal rotation age and carbon price is the same as in the case of a constant carbon price (Tables 6.8 and 6.9), that is, the optimal rotation age decreases as the maximum value of carbon price increases. The *Acacia mangium*'s optimal rotation age is more sensitive to the carbon price than that of *Eucalyptus urophylla*. At the maximum carbon price of 1.2 m. VND/tonne (or 70 USD/tonne), the rotation age of *Acacia mangium* is reduced to 1 year for all ten runs.

Discount Rate

In general, the optimal rotation age is shorter with an increasing discount rate. The speed with which the optimal rotation age shortens with an increasing discount rate is lower compared to the case of a constant carbon price. Moreover, stochastic carbon price lengthens the optimal rotation age. For example, for *Acacia mangium*, at a constant carbon price of 0.34 million VND, all rotation ages are 1 year, while at a stochastic price of 0.3–0.5 million VND, many rotation ages are greater than 1 year. For *Eucalyptus urophylla*, at a constant price of 1.19 million VND, the rotation ages are less than 4 years, while at a stochastic price, they are almost greater than 5 years. This result agrees with the results of Chladna (2007) and Thomson (1992).

Table 6.8 Sensitivity analysis of Faustmann rotation age with stochastic carbon price for *Eucalyptus urophylla*

		T (year)					NPV (m. VND/ha)				
		Discount rate (%)					Discount rate (%)				
Maximum carbon price (m.VND/tonne)	The order of the run (out of ten)	$r = 1$	3	5	8	10	$r = 1$	3	5	8	10
1.2	First run	7	7	6	6	6	238.7	148.6	98.9	62.1	47.6
	Fourth run	10	10	10	10	10	221.1	140.4	95.5	58.6	44.0
	Tenth run	10	10	5	5	5	282.5	176.1	126.3	84.8	68.1
0.5	First run	9	9	9	6	6	122.2	73.5	46.9	27.0	20.2
	Seventh run	12	8	8	8	8	124.8	76.0	52.4	32.5	24.4
	Tenth run	9	11	6	6	6	113.8	66.2	44.4	27.7	21.0
0.3	First run	8	8	8	8	7	89.0	52.8	33.1	17.5	12.2
	Third run	14	14	11	7	9	101.4	56.7	35.1	19.2	13.2
	Fifth run	10	11	11	6	6	103.5	58.7	35.8	18.5	13.1
	Tenth run	10	10	8	8	8	96.3	56.2	37.2	21.9	15.9

Table 6.9 Sensitivity analysis of Faustmann rotation age with stochastic carbon price for *Acacia mangium*

Maximum carbon price (m.VND/tonne)	The order of the run (out of ten)	*T* (year)					NPV (m. VND/ha)				
		Discount rate (%)					Discount rate (%)				
		1	3	5	8	10	1	3	5	8	10
1.2	First run	1	1	1	1	1	1,159.2	791.7	583.0	411.1	343.9
	Fifth run	1	1	1	1	1	1,055.2	740.7	555.2	395.3	329.8
	Eighth run	1	1	1	1	1	1,067.2	700.7	498.9	339.0	278.4
0.5	First run	1	1	2	2	2	302.3	199.2	143.5	103.2	86.9
	Third run	2	2	1	1	1	195.9	126.8	92.9	66.2	56.4
	Ninth run	1	1	1	1	1	356.5	232.6	161.8	103.3	80.4
0.3	First run	14	10	5	5	5	124.5	72.1	50.8	32.7	25.3
	Second run	11	9	9	5	5	139.7	84.5	55.6	34.3	26.7
	Fourth run	1	1	1	1	1	1,159.2	791.7	583.0	411.1	343.9
	Fifth run	1	1	1	1	1	1,055.2	740.7	555.2	395.3	329.8

Carbon Payment Scheme

The project also investigated the impacts of various carbon payment schemes on the optimal management strategy (Tables 6.10 and 6.11). Carbon value can be paid to forest owners at the beginning of the rotation (with an expectation of rotation age) every year or at the end of the rotation. It was found that a lump sum payment at the beginning of the rotation lengthens the optimal rotation age and increases the NPV compared to an annual payment and payment at the end of the rotation. Interestingly, the optimal rotation age increases with a rise in the discount rate and a carbon price of 1.19 million VND (70 USD/tonne). Intuitively, this result follows from having the payment at the beginning of the rotation in combination with a high carbon price, making the carbon benefit greater than the timber value and worthwhile lengthening the rotation age.

Table 6.10 Sensitivity analysis of Faustmann rotation age with carbon payment scheme for *Eucalyptus urophylla*

Carbon price (m VND/ tonne)	Discount rate (%)	Ending payment T (year)	Ending payment NPV (m. VND/ha)	Annual payment T	Annual payment NPV	Beginning payment T	Beginning payment NPV
0.051	1	10	64.44	10	65.15	10	65.61
	5	9	21.24	10	22.79	10	23.95
	8	9	10.35	9	11.94	9	13.09
	10	8	6.42	8	7.90	9	9.08
0.17	1	10	93.54	10	95.89	10	97.43
	5	9	33.75	9	38.71	10	42.13
	8	8	18.43	8	23.33	9	27.07
	10	7	12.91	8	17.63	9	21.36
0.34	1	9	135.70	9	139.93	10	142.90
	5	7	52.75	8	61.63	9	68.37
	8	6	31.22	7	39.99	9	47.04
	10	6	23.20	7	31.78	10	38.93
1.19	1	4	385.18	4	393.22	5	397.99
	5	3	174.62	4	191.27	5	202.53
	8	3	115.12	4	132.22	37	157.95
	10	2	92.84	4	109.35	41	157.83

Table 6.11 Sensitivity analysis of Faustmann rotation age with carbon payment scheme for *Acacia mangium*

Carbon price (m VND/ tonne)	Discount rate (%)	Ending payment T (year)	Ending payment NPV (m. VND/ ha)	Annual payment T	Annual payment NPV	Beginning payment T	Beginning payment NPV
0.051	1	17	75.65	17	77.61	17	78.30
	5	10	24.11	10	27.90	13	29.10
	8	9	11.71	10	15.73	10	16.90
	10	9	7.32	9	11.31	10	12.43
0.17	1	6	123.87	9	128.57	9	129.77
	5	4	51.29	5	61.05	6	62.71
	8	3	31.84	5	41.73	5	43.43
	10	3	24.81	4	34.36	5	36.04
0.34	1	1	533.47	1	533.47	1	533.47
	5	1	258.31	1	258.31	1	258.31
	8	1	178.04	1	178.04	1	178.04
	10	1	146.97	1	146.97	1	146.97
1.19	1	1	2,924.46	1	2,924.46	1	2,924.46
	5	1	1,416.04	1	1,416.04	1	1,416.04
	8	1	976.01	1	976.01	1	976.01
	10	1	805.67	1	805.67	1	805.67

Table 6.12 Sensitivity analysis of Faustmann carbon rotation age with planting cost subsidy *Eucalyptus urophylla*

Discount rate (%)	Subsidy level (m. VND/ha)	*T* (year)	NPV (m. VND/ha)
1	6.85	9	94.44
	4	10	81.00
	2	10	72.72
	1	10	68.58
	0.5	10	66.51
5	6.85	7	40.44
	4	8	31.76
	2	9	26.47
	1	9	23.85
	0.5	9	22.55
8	6.85	6	25.91
	4	7	18.96
	2	8	14.54
	1	8	12.39
	0.5	9	11.33
10	6.85	6	20.38
	4	7	14.21
	2	8	10.15
	1	8	8.29
	0.5	8	7.35

Planting Cost Subsidy

A sensitivity analysis of the carbon rotation age with a planting cost subsidy was also carried out, and the results are shown in Tables 6.12 and 6.13. Since planting costs for both species are around six million VND, the study used a subsidy level of between 0.5 and 6.77/6.85 (full subsidy) million VND per ha per rotation. It is shown that the larger the planting cost subsidy, the shorter the rotation age. In particular, for *Eucalyptus urophylla*, a full subsidy always makes the rotation age shorter, and a subsidy of 0.5 m VND shortens the rotation only at a 5 % discount rate. Moreover, at a 5 % discount rate, all subsidy levels induce forest owners to shorten the rotation age. The NPV is higher with the lower discount rate and the higher subsidy level.

Timber Price

Timber of different sizes is sold at different prices. The study thus employed a nonconstant timber price and assumed that harvest size is strongly correlated with timber age. Using survey data, the study used real timber prices (*P*) of 0.37 m. VND/m^3 for *Eucalyptus urophylla* and 0.33 for *Acacia mangium* as the price for

Table 6.13 Sensitivity analysis of Faustmann carbon rotation age with planting cost subsidy *Acacia mangium*

Discount rate (%)	Subsidy level (m. VND/ha)	*T* (year)	NPV (m. VND/ha)
1	6.77	10	99.60
	4	13	88.38
	2	13	81.72
	1	13	78.39
	0.5	17	76.92
5	6.77	6	42.66
	4	9	33.90
	2	10	28.84
	1	10	26.47
	0.5	10	25.29
8	6.77	5	27.55
	4	7	20.15
	2	8	15.69
	1	9	13.68
	0.5	9	12.69
10	6.77	5	21.85
	4	6	15.14
	2	8	11.02
	1	8	9.16
	0.5	8	8.22

timber at age 7 years. If timber harvest age (T) is smaller than 5 years, timber price equals $P_t = P\text{-}0.15$. If timber age is between 5 and 6 years, timber price equals $P_t = P\text{-}0.05^*(7\text{-}T)$. If timber age is from 7 to 14 years, timber price equals $P_t = P + 0.1^*(T\text{-}7)$. If timber age is greater than 14 years, then $P_t = P + 0.8$. The real timber price varied from 0.1 to 3 m. VND/m^3 for a sensitivity analysis.

For *Eucalyptus urophylla*, the study found that when the real timber price is low, it is optimal to grow large-size timber (i.e. the rotation age is around 13–15 years) at all discount rates. Moreover, at the real timber price (0.37 m. VND) or at the price of 0.1 m. VND, the rotation age remains the same for all discount rates. The higher timber price lowers the rotation age. Only at an 8 % discount rate and at a three million VND timber price is it optimal to grow small-sized timber (i.e. 8 years). Assuming a nonconstant timber price makes the rotation age significantly longer (i.e. almost from 13 to 15 years) than with the constant-price Faustmann solution (i.e. from 8 to 10 years). The NPV increases with timber price and decreases with the discount rate. The relationships between the optimal rotation/the NPV and timber price for *Acacia mangium* are similar to those for *Eucalyptus urophylla* (Tables 6.14 and 6.15).

Table 6.14 Sensitivity analysis of Faustmann carbon rotation age with timber price where timber price is nonconstant for *Eucalyptus urophylla*

Discount rate (%)	Timber price (m. VND/m^3)	*T* (year)	NPV (m. VND/ha)
1	0.1	15	140
	0.3	15	174
	0.37	14	186
	0.5	14	210
	1	14	302
	3	10	700
5	0.1	15	46
	0.3	14	58
	0.37	14	63
	0.5	14	72
	1	13	107
	3	9	280
8	0.1	15	22
	0.3	14	29
	0.37	14	32
	0.5	13	37
	1	11	59
	3	**8**	168

Table 6.15 Sensitivity analysis of Faustmann carbon rotation age with timber price where timber price is nonconstant for *Acacia mangium*

Discount rate (%)	Timber price (m. VND/m^3)	*T* (year)	NPV (m. VND/ha)
1	0.2	17	232.52
	0.33	17	262.95
	0.5	17	302.76
	2	17	653.96
	3	17	888.10
5	0.2	15	79.55
	0.33	15	90.68
	0.5	15	105.22
	2	15	233.58
	3	14	320.94
8	0.2	15	40.61
	0.33	15	46.72
	0.5	15	54.71
	2	13	127.08
	3	11	180.91
10	0.2	15	26.67
	0.33	15	30.98
	0.5	15	36.62
	2	11	91.49
	3	9	133.81

Table 6.16 Sensitivity analysis of Faustmann carbon rotation age when timber price is nonconstant and increasing every year for *Eucalyptus urophylla*

Price increasing/year		1 %		10 %	
Timber price (m. VND/m^3)	r (%)	*T*	NPV	*T* (year)	NPV (m. VND/ha)
0.2	1	15	218	18	6,501
	5	15	69	15	1,257
	8	14	33	15	447
	10	14	21	15	242
0.37	1	14	258	18	7,608
	5	14	83	14	1,487
	8	14	41	14	526
	10	13	27	14	285
0.5	1	14	290	18	8,454
	5	14	94	14	1,669
	8	14	47	14	591
	10	13	31	14	321
2	1	13	671	13	18,438
	5	10	239	14	3,766
	8	9	135	14	1,340
	10	9	99	13	735
3	1	10	940	13	25,534
	5	9	348	13	5,191
	8	9	202	13	1,855
	10	8	150	10	1,032

The study also computed the optimal rotation age when the timber price increases every year at 1 % and 10 %. For *Eucalyptus urophylla*, compared with the carbon rotation age with a constant price, the rotation age is longer at every discount rate, increasing from 8–10 years to 13–14 years and to 14–18 years at 1 % and 10 %, respectively. With the timber price increasing at 1 % every year, the rotation age decreases as the discount rate increases. However, with a 10 % annual price increase, this rule seems to be no longer valid. In particular, at the timber price of two million VND/m^3, the rotation age is 13–14-13 years as the discount rate increases (Table 6.16). This unusual relationship between the discount rate and the rotation age occurs not only with this carbon sequestration function but also with the other carbon sequestration function and for *Acacia mangium* as well. Interestingly, at a 10 % increase, the optimal rotation of *Acacia mangium* is no longer sensitive to the timber price. At a 1 % discount rate, it is optimal to keep trees until the end of the planning horizon (Table 6.17).

Table 6.17 Sensitivity analysis of Faustmann carbon rotation age when timber price is nonconstant and increasing every year for *Acacia mangium*

Price increasing/year		1 %		10 %	
Timber price (m. VND/m^3)	r (%)	*T*	NPV	*T* (year)	NPV (m. VND/ha)
0.2	1	17	322	50	14,027
	5	15	103	25	2,107
	8	15	52	26	654
	10	15	34	17	346
0.33	1	17	364	50	15,850
	5	15	117	25	2,381
	8	15	59	26	740
	10	15	39	17	392
0.5	1	17	419	50	18,235
	5	15	136	25	2,740
	8	15	69	26	853
	10	15	46	17	451
2	1	17	904	50	39,274
	5	15	300	25	5,909
	8	14	157	26	1,843
	10	11	110	17	980
3	1	17	1,227	50	53,300
	5	14	410	25	8,021
	8	13	219	26	2,504
	10	10	159	17	1,332

Conclusions and Policy Implications

This research applied the Faustmann formula to find the optimal rotation age and the net present value when carbon sequestration is included, for *Eucalyptus urophylla* and *Acacia mangium* in Yen Bai province, Vietnam.

The survey results show that the actual tree cutting age is 5 years. When only timber value is considered, the optimal rotation age of *Eucalyptus urophylla* is 10 and 9 years for households and forest enterprises, respectively, at a 5 % discount rate. For *Acacia mangium*, the optimal rotation age is 13 years for both households and forest enterprises. The net present value is 16 million VND per ha for households for both *Eucalyptus urophylla* and *Acacia mangium*. For forest enterprises, the NPV is 57 and 62 million VND/ha for *Eucalyptus urophylla* and *Acacia mangium*, respectively.

Adding carbon values makes the optimal rotation age slightly shorter and the net present value higher. Similarly, taking into account the spatial arrangement of forest stands shortens the rotation age and improves the net present value. Nevertheless, the rotation age including carbon or forest level is still well above the actual age. The optimal rotation age decreases with an increasing carbon price, planting subsidy level, constant timber price and discount rate. The optimal rotation age increases with a rise in timber price, the level of economies to planting scale and the stochastic carbon price. The optimal rotation age is sensitive to the particular carbon payment scheme.

The government could use a lump sum payment of carbon benefits at the beginning of the rotation to encourage forest owners to lengthen the rotation age. Survey results show that households equally prefer two payment schemes, with payment either at the beginning or annually. Moreover, households would agree to lengthen rotation age with a subsidy of seven million VND per ha per year or 21 million VND per ha per rotation and with a commitment to keep their forest until 11 years. These figures show that the optimal management strategy can be achieved, for example, at carbon price of 3 USD (0.051 million VND) per metric tonnes with a 5 % discount rate. Almost all households agreed to engage in a carbon pooling project, and one fourth of them believed that there are no obstacles to implementing such a project.

Given that the majority of households cut their trees sooner based on their family's financial status, a planting cost subsidy would help to narrow the actual and the optimal rotation age. Intuitively, instead of investing money to plant trees, households can use it to meet their living costs and, hence, can delay their harvest. At a 5 % discount rate, a full subsidy can reduce rotation age from 10 to 6 years.

If the assumption of constant timber price is relaxed, the study found that it is optimal to grow large-size timber. This is consistent with the fact that in Vietnam, the wood-processing industry is growing rapidly, and currently, most large-size timber is imported. Economies of scale in planting show that it is more profitable to produce at a large scale to reduce production costs.

Finally, the survey data shows that households are financially unable and unwilling to delay their harvest till 13–15 years to become large-size timber producers. Therefore, cooperation or joint venture among households, forest enterprises and wood-processing companies could bring a win-win solution. In such an arrangement, households can receive technical and capital support and achieve the benefits from a large-scale production, while wood-processing companies can take advantage of a cheap timber supply from domestic sources.

References

Bui DT, Hong BN (2006) Payments for environmental services in Vietnam: assessing an economic approach to sustainable forest management. Economy and Environment Program for Southeast Asia, Singapore

Chladna Z (2007) Determination of optimal rotation period under stochastic wood and carbon prices. For Policy Econ 9:1031–1045

Department of Forestry (2006) Vietnam's joining WTO: opportunities and challenging for Vietnam's forest enterprises, News on Five Million Hectare Program 2 (2006). Ministry of Agriculture and Rural Development, Hanoi

Department of Science and Technology-MARD (2003) Yield table of 14 dominant tree species in productive planted forests in Vietnam, Agriculture, Hanoi

Diaz-Balteiro L, Rodriguez LCE (2006) Optimal rotations on Eucalyptus plantations including carbon sequestration: a comparison of results in Brazil and Spain. For Ecol Manage 229:247–258

Englin J, Callaway J (1993) Global climate change and optimal forest management. Nat Res Model 7:191–202

FAO (2006) Global resources assessment 2005. FAO, Rome

Forest Protection Department (2007) Forest statistics. Ministry of Agriculture and Rural Development, Hanoi

Faustmann M (1849) On the determination of the value which forest land and immature stands possess for forestry. J For Econ 1(1995):7–44

Gutrich J, Howarth RB (2007) Carbon sequestration and the optimal management of New Hampshire timber stands. Ecol Econ 62:441–450

MONRE (2005) Project No. 264 of planting trees in Thai Nguyen goes bankrupt, Retrieved 30 July 2009. http://www.nea.gov.vn/thongtinmt/noidung/btn_4_7_05.htm, 2005

Nghiem THN (2011) Optimal forest management for carbon sequestration and biodiversity maintenance. PhD thesis, Massey University, Manawatu. http://mro.massey.ac.nz/handle/10179/2669

Nghiem N (2013) Biodiversity conservation attitudes and policy tools for promoting biodiversity in tropical planted forests. Biodivers Conserv 22:373–403. doi:10.1007/s10531-012-0418-8

Nghiem N (2014) Optimal rotation age for carbon sequestration and biodiversity conservation in Vietnam. For Pol Econ 38:56–64. doi:http://dx.doi.org/10.1016/j.forpol.2013.04.001

Son NH, Nam DH (2005) Characteristics of growth and technologically exploitable age of hybrid acacia planted in Eastern region of South Vietnam. Vietnam J Agric Rural Dev 14:63–66

Nguyen NB, Nguyen VT, Bui CN, Trinh QT (2006) Why do farmers choose to harvest small-sized timber?–A survey in YenBai Province. Northern Vietnam. Economy and Environment Program for Southeast Asia, Singapore

Pawson SM, Brockerhoff EG, Norton DA, Didham RK (2006) Clear-fell harvest impacts on biodiversity: past research and the search for harvest size thresholds. Canadian J For Res-Revue Canadienne De Recherche Forestiere 36:1035–1046

Tavoni M, Sohngen B, Bosetti V (2007) Forestry and the carbon market response to stabilize climate. Energy Policy 35:5346–5353. doi:http://dx.doi.org/10.1016/j.enpol.2006.01.036

Thomson TA (1992) Optimal forest rotation when stumpage prices follow a diffusion process. Land Econ 68:329–342

van Kooten GC, Binkley CS, Delcourt G (1995) Effect of carbon taxes and subsidies on optimal forest rotation age and supply of carbon services. Am J Agricu Econ 77:365–374

Vietnamese Government (1999) Decree of the Government on allocation of forest lands to households, individuals and organizations for long-term forestry purposes, Decree No 163/1999/QD-C, Dated 16 Nov 1999, Hanoi

Vietnamese Government (2005a) Decree of the Government on contracting agricultural lands, productive forest lands, and aquatic lands of state agricultural enterprises and state forest enterprises, Decree No 135/2005/QD-CP, Dated 08 Nov 2005, Hanoi

Vietnamese Government (2005b) Decision of Head of Ministry of Agriculture and Rural Development on Regulations on harvesting timber and non-timber products, Decision No. 40/2005/QD-BNN, Dated 7 July 2005, Hanoi

Vietnamese Government(2007a) Decision of the Prime Minister on some policies to develop productive planted forests in the period 2007–2015, Decision No 147/2007/QD-TT, Dated 10 Sept 2007, Hanoi

Vietnamese Government (2007b) Decision of the Prime Minister on the amendments of some policies regarding Five Million Hectare Program, Decision No 100/2007/QD-TTg, Dated 06 July 2007, Hanoi

Vo DH, Dang TT, Nguyen HT, Nguyen VB, Dang TD (2009) Productivity and carbon sequestration of some major planted forest in Vietnam, Agriculture, Hanoi

Yen Bai Forestry Department (2008) Report on forest status. Yen Bai Forestry Department, Yen Bai

Chapter 7
Rubber Plantation Development in Cambodia: A Cost-Benefit Analysis

Dararath Yem, Neth Top, and Vuthy Lic

Abstract This study focuses on several new policy instruments introduced by the Government of Cambodia to encourage changes in land use, especially a shift away from existing uses to smallholder rubber plantations. A cost-benefit analysis is conducted for two land conversion schemes: first, the conversion of forestland to large-scale rubber plantations in Tumring Commune, Sandan District, Kampong Thom Province, and second, the conversion of crop production to smallholder rubber plantations in several districts of Kampong Cham Province. Five specific options were evaluated. The results of the cost-benefit analysis show clearly that the conversion from crop production (maize, soybean, cassava, and cashew) to small-holder rubber plantation provides the largest benefit to farmers involved in those conversion schemes. The conversion of forestland into large-scale rubber plantation ranks last in economic terms. Sensitivity analyses demonstrate that despite varying key basic assumptions, the ranking of all crops and forest conversion schemes remains unchanged.

Keywords Cambodia • Land use • Cost-benefit analysis • Sensitivity analysis • Forest conversion

Introduction

Background

The available land for rubber plantation in Cambodia is located in Kampong Cham, Kratie, Kampong Thom, Ratanakiri, Mondulkiri, Battambang, Preah Vihear, and Pailin Provinces. It covers approximately 900,000 ha, where more than 700,000 ha is basaltic red soil (MAFF 2006). Of the total area of basaltic red soil in Cambodia, 186,600 ha is located in Lower Mekong including Kampong Cham and Kratie Provinces; 520,000 ha in the northeast plateau including Ratanakiri Province

D. Yem (✉) • N. Top • V. Lic
Environment and Natural Resources Unit, Cambodia Development Research Institute (CDRI), Phnom Penh, Cambodia
e-mail: dararath@online.com.kh; licvuthy@yahoo.com

D. James, H.A. Francisco (eds.), *Cost-Benefit Studies of Natural Resource Management in Southeast Asia*, DOI 10.1007/978-981-287-393-4_7

(180,000 ha), Mondulkiri (320,000 ha), and Kratie (20,000 ha); and 10,000 ha in mountain areas. A realistic estimate of the potential area for rubber cultivation in Cambodia for the north and east areas of the country is approximately 330,000 ha.

As of 2006, the area under rubber plantation was 63,000 ha. The plantations can be divided into three categories of ownership: state, household-owned, and private-industrial plantations. The state-owned plantations are mainly located in Kampong Cham Province, comprising over 63 % of the total rubber plantation land. They are controlled by seven state companies: Chup, Peam Cheang, Krek, Memot, Snuol, Chamcar Andong, and Boeung Ket Rubber Plantations. Privately owned plantations covered around 17 % (Khun 2006). Household-owned plantations – or to use another term "smallholder rubber plantations" – commenced in 1990 (AFD 2006). The areas under smallholder plantation have increased rapidly from 10,000 ha in 1995 to 18,600 ha in 2006 (GDRP 2006). AFD (2006) projections indicate that areas under smallholder rubber plantations would increase to 35,000 ha by year 2010.

In addition to existing private-industrial plantations, more privately owned plantations are granted under the economic land concession (ELC) scheme, initiated by the Royal Government of Cambodia (RGC) in 1992. Such concessions comprise agro-industrial plantations, including cash crops (palm oil, cashew nuts, cassava, bean, sugar cane, rice, and corn), fast-growing trees (acacia, eucalyptus, pine), and other valuable trees such as rubber and teak. Since then, about 907,000 ha of land have been approved for development under 50 concessions. Of the total land area, about 13 % (approximately 120,000 ha) was granted partly for growing rubber trees. Such plantation development seems to be one of the main crops currently planned to be cultivated. However, only a limited number of plantations are actually in progress, while for others, it remains unknown whether rubber trees have been planted or not (MAFF 2006).

Most of the rubber plantation area in Cambodia has been converted from forestland. Large areas of forestland have been cleared to develop rubber plantations. Based on MAFF (2006), large parts of many concessions fall within areas covered by natural forest – as has been found by tracking the coordinates of concessions provided in the agreement contracts using the digital land-use map produced by the Forestry Administration in 1997 – meaning that these forests are subject to clearance for the purposes of establishing rubber plantations. In 2001, for example, the Royal Government of Cambodia canceled 6,200 ha of forest concessions of three logging companies in Tumring Commune, Kampong Thom Province. Of the total canceled area, 912 ha were given to local people, 929 ha were reserved for smallholder rubber production, and 4,359 ha were given to Chup Rubber Plantations for rubber development. Chup Rubber Plantation, moreover, also hired the adjacent forest concessionaires to log 1,200 ha to develop rubber plantations. The development of rubber plantations is thus considered one of the major causes of deforestation in developing countries (Liu et al. 2006).

Jiang and Wang (2003) claimed that rubber plantations play an important role similar to natural forests. Their functions have been extensively studied in terms of rainfall, water runoff, evaporation and transpiration, and soil moisture. The same study showed that annual rainfall, which is intercepted by plantations, accounted for 11 and 63 % for evaporation and transpiration (water returned to the atmosphere from plant cells and soil moisture). Further, it was shown that plantations decrease soil erosion by reducing rainwater runoff and hence can play a very important role in protecting watersheds. Another indirect benefit of rubber plantations is carbon sequestration. According to Khun (2006), rubber plantations at the age of 25 can sequester 525 tons of carbon per hectare.

The present study takes a closer look at the economic aspects of land conversion to rubber plantations by analyzing the economic costs and benefits of conversion scenarios at different locations. The government declaration No. 03 Rbk dated 08/08/00 encourages farmers to plant rubber trees on their small-scale plantations with credit support and land tenure policy. The policy aims to push rubber development as an agro-industry not only on farmers' lands but also on private lands especially small-scale rubber plantation. The study endeavors to contribute to policy discussion on the potential role of rubber development in poverty alleviation and economic development in Cambodia.

Aims of the Study

The general objective of this study was to analyze trade-offs related to rubber plantation development in Cambodia based on a cost-benefit analysis (CBA) framework. More specifically, the study compares the direct and indirect values of existing land uses with those of rubber plantations based on estimated returns from established rubber plantations of different ages, in Kampong Cham and Kampong Thom Provinces.

Methodology

Data Collection

The research team collected secondary information related to economic land concessions, forestland, rubber plantations, and others prior to fieldwork. They also interviewed key informants from relevant institutions such as the General Directorate of Rubber Plantation, the Rubber Research Institute of Cambodia, the Forestry Administration, and the National Institute of Statistics.

Information on standing stock, timber productivity, non-timber forest products (NTFPs), yield of latex, latex price, and the environmental functions of forests and rubber plantations such as carbon sequestration, soil erosion control, and watershed protection was obtained from published literature (e.g., Hansen and Top 2006; Khun 2006; Yamashita et al. 1999; Balsiger et al. 2000).

The fieldwork comprised three phases. Phase I covered the selection of sites, general observation, and positioning of the selected plantations using GPS, etc. This kind of information was used in identifying the actual location of plantations and previous types of land use before the arrival of plantations, using the land-use map of 1998 produced by the Forestry Administration in 1999 and ArcView 8.3 (ESRI, Inc.) software application. Apart from natural forests, additional baseline land uses were identified. Phase II involved the pretesting of a questionnaire and refinement of the survey instrument. Phase III was the field survey of households and plantation owners, farmers' key informant interviews, and Rapid Rural Appraisal (RRA).

The household survey collected information on changes in the livelihoods of people in the communes before and after the establishment of rubber plantations. The information gathered includes the income generated from natural forests and rubber, the contribution of rubber plantations in improving livelihoods, the direct and indirect values of rubber plantations, reasons why local people decided whether or not to plant rubber, and past and current price of latex. Information on people's perceptions of land use changes and the impacts of land conversion on local livelihoods was also collected. Three villages located within and/or surrounding selected plantations were chosen for the survey. Approximately 10 % of the total households within each village were randomly selected for interview.

The plantation holder interviews took place parallel to the household survey. Plantation holders included owners of both small- and large-scale plantations. Around 20 small-scale and 4 large-scale plantations located in different places were selected for the study. Large-scale plantations comprised two plantations, each selected from industrial and state-owned plantations. The respondents were asked about the history of rubber development, costs of establishment and maintenance of rubber plantations, expected income from production of latex and timber over a production cycle, distribution of income originating from rubber plantations, location of rubber plantations in relation to preferred soil type, or originality of land-use type (e.g., forest or barren land). Key informants, including chiefs of villages or communes and officials and workers at the plantations, also provided similar information related to rubber development in the area and the effects of this on villagers' livelihoods.

Specification of Land-Use Options

The study considered the five following land-use conversion schemes:

- Option 1: Conversion from forest land to large-scale rubber plantation
- Option 2: Conversion from cassava production to smallholder rubber plantation
- Option 3: Conversion from soybean production to smallholder rubber plantation
- Option 4: Conversion from maize production to smallholder rubber plantation
- Option 5: Conversion from cashew production to smallholder rubber plantation

Benefits and costs of each option were estimated and the results compared. Sensitivity analysis was also carried out. Details of the analysis appear below.

Option 1: Conversion from Forestland to Large-Scale Rubber Plantation

Baseline Land Use: Forest Conservation

Under forest conservation land use, it was assumed that no changes are made to the current forest cover. Logging is banned, and the local population is freely allowed to collect NTFPs. The major category of net benefits comprises net returns associated with environmental benefits, especially biodiversity conservation and carbon sequestration. These values were taken from existing studies. The present value of net economic benefits of forest conservation was estimated over a time period of 25 years, at a 10 % discount rate.

Benefits from Forest Conservation

The benefits of forest conservation (see Table 7.1) were divided into two categories:

- *Direct revenue from NTFPs.* The value of NTFPs of USD 375 per ha per year was extracted from Hansen and Top (2006).
- *Environmental benefits.* These values were extracted from previous studies especially: (1) Hansen and Top (2006) value of water conservation of USD 70 per ha per year, value of soil conservation of USD 60 per ha per year, and value of carbon sequestration of USD 759 per ha per year with an increment of 2 % per year, and (2) Bann (1998) values of biodiversity conservation of 300 per ha per year.

Cost of Forest Conservation

The cost of forest conservation (see Table 7.2) was divided into the following items:

- *Cost of protection and silviculture.* The value of forest area protection and silviculture was extracted from a previous study of USD 25 per ha per year (Hansen and Top 2006).
- *Capital investment for NTFPs collection.* This value was extracted from an existing report (CBNRM-LI 2008). These costs involve the purchasing of materials used for timber and non-timber collection especially ox, oxcart, knife, ax, line, shoe cloth, and mosquito net. The costs would be incurred only once in year one. It was assumed that the ox and oxcart would last for the whole project period of 25 years. The knife, ax, line, shoe cloth, and mosquito net

Table 7.1 Estimated benefits accruing from forest conservation (Hansen and Top 2006 except (*): Bann 1998)

Year	PV	Year 1	Year 2	Year 3	Year 4	Year 5	...	Year 24	Year 25
Value of NTFPs products	3,404	375	375	375	375	375	...	375	375
Value of water conservation	635	70	70	70	70	70	...	70	70
Value of soil conservation	545	60	60	60	60	60	...	60	60
Value of biodiversity *	2,723	300	300	300	300	300	...	300	300
Carbon sequestration value	8,379	759	781	803	825	847	...	1,265	1,287
Total benefits	15,686	1,564	1,586	1,608	1,630	1,652	...	2,070	2,092

Notes: All values are in USD per ha

Table 7.2 Estimated costs incurred in forest conservation (Hansen and Top 2006; CBNRM-LI 2008)

Year	PV	Year 1	Year 2	Year 3	Year 4	Year 5	...	Year 23	Year 24	Year 25
Protection and silviculture (1)	227	25	25	25	25	25	...	25	25	25
Cost of NTFPs collection (2)	702	460	30	40	30	40	...	40	30	40
Total costs	929	485	55	65	55	65	...	65	55	65

Notes: All values are in USD per ha

would be replaced every 2 years. The costs of timber and non-timber product collection are included in the costs of purchasing food as well as the cost to cover medical expenses to treat illness incurring the stay in the forest for NTFP collection.

Alternative Land Use: Large-Scale Rubber Plantation

In the large-scale rubber plantation scheme, it was assumed that the current forest cover area will be cleared and transformed into large-scale rubber plantations. The benefit accrues only in the sixth year after planting rubber trees. The present value of net economic benefits in the rubber plantation was calculated using a period of 25 years and a 10 % discount rate. No attempt was made to estimate the economic value of the change in people's livelihood, timber products when cut, wildlife trade, environmental damages, and all indirect benefits that could accrue especially social infrastructure (road, school, hospital) or credits from the Clean Development Mechanism.

Table 7.3 Estimated benefits accruing from large-scale rubber plantation (Hansen and Top 2006)

Year	PV	Year 1	…	Year 6	Year 7	Year 8	…	Year 23	Year 24	Year 25
Value of dried rubber	11,686	0	…	0	396	991	…	3,964	3,964	3,964
Value of rubber by-product	23	0	…	0	1	2	…	8	8	8
Value of rubber tree	923	0	…	0	0	0	…	0	0	10,000
Carbon sequestration value	7,153	0	…	659	766	872	…	2,472	2,579	2,685
Total benefits	**19,785**	**0**	…	**659**	**1,163**	**1,865**	…	**6,443**	**6,550**	**16,657**

Notes: PV, present value. All values are in USD per ha

Benefits of Large-Scale Rubber Plantation

The benefits accruing from large-scale rubber plantations are generated from selling dried rubber products, their by-products, old rubber trees at 25th year, and the indirect value of carbon sequestration (see Table 7.3). There is also the possibility of obtaining revenue from renting the lands for intercrop production during the first 3 years, but the study does not take into account this benefit, as it is problematic. Some large-scale rubber plantation owners encourage their workers to undertake orchard crop production on the land available without charge, as a bonus for them.

The price of dried rubber products at international market (SMR-L) varies monthly and represents USD 3,290 per ton on June 2008. The study took into account the average rate in 2007 of USD 2,331 per ton for the economic calculations. In line with the results of the field survey conducted in 2007, the study assumed that the price of dried rubber products at farm gate is equal to 85 % of the rate in the international market, equivalent to USD 1,982 per ton. Dried rubber production varies from 0.72 ton per ha in the 6th year to a peak of 1.80 ton per ha in the 16th year and declines to 0.79 ton per ha in the 25th year. The rubber by-products represent 0.2 % of total dried rubber. The revenue from selling the old rubber tree was estimated at USD 3,300 per ha.

Costs of Large-Scale Rubber Plantation

Table 7.4 shows detailed costs incurred in large-scale rubber plantation schemes. The costs are divided into six components comprising: (1) maintenance of rubber trees, (2) tapping and harvesting, (3) material and equipment, (4) manufacturing and coagulating, (5) taxes, and (6) overhead expenses. The average annual total cost for rubber plantations is USD 628 per ha with minimum cost of USD 226 per ha in the second year and maximum of USD 809 per ha at the 11th year.

Option 2: Conversion from Agriculture to Smallholder Rubber Plantation

Baseline Land Use: Agriculture

Benefits of Cassava Production The benefits of cassava production are generated from selling cassava roots for the average price of USD 20 a ton. For economic analysis, the study assumed cassava production with an average yield of 12.06 ton/ha (ACI 2005). Thus, the gross benefit of one-hectare cassava farm is constant over the year and represents USD 254 per year.

Costs of Cassava Production The costs are divided into (1) hiring tractor or bullock for land preparation; (2) labor costs covering land preparation, seeding,

Table 7.4 Costs incurred in large-scale rubber plantation (Hansen and Top 2006)

Year	1	2	3	4	5	6	7	8	9	10	11	12	13	14	15	16	17	18	19	20	21	22	23	24	25
Maintenance of rubber trees																									
Land clearing	74	7	4	4	3	4	2	3	4	5	5	9	5	6	6	6	7	4	5	5	9	5	6	6	6
Planting	103	11	0	0	0	0	0	0	0	0	0	0	0	0	0	0	0	0	0	0	0	0	0	0	0
Maintenance	107	105	107	61	55	55	41	51	46	50	47	43	46	43	45	45	44	46	50	47	43	46	43	45	45
Fertilizing	63	63	101	133	172	210	0	0	0	0	0	0	2	1	1	1	1	0	0	0	0	2	1	1	1
Maladies treatment	0	0	0	11	11	17	10	22	20	12	10	14	15	15	14	14	14	20	12	10	14	15	15	14	14
Subtotal	**347**	**185**	**211**	**208**	**242**	**285**	**53**	**76**	**70**	**67**	**62**	**66**	**67**	**65**	**67**	**66**	**66**	**70**	**67**	**62**	**66**	**67**	**65**	**67**	**66**
Tapping and harvesting																									
Tapping wages	0	0	0	0	0	0	83	85	87	114	113	115	114	112	113	112	114	87	114	113	115	114	112	113	112
Tapping bonus	0	0	0	0	0	0	96	92	83	77	80	69	66	65	66	63	62	83	77	80	69	66	65	66	63
Chemical using	0	0	0	0	0	0	33	33	44	47	49	54	54	50	48	48	46	44	47	49	54	54	50	48	48
Transport of latex	0	0	0	0	0	0	15	16	19	21	24	24	25	26	27	25	26	19	21	24	24	25	26	27	25
Subtotal	**0**	**0**	**0**	**0**	**0**	**0**	**226**	**231**	**233**	**260**	**266**	**261**	**258**	**254**	**254**	**249**	**248**	**233**	**260**	**266**	**261**	**258**	**254**	**254**	**249**
Material/equipments																									
Material for tappers	0	0	0	0	0	0	7	7	7	4	4	4	4	4	4	4	4	7	4	4	4	4	4	4	4
Material for rubber trees	0	0	0	0	0	0	13	13	15	18	16	15	15	14	15	18	18	15	18	16	15	15	14	15	18
Tanks/harvested equipments	0	0	0	0	0	0	0	0	0	1	1	0	1	0	0	0	0	0	1	1	0	1	0	0	0
Subtotal	**0**	**0**	**0**	**0**	**0**	**0**	**21**	**20**	**22**	**23**	**21**	**19**	**19**	**19**	**20**	**22**	**22**	**22**	**23**	**21**	**19**	**19**	**19**	**20**	**22**
Manufacturing/ coagulating																									
Wages expenses	0	0	0	0	0	0	9	8	6	5	3	9	8	8	6	7	7	6	5	3	9	8	8	6	7
Bonus expenses	0	0	0	0	0	0	0	0	0	0	0	0	0	0	0	0	0	0	0	0	0	0	0	0	0
Oil expenses	5		5	5	4	6	14	15	15	17	17	19	16	16	17	16	15	15	17	17	19	16	16	17	16
Maintenance of machineries	5	5	6	5	5	5	6	6	5	5	7	10	9	8	8	8	8	5	5	7	10	9	8	8	8

(continued)

Table 7.4 (continued)

Year	1	2	3	4	5	6	7	8	9	10	11	12	13	14	15	16	17	18	19	20	21	22	23	24	25
Electricities/ Water	11	11	11	11	12	12	12	28	35	46	48	64	63	64	63	64	63	35	46	48	64	63	64	63	64
Chemical using expenses	0	0	2	3	7	12	14	13	13	14	14	7	8	9	9	9	9	13	14	14	7	8	9	9	9
Wrapping expenses	0	0	0	0	0	0	40	42	39	42	35	36	35	35	35	35	35	39	42	35	36	35	35	35	35
Transport expenses	0	0	0	0	0	0	1	1	2	2	2	2	3	3	3	4	4	2	2	2	2	3	3	3	4
Warehousing expenses	0	0	0	0	0	0	0	0	0	0	0	0	0	0	0	0	0	0	0	0	0	0	0	0	0
Subtotal	**21**	**21**	**24**	**25**	**27**	**34**	**95**	**114**	**116**	**131**	**126**	**148**	**143**	**143**	**143**	**143**	**141**	**116**	**131**	**126**	**148**	**143**	**143**	**143**	**143**
Taxes																									
Export taxes	0	0	0	0	0	0	18	27	27	12	86	143	135	138	140	142	143	27	12	86	143	135	138	140	142
Export fees	0	0	0	0	0	0	17	20	13	14	27	19	17	18	17	16	17	13	14	27	19	17	18	17	16
Subtotal	**0**	**0**	**0**	**0**	**0**	**0**	**35**	**47**	**41**	**26**	**113**	**162**	**153**	**157**	**158**	**158**	**161**	**41**	**26**	**113**	**162**	**153**	**157**	**158**	**158**
Overhead Expenses	36	20	22	22	25	30	22	160	168	154	151	154	149	149	151	148	147	168	154	151	154	149	149	151	148
Total costs	**404**	**226**	**258**	**255**	**294**	**349**	**452**	**649**	**650**	**660**	**739**	**809**	**789**	**787**	**792**	**786**	**785**	**650**	**660**	**739**	**809**	**789**	**787**	**792**	**786**

Notes: All values are in USD per ha of rubber plantation. (1) Land clearing, route, bridging, stream, Setting up poles, digging holes, planting. Source: (Hansen and Top 2006)

weeding, and harvesting; (3) input components including purchase of seed; and (4) transport costs. The costs are considered constant over time and represent USD 98 per ha per year.

Benefits of Soybean Production The benefits of the soybean production are generated from selling soybeans for the average price of USD 327 a ton. The study assumes wet season soybean production with an average yield of 1.217 ton/ha (ACI 2005). Thus, the gross benefit of a 1-ha soybean farm is considered constant over the year and represents USD 397 per year.

Costs of Soybean Production The costs are divided into (1) labor costs covering plowing, seeding, weeding, harvesting, and graining; (2) input components including purchase of seed; and (3) transport costs. The costs are considered to be constant over time and represent USD 311 per ha per year.

Benefits of Maize Production The benefits of the maize production are generated from selling maize at the average price of USD 141 a ton. The study assumed maize production with an average yield of 2.71 ton/ha (ACI 2005). Thus, the gross benefit of 1-ha maize farm is considered constant over a year and representing USD 381 per year.

Costs of Maize Production Those costs are divided into (1) labor costs covering plowing, seeding, weeding, and harvesting; (2) input components including purchase of seed; and (3) transport costs. The costs are considered to be constant over time and represent USD 316 per ha per year.

Benefits of Cashew Production Cashew produces a crop only in 3rd year after the planting. The benefit accruing from cashew production is generated from selling of raw cashew nut with the price of USD 730 a ton. Cashew nut production varies from 300 kg per ha in 3rd to 5th year to the peak of 1,500 kg per ha from 11th to 15th year, falling to 1,100 kg per ha in 17th to 25th year and from 6th to 10th year.

Costs of Cashew Production The study divided the cost items for cashews into four major components, shown in Table 7.5. Land preparation includes the costs to cover, planting materials, land clearing, digging planting holes, and planting. These costs accrue only in year 1 of starting the cassava production. Fertilizers include the cost to cover the purchase of chemical fertilizer (NPK) and manure. This cost accrues every year. Operation and maintenance costs include the costs of fertilizer application and management and infrastructure. These costs accrue every year. Harvesting covers the costs to hire labor during harvesting. It accrues only in the fourth year after plantation. The total costs differ from 1 year to another.

Alternative Land Use: Smallholder Rubber Plantation

The study assumed that benefits accrue only in the sixth year after planting the rubber trees. The present value of net economic benefits was calculated for a period of 25 years and a 10 % discount rate. No attempt was made to estimate the

Table 7.5 Costs incurred in cashew production (Hansen and Top 2006; EIC 2006)

Year	PV	Year 1	Year 2	Year 3	…	Year 6	Year 7	Year 8	…	Year 23	Year 24	Year 25
Land preparation	104	114	0	0	…	0	0	0	…	0	0	0
Fertilizers (NPK manure)	2,126	425	213	213	…	213	213	213	…	213	213	213
Operation and maintenance	536	95	55	55	…	55	55	55	…	55	55	55
Harvesting	82	0	0	0	…	10	10	10	…	15	15	15
Total costs	**2,848**	**634**	**268**	**268**	…	**278**	**278**	**278**	…	**283**	**283**	**283**

Notes: PV, present value. All values are in USD per ha

Table 7.6 Estimated benefits accruing from the smallholder rubber plantation (EIC 2006)

Year	PV	Year 1	…	Year 6	Year 7	Year 8	…	Year 23	Year 24	Year 25
Value of coagulum latex	9,462	0	…	875	1,250	1,625	…	2,250	2,250	2,250
Value of intercropping	622	250	…	0	0	0	…	0	0	0
Value of rubber tree	305	0	…	0	0	0	…	0	0	3,300
Total benefits	10,388	250	…	875	1,250	1,625	…	2,250	2,250	5,550

Notes: PV, present value. All values are in USD per ha

economic value of the change in people's livelihood, environmental damages, carbon sequestration, and the opportunity of a clean development mechanism (CDM) component of the project.

Benefits of Smallholder Rubber Plantation

The benefits accruing in the smallholder rubber plantation scheme (see Table 7.6) are generated from selling (1) coagulum latex with the price of USD 0.63 per kg where coagulum latex production varies from 1,400 kg per ha in 6th year to the peak of 3,600 kg per ha in 25th year; (2) old rubber trees estimated at USD 3,300 per ha; and (3) intercropping of soybean and mung bean (twice a year) of USD 250 per ha per year. This last benefit accrues only in the first 3 years of planting the rubber trees.

Costs of Smallholder Rubber Plantation

For the smallholder rubber plantation, the required investment to transform coagulum latex into dried rubber is high. Thus, the farmer prefers to sell the coagulum latex to an intermediary at the farm gate. Table 7.7 shows detailed costs incurred in smallholder rubber plantations. The costs are divided into labor and other input components. Labor for planting includes land preparation, sticking, digging hole, and gardening. Labor for maintenance includes weeding, fertilizer application, plant replacement, suckering, and treatments. The input components include purchase of rubber tree, fertilizer, chemical products, and harvesting materials.

Results of Cost-Benefit Analysis

NPVs of Land-Use Options

The study assessed (1) the incremental net benefit of converting from forest conservation to large-scale rubber plantations and (2) the incremental net benefit of converting the four crop production schemes to smallholder rubber plantations. Incremental net benefits comprise the difference between the net benefits of rubber plantation schemes (large-scale or smallholder) and the net benefits accruing for conserved forestland or the four crop production schemes.

The present values of net benefits or net costs were estimated over a 25-year period with a discount rate of 10 %. The results are shown in Table 7.8 in which the various options are ranked according to their incremental NPVs. The higher the incremental net benefits, the more economically desirable is the conversion to rubber plantation (either large-scale or smallholder)

Table 7.7 Costs incurred in smallholder rubber plantation (EIC 2006)

	PV	Year 1	Year 2	Year 3	…	Year 6	Year 7	Year 8	…	Year 23	Year 24	Year 25
Labor for planting	217	226	14	0	…	0	0	0	…	0	0	0
Labor for maintenance	580	113	181	130	…	32	32	32	…	0	0	0
Labor for harvesting	1,269	0	0	0	…	240	240	240	…	240	240	240
Rubber tree	307	293	49	0	…	0	0	0	…	0	0	0
Fertilizer	244	40	40	60	…	0	0	0	…	0	0	0
Chemical	73	23	23	18	…	5	5	5	…	0	0	0
Harvesting materials	37	0	0	0	…	66	0	0	…	0	0	0
Total costs	**2,727**	**695**	**307**	**208**	…	**343**	**277**	**277**	…	**240**	**240**	**240**

Notes: PV, present value. All values are in USD per ha

Table 7.8 Present value (PV) of incremental net benefits (USD per ha)

	PV of net benefit	PV of incremental net benefit	BCR	IRR
Base case 1 – forest conservation	14,757	–	0	0 %
Base case 2 – cassava production	1,416	–	0	0 %
Base case 3 – soybean production	785	–	0	0 %
Base case 4 – maize production	584	–	0	0 %
Base case 5 – cashew production	2,270	–	0	0 %
NPV of smallholder rubber plantation				
Option 4 – maize to rubber (smallholder)	7,661	**7,076**	47.9	38 %
Option 3 – soybean to rubber (smallholder)	7,661	**6,875**	72.0	36 %
Option 2 – cassava to rubber (smallholder)	7,661	**6,244**	4.4	32 %
Option 5 – cashew to rubber (smallholder)	7,661	**5,390**	43.7	N/A
Option 1 – forest to rubber (large scale)	15,690	**934**	1.3	11 %

Table 7.8 shows that all crop conversion options yield positive values of incremental net benefits, ranging from USD 934 to 7,076 per ha over a 25-year period with a 10 % discount rate. The benefit cost ratios (BCRs) range from 1.3 for Option 1 (conversion from forest to large-scale rubber plantation) to the highest BCR of 72 for Option 3 (conversion from soybean production to smallholder rubber plantation).

Option 4 (conversion from maize production to smallholder rubber plantation) is the most preferred on economic efficiency grounds and ranks first compared relative to other crop conversion schemes. The incremental net benefit of Option 4 has a positive value of USD 7,076 per ha of farmland with a BCR of nearly 48.

Option 3 (conversion from soybean to smallholder rubber plantation) is the second most preferred on economic efficiency grounds. It ranks second closely behind Option 4 (maize to rubber) with a high BCR of 72.

Option 2 (conversion from cassava to smallholder rubber plantation) is the third most preferred on economic efficiency grounds. It ranks third closely behind Option 3 (soybean to rubber). The BCR of Option 2 is quite low at 4.4.

Option 5 (conversion from cashew to smallholder rubber plantation) is the fourth most preferred on economic efficiency grounds. It ranks fourth closely behind Option 2 (cassava to rubber). The BCR of Option 5 is 43.7.

Option 1 (conversion from forest to large-scale rubber plantation) is the least preferred on economic efficiency grounds. It ranks last, far behind other crop conversion schemes. The BCR of Option 1 is very low at 1.3.

The results of the cost-benefit analysis show clearly that the conversion from crop production (cassava, soybean, maize, and cashew) to smallholder rubber plantations provides large benefits to farmers involved in those conversion schemes. This suggests a need for technical assistance relating to rubber plantations and an extensive follow-up program from a competent authority. This conclusion is consistent with the finding of the AFD project on the establishment of smallholder rubber plantation in six districts of Kampong Cham Province.

Sensitivity Analysis

A sensitivity analysis was conducted to investigate the effects on the present values of incremental net benefits of varying key assumptions relating to the discount rate and project lifetime. The analysis provides a measure of the degree to which these variables can deviate from their estimated values before the preferred options cease to be economically desirable. Different scenarios were tested.

Scenario 1: Change the Discount Rate from 10 % to 15 %

For all crops and forest conversion options, the discount rate was increased from 10 % in the Initial Scenario to 15 % in Scenario 1. The project duration was set at 25 years. Table 7.9 shows that there was no change in the ranking of all crops and forest conversion options. Option 1 would actually result in a negative incremental net benefit of 1,984.

Scenario 2: Reduce Project Lifetime to 15 Years

The effects of reducing the life of the project from 25 to 15 years are shown in Table 7.10. This table shows that again there is no change in the ranking of all crops and forest conversion options. With a reduced lifetime for the project, the conversion from forest to large-scale plantation becomes uneconomical.

Conclusions and Policy Recommendations

The results of the cost-benefit analysis and the sensitivity analyses show that despite variations in the basic assumptions relating to the discount rate, project duration, production costs, and value of crops, the ranking of all crops and forest conversion schemes remains unchanged. It clearly reveals that smallholder rubber plantations represent the economically most desirable land use.

The conversion from forestland to large-scale rubber plantation is a sensitive issue. Most of the benefits accruing from forest conservation relate to biodiversity values and environmental services. Some previous reports, especially those financed by environmental organizations, put the forest value and preservation of

Table 7.9 Scenario 1: PV of incremental net benefits

	PV of net benefit	PV of incremental net benefit	BCR	IRR
Base case 1 – forest conservation	10,191	–	0	0 %
Base case 2 – cassava production	1,009	–	0	0 %
Base case 3 – soybean production	559	–	0	0 %
Base case 4 – maize production	416	–	0	0 %
Base case 5 – cashew production	1,016	–	0	0 %
NPV of smallholder rubber plantation				
Option 4 – maize to rubber (smallholder)	3,944	3,528	5,821	38 %
Option 3 – soybean to rubber (smallholder)	3,944	3,384	96.4	36 %
Option 2 – cassava to rubber (smallholder)	3,944	2,935	3.1	32 %
Option 5 – cashew to rubber (smallholder)	3,944	2,928	51.1	N/A
Option 1 – forest to rubber (large scale)	8,207	−1,984	0.1	11 %

Notes: The values in the PV column are in USD per ha. N/A: not applicable

Table 7.10 Scenario 2: PV of incremental net benefits

	PV of net benefit	PV of incremental net benefit	BCR	IRR
Base case 1 – forest conservation	11,938	–	0	0 %
Base case 2 – cassava production	1,187	–	0	0 %
Base case 3 – soybean production	658	–	0	0 %
Base case 4 – maize production	489	–	0	0 %
Base case 5 – cashew production	1,506	–	0	0 %
NPV of smallholder rubber plantation				
Option 4 – maize to rubber (smallholder)	5,189	4,700	144.0	38 %
Option 3 – soybean to rubber (smallholder)	5,189	4,531	451.3	36 %
Option 2 – cassava to rubber (smallholder)	5,189	4,003	3.5	31 %
Option 5 – cashew to rubber (smallholder)	5,189	3,683	63.2	N/A
Option 1 – forest to rubber (large scale)	9,953	−1,985	0.3	6 %

Notes: The values in the NPV column are in USD per ha. N/A: not applicable

people's livelihoods at the highest level. By contrast, private companies and the government often assert that the conversion of forestland to large-scale rubber plantations would be an ideal means of achieving poverty alleviation targets.

Those benefits would include job creation and other economic benefit from conversion schemes.

Over the past few years, the government has awarded large areas of forestland to private companies under the economic land concession scheme. Some of those companies intend to undertake large-scale cassava production. Extrapolating the results of this study, the conversion from forestland to large-scale cassava production would not be economically viable in the long term.

There is a significant ongoing debate in Cambodia over forest conversion into agricultural industry practice under the economic land concession scheme. This study reconfirms previous studies, finding that the social component has been ignored and the local population generally becomes worse off after the forest conversion scheme. The study recommends that the social component of forest conversion schemes should be considered a high priority. In order that the economic land concession would yield benefits to the local population, the study suggests that civil society and NGOs should be involved in the land compensation procedure. Follow-up programs and technical assistance should be provided to the affected population.

The results of the study indicate that there are significant potential net benefits from large-scale rubber plantations (with a PV of net benefits of USD 15,690 per ha over a 25-year period with a 10 % discount rate) compared to the highest net benefit from other crop production (PVs of USD 2,270 per ha for cashew and USD 1,416 per ha for cassava). Large-scale rubber plantations also generate higher net benefits than conserved forests in economic terms. The PV of forest conservation was estimated at USD 14,757 per ha.

However, the PV of *incremental* net benefits for the conversion from forestlands to large-scale rubber plantation represents the least economically desirable option, with a PV of only USD 934 per ha for the project duration of 25 years, assuming a 10 % discount rate. This is lower than even the worst performing crop conversion, for which the minimum PV of incremental net benefit is USD 5,390 per ha. Indeed, if the discount rate is changed from 10 % to just 12 %, the conversion from forestland to large-scale plantations will not be economically desirable, as the PV of incremental net benefits becomes negative.

Smallholder rubber plantations should be encouraged and promoted in an effort to reduce poverty rates in red soil provinces of Cambodia. Although the development of smallholder rubber plantation requires huge investment in both financial and technical resources, especially for the first 6 years of its establishment, some mechanisms could help farmers to overcome those difficulties.

This study proposes that the government and project developers should adopt various mechanisms to help farmers overcome any difficulties. An appropriate mechanism has been implemented by the AFD project in Kampong Cham Province, representing a successful pilot project, and this model should be extended to other red soil provinces in Cambodia.

The study also demonstrates that cost-benefit analysis could be used as a decision support tool by the Department of Environmental Impact Assessment of the Ministry of Environment to evaluate the economic return of investment projects that involve forestland conversion schemes. Those projects are obliged to produce a report on environmental impact assessment of the relevant investments and submit it to the Ministry of Environment for approval.

References

ACI (Agrifood Consulting International) (2005) Final report for the Cambodian Agrarian Structure Study. Prepared for the Royal Government of Cambodia and the World Bank, Bethesda

AFD (Agence Française de Développement) (2006) Smallholders rubber development in Kampong Cham province. International workshop on rubber development, Vientiane

Balsiger J, Bahdon J, Whiteman A (2000) The utilization, processing and demand for rubber wood as source of wood supply, Working paper No. APFSOS/WP/50. Food and Agriculture Organization, Rome

Bann C (1998) The economic valuation of tropical forest land use options: a manual for researchers. Economy and Environment Program for South East Asia, Singapore

CBNRM-LI (2008) Economic valuation of a natural resource management project: a cost and benefit analysis of the Community Forestry Project in Chumkiri district, Kampot province, Cambodia. CBNRM-Learning Institute, Phnom Penh

EIC (Economic Institute of Cambodia) (2006) Cambodia's export diversification and value addition, Final Report. EIC and UNDP-RCC, Phnom Penh

GDRP (General Directorate of Rubber Plantation) (2006) Natural Rubber Bulletin, Phnom Penh, Cambodia

Hansen K, Top N (2006) Natural forest benefits and economic analysis of natural forest conversion in Cambodia. Cambodia Development Resource Institute, Phnom Penh

Jiang JS, Wang RS (2003) Hydrological eco-service of rubber plantations in Hainan Island and its effect on local economic development. J Environ Sci 15(5):701–709

Khun K (2006) Triple potential of hevea-culture in Cambodia, Masters degree thesis, Kyushu University, Japan

Liu W, Hu H, Ma Y, Li H (2006) Environmental and socioeconomic impacts of increasing rubber plantation in Menglun Township, Southwest China. Mt Res Dev 26(3):245–253

MAFF (2006) Price bulletin for agricultural commodities in 2005. Ministry of Agriculture, Forestry and Fishery, Phnom Penh

Yamashita Y, Watanabe H, Takeda S (1999) Rubber wood: characteristic of its supply and development of its utilization. For Res 70:65–70

Chapter 8
Environmental Consequences of and Pollution Control Options for Pond "Tra" Fish Production in Thotnot District, Can Tho City, Vietnam

Thi Lang Vo, Quang Vinh Ky, and Thi Thanh Truc Ngo

Abstract This study examines the environmental consequences arising from pond "tra" fish breeding in the Mekong Delta. It explores technically and economically feasible wastewater treatment options for bringing water pollution down to an acceptable level in accordance with Vietnam's environmental standards. Results show that the farmers' practice of exchanging water between fishponds and outside water bodies is one of the main causes of surface water pollution in tra fish-breeding areas. Water sample analysis indicated that the COD concentration in tra pond water was 34 mg/l, exceeding the limit of <10 mg/l according to Vietnam's surface water quality standards (TCVN 5942-1995, class A). The pollution load rate in terms of COD per kilogram of fish produced was 0.098.

To cope with the pollution situation more effectively, some technical pollution control options are proposed, of which the trickling filter technology was found to be the most cost-effective. Some policy recommendations to address the environmental pollution caused by pond tra fish production in the Mekong Delta are also given.

Keywords Mekong Delta • Pollution control • Fish production • Cost-effectiveness analysis • Water quality

---

T.L. Vo (✉)
School of Economics & Business Administration, Can Tho University, Can Tho City, Vietnam
e-mail: vtlang@ctu.edu.vn

Q.V. Ky
Climate Change Coordination Office of Can Tho City, Can Tho City, Vietnam
e-mail: kqvinh@ctu.edu.vn; vpbdkhct@cantho.gov.vn

T.T.T. Ngo
Department of Agricultural, Natural Resources and Environmental Economics, Can Tho University, Can Tho City, Vietnam
e-mail: ntttruc@ctu.edu.vn

D. James, H.A. Francisco (eds.), *Cost-Benefit Studies of Natural Resource Management in Southeast Asia*, DOI 10.1007/978-981-287-393-4_8

Introduction

Background

Aquaculture production is a thriving industry in the Mekong Delta (MD) of Vietnam. The industry has grown rapidly in the last two decades. Tra fish (*Pangasius hypophthalmus*) and basa fish (*Pangasius bocourti sauvage*) are two popular catfish species bred in the MD. They can be cultured in floating cages or ponds. With ponds, farmers can breed the fish all year round, whereas with cages/pens, they can only do so for 6–7 months in a year. Thus, pond fish-breeding areas have mushroomed in the delta.

Fish producers discharge waste into local water sources, and this adversely impacts local communities that rely on river water as their main domestic water source. In addition, polluted water discharged from fishponds can cause disease in fish. Due to the tides, a great amount of wastewater remains on site, subsumed in nearby water channels, lacking time to go far and so, ends up being eventually pumped back into fishponds—this creates disease in the fish and increases antibiotic costs in breeding them. Furthermore, poor water quality can change the color of tra fish meat from white to yellow, and this may cause the produce to fall below quality standards for domestic consumption and export. The deterioration of surface water conditions associated with catfish farming can be the result of many factors. Increased fish-farming areas, intensive farming with high densities of fish, excessive use of feed, overuse of antibiotics and other chemicals, and lack of wastewater treatment systems are among them.

The 2003 Can Tho Statistical Yearbook reports more than 600,000 ha of surface water in the MD being used for aquaculture. At a depth of 1 m, the water volume for such acreage is around six billion cubic meters. If fish breeders substitute one-fourth of this volume of (pond) water per day using water from canals or rivers, one and a half billion cubic meters of wastewater per day or more than 540 billion cubic meters per year will be dumped into water bodies. As the COD concentration in wastewater from tra and basa ponds is around 118 mg/l (Ky 2004), then the annual flow of the Mekong River (500 billion cubic meters/year) cannot dilute it enough to meet the COD concentration limits in surface water (less than 10 mg/l for Class A) in accordance with TCVN 5942-1995. Given the situation, fish producers as well as the government need to take stronger action to reduce the adverse impacts of tra fish farming on the environment.

Aims of the Study

The general objective of this study was to provide an economic assessment of the environmental consequences of and pollution control options for pond tra fish production in the MD. The study considered only the options for commercial farms selling to export companies.

The specific objectives were:

- To identify the environmental problems associated with the dumping of waste from pond tra fish production
- To identify cost-effective technological solutions
- To recommend policy options to address the water pollution problems caused by pond tra fish production

Methodology

Study Site Selection and Description of Study Area

Tra fish production is found in many provinces in the MD like An Giang, Dong Thap, Can Tho, and Vinh Long, but Thotnot District of Can Tho City is the biggest producer; hence, it was chosen as the study area. Thotnot had the biggest area under tra fish farming—318 ha of tra and basa fish out of a total of 671 ha under tra fish farming in the whole city (Can Tho City Statistical Yearbook 2004). Thotnot is also located near the Can Tho City Environmental Monitoring Station making wastewater sample analysis easier. Thotnot District has eight communes, of which two raise the most tra fish, namely, Thoi Thuan and Tan Loc. These two were among those selected for a household survey, focus group discussions, and water sampling.

Thotnot District lies along the Hau River. It has one town and seven communes. Its natural land area is 17,110.08 ha, with a population of 192,327 inhabitants and a population density of 1,124 persons/km^2 (Can Tho Statistical Yearbook 2004). Aquaculture areas in the past years have shown an upward trend, increasing from 209 ha in 2000 to 393 ha in 2004 and 484.4 ha in 2005, with the pond tra fish as the main species bred for export. The industry has flourished dramatically in the past 3 years since export demand rose in 2003. Tra fish is raised in all Thotnot communes that are situated along canals and the Hau River, but the largest production areas are in the Thoi Thuan (154 ha) and Tan Loc communes (136 ha).

Data Collection

Secondary data relating to tra fish production and its environmental consequences was collected from the Cantho Service of Agriculture and Rural Development, Can Tho University, and local management agencies.

To get general information on the opinions and perceptions of local communities about tra fish farming and its environmental consequences, three focus group discussions were held with three groups of people in Thoi Thuan Commune. The first FGD consisted of 13 local leaders and heads of local organizations such as women's unions and farmer associations. The second FGD comprised seven tra fish breeders, while the third consisted of five farmers who did not raise tra fish and lived along canals from which fish breeders took water to flush their ponds.

Households living along the Hau River and canals (of different levels) were randomly chosen for interview. The total number of surveyed fish farmers was 131 (31 from Tan Loc Commune, 90 from Thoi Thuan, and 10 from the two nearby communes). The data collected from the household survey comprised socioeconomic characteristics of the fish farmers, tra fish-breeding practices, production costs, the advantages and disadvantages of fish breeding, pond wastewater treatment practices, and the farmers' awareness of the environmental impacts of their activities and their responses to possible state regulations.

Water sampling was conducted once a week from five selected fish-farming households. A total of 178 water samples were collected from the households, and eight water samples were taken from the Hau River as control samples. Water quality analysis was done for five parameters: pH, SS, DO, COD, and NH_3-N. The water sample collectors were from the Environmental Monitoring Station of Can Tho City, and the samples were analyzed at the station's laboratory. The pH was analyzed with analytical equipment pH 540-GLP, SS with LF197, DO with YSI-5000, COD with DR 4000, and NH_3-N with DR 4000.

The inlet and outlet water quality analysis results were used to estimate the pollution loads and pollution load rates as well as compared with TCVN 5942-1995 (surface water quality standards). With the water quality analysis results, technical experts designed three technologies capable of treating the organic pollutants, and then economic calculations were made.

Cost-Effectiveness Analysis (CEA)

The CEA was done for proposed technical options to treat the COD effluents generated by pond tra fish production. The CEA is an analytical tool used by economists to evaluate environmental decisions (Field and Olewiler 2005). When there are several ways to attain a certain objective, CEA gives the costs of the various alternatives. The most cost-effective option is the one that achieves the given objective at the lowest cost among all possible options. Although a full cost-benefit analysis (CBA) is a superior tool, due to data limitations, especially difficulties in measuring the benefits (avoided health impacts), a CEA was used for this study.

To identify cost-effective technological options to reduce the water pollution caused by tra fish farming, the abatement cost per unit of pollutant of each control option was estimated and compared. Two measures were used: the abatement cost per kilogram of COD and the COD abatement cost per kilogram of fish growth.

The procedures involved in the calculations are listed below.

Calculating the Abatement Cost per Kilogram of COD

- Determining the appropriate wastewater treatment technology
- Determining the lifespan of the structure (years) (depending on type of materials used)
- Determining the total capacity of the structure (L) [= lifespan (years) abatement capacity per year (kg of COD abated/year)]
- Determining the investment cost of the structure (C_1)
- Determining the discounted flow of annual operation and maintenance (O&M) costs for the structure's lifespan (C_2)
- Calculating the abatement cost per kg of COD using the following equation:

$$C = (C_1 + C_2)/L$$

Calculating the COD abatement cost per kilogram of fish growth:
COD abatement cost per kg of fish growth = C* PLR

Technical Options to Reduce COD Loads

Fish Farmers' Pollution Abatement Practices

To reduce the pollution in tra fishpond water, fish farmers in Thotnot District usually replace pond water with river water to dilute the pollutant concentrations in fishponds. During the tra fish production process, uneaten feed, especially home-processed feed, and excrement pollute the pond water, which can cause disease in the fish. All the respondents reported that to cope with this problem, they regularly flushed the fishponds with water from external sources such as rivers and canals, as advised by technical agencies. This measure helped improve the quality of the pond water but caused problems for the external water bodies. Some fish farmers also use lagoons to reduce water pollution (10 %). However, in Thotnot District, this method is not common because of insufficient land.

Fish farmers used a variety of chemicals to deal with pond water quality problems and fish disease. In addition, they often used the method of changing pond water with outside water to reduce water pollution in the ponds as well as to prevent disease. According to the survey results, common tra fish diseases affected fish organs like the liver (77 %), kidney (70 %), gill (41 %), skin (63 %), head (25 %), and swim bladder and intestines (3 %). The farmers interviewed said that these diseases occurred due to weather changes (49 %), polluted water sources (72 %), parasites and bacteria in water (17 %), and home-processed feed (5 %). When farmers saw their fish get sick, they usually took the fish to the local aquaculture station or a veterinary clinic for them to be diagnosed and then bought drugs for treatment. These medicines were either mixed with feed or water, and the mixture was then dumped into the ponds. In addition, the farmers also resorted to exchanging greater quantities of pond water with outside water to dilute the concentration of organic pollutants.

Pond sediment removal is a common practice in tra fish production. Almost all of the farmers (96.9 %) said that they had their ponds dredged several times, once after harvesting the fish and preparing the ponds for the new stock, and two to three times during the culturing process. The dredged sediment was dumped in several places like dikes, orchards, containment ponds, or water channels. Besides creating water pollution, dumping was contributing to the reduction of canal depth, which threatened the navigation of vessels and the transportation of goods in the area and created/increased canal dredging costs. To deal with this problem, the district and commune authorities passed a regulation to ban the dumping of pond sediment into water channels. However, only 24.4 % of the respondents said that they were aware of the regulation.

Proposed Pollution Abatement Technologies

For this study, three technical control options were evaluated: the aeration system, the trickling filter, and the constructed wetlands. The aeration system is being applied in aqua-product processing plants (Meko and Binh An in Can Tho Industrial Zone), while the trickling filter system is installed at the Ha Tien II Cement Factory in Can Tho City to treat domestic wastewater from workers' activities and kitchens. Constructed wetlands have so far not yet been applied in Can Tho City, but this system is going to be installed to treat wastewater from the students' dormitory at Can Tho University. The three options are explained and illustrated below.

Aeration System

In this system (see Fig. 8.1), wastewater from fishponds will be pumped into a temporary tank (1). Then, the water will be pumped into an aeration tank (2). After a specific time for aeration, the water will flow into a sediment tank (3), then to a sterilizing tank (4), and finally return to the fishponds or flow out to local surface water bodies. After 3–6 months of operation, there will be some sludge settling at the bottom of the biofilter tower. This sludge must be pumped to the sludge-storing tank (5) for other uses.

Trickling Filter System

With this system (see Fig. 8.2), wastewater from the fishponds will be pumped into a temporary tank (1). Then, the water will be pumped to the top of the biofilter tower and sprayed for leaching through the filter material to the sediment tank at the

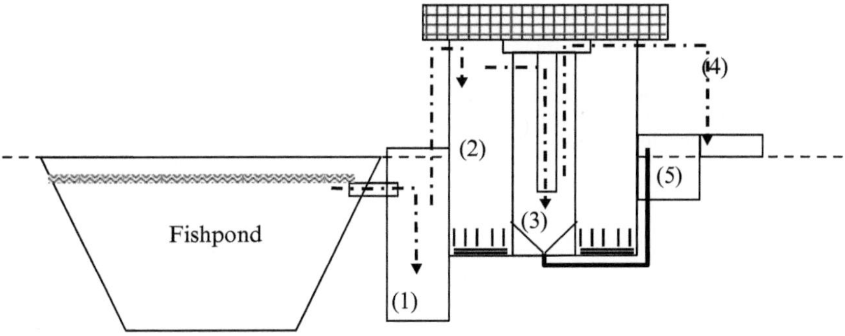

Key
- - · ▶ : Wastewater flowing route
(1) : Temporary tank
(2) : Aeration tank
(3) : Sediment tank
(4) : Sterilizing tank
(5) : Sludge-storing tank

Fig. 8.1 Aeration system

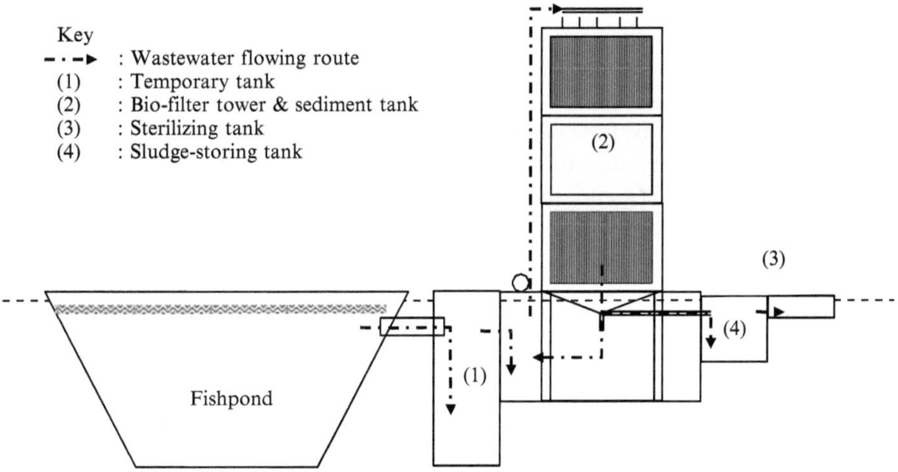

Fig. 8.2 Trickling filter system

bottom (2). After a few selected cycles of the biofiltering process, the water will flow into a sterilizing tank (3) and then return to the fishponds or flow out to local surface water bodies. After 3–6 months of operation, there will be some sludge settling at the bottom of the biofilter tower. This sludge should be pumped into the sludge-storing tank (4) for other uses.

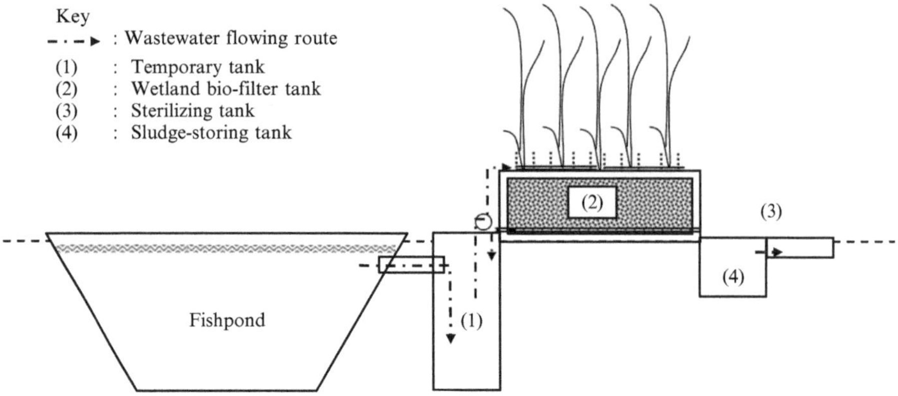

Fig. 8.3 Wetland system

Wetland System

With a wetland system (see Fig. 8.3), wastewater from the fishponds will be pumped into a temporary tank (1). Then, the water will be pumped to eject about 50 cm over the top of the wetland biofilter tank (2) to fall and leach through the filter material. It will then flow to the temporary tank (1) again through a collecting pipe system at the bottom of the biofilter tank. After a few selected cycles of the biofiltering process, the water will be poured into a sterilizing tank (3) and then return to the fishponds or flow out to local surface water bodies. The workers must check the system every day for any operational problems. After 3–6 months of operations, there will be some sludge settling at the top of the wetland biofilter tank. The filter must be cleaned and the sludge transferred to the sludge-storing tank (4) for other uses.

Scientific Results

Pollution Load Calculations for Tra Fish Production in Thotnot District

The calculation of pollution load was based on the COD parameter or the amount of oxygen needed to oxidize the entire organic matter in a fishpond (in this case). The more COD there is in the water, the more the amount of organic matter in it, that is, the water is more polluted.

Due to time constraints, the water analysis stage barely covered 5 months from September 2006 to January 2007. In addition, it was hard to control the private fish raisers' culturing schedules so the research team chose the ponds which had fish in the development stage, which was the fastest growing stage in the tra fish

Table 8.1 Hau River water quality (Tan Loc Islet September 2006 to January 2007)

	Tested parameters				
Date	pH	SS	DO	COD	NH_3-N
TCVN 5942-1995 Class A	6–8.5	≤20 mg/l	≥6 mg/l	<10 mg/l	≤0.05 mg/l
01/09/2006	7.91	76	5.2	4.0	0.050
23/09/2006	7.16	51	3.6	14.0	0.060
14/10/2006	7.09	115	5.2	11.0	0.100
28/10/2006	7.19	112	3.2	6.0	0.120
18/11/2006	6.38	58	2.9	9.0	0.247
16/12/2006	6.65	87	3.1	12.0	0.147
13/01/2007	7.16	61	3.2	10.0	0.098
27/01/2007	6.85	22	3.3	5.1	0.163

Water sample analysis results from the Can Tho Environmental Monitoring Station

development cycle. This is the stage during which fish have a weight of 10–800 g each. Stages where the weight is less than 10 g or above 800 g are considered as very slow growth rate stages.

River Water Quality (Hau River)

There were eight water samples taken from the middle of the Hau River at different times during the study period. Table 8.1 shows the results of the Hau River water analysis.

- pH values varied from 6.38 to 7.91, most of which were within Vietnamese environmental standards (TCVN 5942-1995). Specifically, the 18/11/2006 water sample pH had the lowest value of 6.38, violating the TCVN standard of 6.5, but the violation level was not high and was still within normal range compared to previous Hau River water monitoring results.
- Suspended solids (SS) in the samples varied according to sampling times with the highest value of 115 mg/l in October 2006, thereafter decreasing until the end of January 2007 with the lowest value of 22 mg/l. Hence, at the beginning of the study stage, the amount of SS was high due to the flood season with abundant silt, and at the end of the study period, the amount of SS was low because of the start of the dry season in the MD.
- Dissolved oxygen (DO) amount in the water varied from 2.9 to 5.2, usually hovering between 3.1 and 3.6 mg/l. The DO values of the Hau River water were lower than that of TCVN 5942-1995, class A (>6 mg/l), showing that the water was being polluted by organic matter. Low DO levels affect aquatic life.

- Organic content, represented by the COD parameter, varied from 4 to 14. The COD values were greater than the allowed rate of TCVN 5942-1995, class A ($<$10 mg/l) four times.
- The NH_3-N content in the samples often exceeded the TCVN 5942-1995, class A standard of 0.05 mg/l. It showed a tendency to increase significantly at the end of 2006 and at the end of the study period. In September 2006, the NH_3-N content in the water was 0.05 mg/l, equal to the allowed level. In November 2006, NH_3-N had the highest concentration of 0.247 mg/l, exceeding the permissible limit by nearly five times.

In summary, the Hau River water had better quality at the beginning of the study period than at the end. It can be concluded that the Hau River water was polluted due to organic matter, more so in the late months of 2006.

Water Quality in Thotnot District

The inlet and outlet water quality analysis results were compared to the specifications in TCVN 6774:2000 on Freshwater Quality Guidelines for the Protection of Aquatic Life (pH of 6.5–8.5, DO of 5 mg/l, and SS of $\leq$100 mg/l) and 28 TCN 176:2002 on aquaculture standards (COD of $<$10 mg/l and NH_3-N of $<$1 mg/l). In general, pH values of inlet and outlet water at the five fishponds were still within TCVN 6774:2000. The pH values of the outlet water tended to be lower than those of the inlet water. SS measured at the five ponds varied from 28 to 303 mg/l, while the allowed level in TCVN 6774:2000 is equal to or less than 100 mg/l.

- The DO values in inlet and outlet water were much lower than the allowed standard in TCVN 6774:2000 (5 mg/l). DO amounts in the outlet water were far less than those in the inlet water, varying from 0.8 to 6.8 but generally less than 4.1. Perhaps too high stocking rates (33–120 fingerlings/m^2) coupled with uneaten feed and fish feces contributed to this. Due to high population densities and lack of oxygen, fish loss rates were also high at 19–50 % of initial stocking rates.
- As for COD values, the water sample analysis results showed that the COD content in the outlet water was much higher than that in the inlet water and often exceeded the allowed COD standard ($<$10 mg/l). The difference in COD contents between inlet and outlet water samples is the COD amount generated from tra fish production, which is subsequently released in the wastewater into the outside water environment.
- NH_3-N in tra fish culture tended to increase with tra fish farming over time. The NH_3-N content in the inlet water was within the permitted range, but in the outlet water, it was often beyond 1 mg/l.

Table 8.2 Summary of the wastewater analysis results for the five fish-farming households

Households	Fish growth (kg)	Pond volume (m^3)	Wastewater volume exchanged (m^3)	COD load (kg)	PLR (kg COD/kg of fish growth)
1	416,500	32,000	925,290	45,126	0.108
2	56,000	4,800	66,000	4,625	0.083
3	40,800	4,000	60,125	3,933	0.096
4	157,000	25,000	452,500	13,601	0.087
5	173,600	36,000	954,000	15,835	0.091
Total/average	843,900	101,800	2,457,915	83,120	0.098

Source: Calculated by the technical expert

Pollution Load Rate (PLR) Results

The PLR is used in the calculation of COD abatement costs per kilogram of fish. Table 8.2 shows the summary of the water sample analysis and PLR results for the five households.

The average PLR for the five fish-breeding households was estimated at 0.098 kg COD/kg of fish growth, that is, one kilogram of fish yielded 98 g of COD. On the other hand, the average COD amount in tra pond wastewater effluents that needed treating was 0.034 kg/m^3 or 34 mg/l. This was the additional COD quantity generated from breeding tra fish beyond the COD amount available in the input water. This concentration is small compared to that of other industrial activities like processing aquaculture products where the COD amount may reach over 1,000 mg/l of wastewater. However, in tra fish production, the wastewater volume dumped into the environment is huge, and so it has caused significant pollution to the local water resources. If this situation continues unchecked, tra fish production could become unsustainable in the near future.

Economic Results

Results of Cost-Effectiveness Analysis (CEA)

The assumptions for the CEA analysis were as follows:

- Pond water surface area: 5,000 m^2 (average pond area in Thotnot District)
- Pond volume: 20,000 m^3 (pond water depth of 4 m)
- Daily discharged wastewater: 6,000 m^3 (average discharge rate of 30 %)
- COD concentration (needing treatment) per cubic meter of wastewater: 0.034 kg/m^3
- Daily COD load (needing treatment): about 200 kg (=6,000 m^3 * 0.034 kg/m^3)

Table 8.3 Estimated costs of the three wastewater treatment systems for a pond 5,000 m^2 in size (unit: '000 VND)

Costs	Aeration system	Trickling filter system	Constructed wetland system
1. Investment costs	**1,871,581**	**975,658**	**812,158**
Construction area (m^2)	915	1,300	1,300
Land	137,250	195,000	195,000
Materials (iron, cement, etc.)	972,281	242,125	191,169
Pumps:			
Wastewater	69,943	69,943	69,943
Sludge	28,876	14,438	9,625
Aeration or filter matter	63,000	234,000	144,000
Electric and mechanical system	200,000	40,000	60,000
Labor for construction	400,231	180,152	142,421
2. Annual operating costs	**151,728**	**89,040**	**138,000**
Operating labor	36,000	36,000	72,000
Electric power	109,728	51,840	64,800
Maintenance cost	6,000	1,200	1,200
3. Present value (@ 10 %)	Lifespan: 30 years	Lifespan: 30 years	Lifespan: 10 years
Investment cost	1,871,581	975,658	812,158
Total operating costs	1,430,327	839,372	847,950
Total present value	**3,301,908**	**1,815,030**	**1,660,108**
4. COD amount that needs treating (kg)	2,190,000	2,190,000	730,000
5. Treatment cost/kg of COD	**1.508**	**0.829**	**2.274**
6. PLR (kg COD/kg fish produced)	0.098	0.098	0.098
7. Treatment cost/kg of fish produced (item 5*PLR)	**0.148**	**0.081**	**0.223**

Source: Calculated by the technical expert and authors
Note: 1 USD ≈ VND 16,000

- Pollution load rate (PLR): 0.098 kg of COD/kg of tra fish growth
- Lifespan of aeration and trickling filter systems: 30 years
- Lifespan of constructed wetland system: 10 years

Table 8.3 shows the estimated costs of the three wastewater treatment systems. For a pond 5,000 m^2 in size, the land area needed for the aeration system is 915 m^2; for the trickling filter system, 1,300 m^2; and for the constructed wetland system, 1,300 m^2. The investment cost of the aeration system is the highest, while that of the constructed wetland is the lowest. The operating cost of the aeration system is also the highest and that of the trickling filter system is the lowest. Of the three systems, trickling filters have the least treatment costs per kilogram of COD abated and per kilogram of fish growth. Therefore, the trickling filter system appears to be the most cost-effective of all. These calculations were based on current input prices. If the input prices increase, the COD abatement costs will be higher.

Comparison of the Three Options: Other Criteria

Table 8.4 shows the comparison of some of the main characteristics of the three options with common advantages as well as disadvantages. It appears that trickling filters have the most preferred characteristics.

Responses of Farmers to Proposed Technical Options

Engagement Process

After the calculations were done, a seminar was organized in Thotnot District to collect the farmers' opinions of the proposed technologies as well as other information. Twenty-two people attended the meeting, of which 14 were fish-breeding farmers and the rest were local officials of related agencies. At the meeting, the study results of the environmental consequences of tra fish production were presented first, and then the proposed technical options with their associated costs were described by the technical expert. After that, the study team asked the participants to assess the technical and financial feasibility of the options. Local officials were interested, but the fish farmers were reluctant to consider the prospect of having to treat their wastewater with costly technologies.

Views of Farmers

The fish farmers' views on the proposed options were as follows:

- The options depended heavily on the state's electricity supply which could not even meet national demands—this was a big constraint.
- The options were only suitable for concentrated fish culture planned zones, not for separate fish-breeding households, due to the lack of extra land on which to build the treatment structures. Moreover, if only some households applied these systems while the others did not, the former would still face the problem of unclean input water if they supplemented pond water with water from the local canals which were already polluted not only by tra fish effluents but also by other polluting sources.
- The state should invest in a pilot project involving these options for a few years for farmers to see the results before they made any decisions.
- Due to the relatively big investments required, if these systems could be funded by some organization or cooperative, then the fish farmers would be willing to pay this organization for clean input water and treatment fees.

Table 8.4 Comparison of some financial and technical characteristics of the three options

Characteristics	Aeration system	Trickling filter system	Constructed wetland system
1. Investment costs	High	Low	Lowest
2. Annual operating costs	High	Low	Medium
3. Electric power	High	Lowest	Low
4. COD abatement costs	Medium	Low	High
5. COD treatment capacity	95–98 %	90 %	75–80 %
6. Lifetime	30 years	30 years	10 years
7. Advantages	High treatment capacity	Rather high treatment capacity	Biological process
	Automated, easy to operate	Automated, easier to operate	Automated, easiest to operate
	Effective in treating high concentrations of organic pollutants	Effective in treating high concentrations of organic pollutants	
	Reliable performance	Simple, reliable	
	Biological process	Biological process	
	Can reuse treated wastewater	Can reuse treated wastewater	
		Durable process elements	
		Low power requirements	
8. Disadvantages	High investment	High investment	High investment
	Needs more land (18 % of pond area), which is a constraint in the study area	Needs more land (26 %), which is a constraint in the area	High labor requirement to run the system
	Consumes a lot of electric power, relying on state power supply, which is big constraint	Consumes electric power, relying on state power supply, which is big constraint	Needs more land (26 %), which is a constraint in the area
	Needs technicians in case the system goes out of order	Needs technicians in case the system goes out of order	Only treats wastewater with low pollutant concentrations and must go through several rounds
			High COD abatement costs
			Consumes electric power, relying on state power supply, which is big constraint
			Needs technicians in case the system goes out of order

- The profits from the tra fish industry were shared among many stakeholders such as input suppliers, so why should the fish farmers be the only one to bear the environmental costs?
- Tra fish prices often fluctuated, depending on export prices, that is, on international tra fish demand. If fish prices go up, then so will input prices. However, if fish prices go down, input prices may not follow suit, which may lead to financial losses for the farmers. Since the farmers were struggling with increasing operating costs, they could not afford any significant investment in wastewater treatment systems. Therefore, they were more inclined to resort to chemical treatment since they thought that this was less costly.

Among the five surveyed fish-breeding households, one had installed an aeration system in his pond to provide oxygen for his fish. He reported to the group that although the COD amount and the number of dead fish in his pond had decreased, so did the fish growth rates. Since he had spent a significant amount of money on this treatment system, he could not consider any alternative treatment.

In response to the farmers' opinions, the technical expert explained that applying aerators directly in ponds reduced the COD content and increased oxygen, but it also increased ammonia (NH_3) which could reduce fish growth or kill fish if levels rose too high. Therefore, aerators should be placed outside the pond. In regard to chemical treatment, the technical expert revealed that his agency had tested some chemicals to treat wastewater in fishponds, but the results showed that their performance was not as good as the three options proposed (they only treated about 70 % of the COD). Moreover, the prices of the chemicals were very high since they were imported from the United States. His view was that treating pond water with chemicals increased the chemical quantity in ponds while not strictly solving the pollution problem. As for input water problems, he said that fish farmers could reuse the treated water instead of releasing it into water channels or they could use the systems to treat the input water. The technical expert added that while treating wastewater would raise fish production costs, the farmers would save on fuel expenditure for pond flushing at the same time.

In all, farmers realized the consequences of tra fish production on local water quality, but due to land and financial constraints, they responded poorly to the proposed technical options. As the above fish farmers were not eager to receive these technologies due to their personal constraints, option ranking was necessary to make a more objective judgment on the three technologies. Therefore, the same fish farmers were invited to another meeting and asked to rank the options using the preference ranking method. This method of ranking is not complex. People are asked to choose between each pair of options in turn, using a two-way choice matrix. The option that appears in the two-way choice matrix the most often is given the highest rank. People were also asked to give reasons for their choices. Table 8.5 lists the ranking results.

Table 8.5 shows that 36 % of the participants chose the aeration system as their most preferred option, 50 % chose trickling filters, none chose wetlands, and 14 %

Table 8.5 Results of preference ranking of options

Most preferred option	Number	Percentage
1. Aeration system	5	36
2. Trickling filter system	7	50
3. Constructed wetland system	0	0
4. No ranking	2	14
Total	14	100

Source: synthesized from individual fish farmers' matrices

refused to rank the options. Their reasons for choosing or not choosing an option are given below.

Choosing the Aeration System

- High treatment capacity.
- Structure of high investment signifies high quality.
- The COD abatement cost is not significant compared to the fish unit production cost.

Choosing the Trickling Filter System

- The COD abatement cost is the lowest.
- Effective treatment.
- Saves money in not having to aerate the wastewater.

Not Choosing the Constructed Wetland System

- Low treatment capacity.
- Needs a lot of labor to operate and clean the system, but labor price is high and labor supply is short in the region.
- The COD abatement cost is the highest.

Refusing to Rank the Options (Not Trusting in Their Feasibility)

- Need to see the performance of the systems before making judgments on them.
- The structures required more land area for construction (18–26 %) than they expected (10 %).
- Input water quality is important to fish health, but these options place more focus on treating effluents, not adequately solving their input water concerns.
- Their ponds are located far from the Hau River, so water supply is not abundant, and there is river water competition among fish producers in this area. Therefore,

they do not think that tra fish farming in these areas will last long, due to the increasingly polluted input water.
- Governments should strictly address other large polluting sources like alcohol production, aqua-product processing activities, etc. first.
- Do not believe that the power demand will be met.

Conclusions and Recommendations

Tra fish production in Thotnot District in particular and the MD in general has brought many benefits to fish-breeding farmers as well as to the nation, namely, by improving the incomes of the farmers, providing nutritious food to the population, and earning foreign income for the country through export. Fish producers in Thotnot District earn profits of about VND 2–2.5 thousand/kg of fish.

However, tra fish culture has caused water pollution. The study uncovered the following problems: (a) earth removal from fishpond digging and periodic dumping of pond sediment into external water bodies have polluted these bodies and reduced their depth, affecting water quality and navigation, and (b) wastewater from tra fishponds has been released into rivers and canals without proper treatment. The organic content in the wastewater was high. The average COD quantity released from tra fish raising processes into outside water bodies was about 34 mg/l of wastewater or 98 g/kg of fish produced, and (c) high stocking rates, coupled with unhygienic homemade feed, polluted input water, abuse of antibiotics, etc., have led to high fish losses due to lack of dissolved oxygen in the water.

To reduce the pollution in tra fishponds, fish breeders often exchange some of the water with freshwater from outside sources as a means of diluting the pollutant concentration in the ponds in order to prevent fish loss. This practice is of grave concern to the local government as well as the people. It also contravenes Vietnam's Law on Environmental Protection 2005 which stipulates regulations for production/business activities, aquaculture, etc., such as treating wastewater before discharging it into the environment. If this situation is not arrested, the areas breeding tra fish will soon lack clean water for the local people's needs, and the tra fish-farming industry will also suffer great losses due to polluted water.

Currently, some fish-farming households use additional ponds to hold the wastewater for some time before releasing it into public water bodies, but these ponds are small and the water retention time allocated by the farmers is too short, so this measure is not effective. To treat effluents more effectively, the study team proposed three possible technical options to reduce water pollution from tra fish farming: (a) the aeration system, (b) the trickling filter system, and (c) the constructed wetland system. Among these, trickling filters were the most cost-effective option. The treatment cost per kilogram of COD estimated was VND 1.51 thousand ($\approx$USD 0.09) for the aeration system, VND 0.83 thousand ($\approx$USD0.05) for the trickling filter system, and VND 2.27 thousand ($\approx$USD 0.14) for constructed wetlands, much greater than the environmental protection fee of

VND 0.3 thousand/kg of COD in wastewater presently imposed by the government on industrial as well as domestic wastewater as specified in Decree 67/2003/NĐ--CP. Industrial wastewater refers to water discharged into the environment from industrial production establishments and agricultural, forestry, and aqua-product processing establishments.

The treatment costs per kilogram of fish produced were VND 0.148 thousand for the aeration system, VND 0.081 thousand for trickling filters, and VND 0.223 thousand for constructed wetlands, equivalent to 7.5 %, 4 %, and 11 % of fish production profits, respectively. All three options needed extra land (about 18–26 % of pond water surface area), adequate power supply, and considerable investment, which are all significant constraints to fish producers. To assess the social acceptability of the three options, two focus group discussions were organized. The first provided general judgments, and the second ranked the options through the preference ranking method. Fifty percent (50 %) chose the most cost-effective option (trickling filters), 36 % chose aeration systems, no one chose constructed wetlands, and 14 % refused to rank. All of them suggested that the state install these systems as a pilot project on a certain plot to demonstrate their performance, and if they proved to be good, then a collective organization should be set up to manage wastewater treatment in the area, and the farmers would be willing to pay for clean input water as well as tra fish wastewater treatment fees.

To continue economic development without sacrificing the natural environment, national laws as well as other government action are needed to address the externalities of economic activities. From the above conclusions, we can see that the dumping of great quantities of tra fish waste into public water channels by fish farmers has considerably jeopardized the health of the fish being reared as well as the domestic water supply of the local people. The Ministry of Fisheries has laid down specific guidelines for pond construction and land requirements for the storing of pond sediment and wastewater. However, in reality, fish farmers do not abide by the latter and still discharge untreated wastewater into local waters. This study proposed three wastewater treatment technologies to treat wastewater more effectively than settling pools, but the COD treatment costs were rather high. Some policy recommendations that would help in reducing water pollution caused by tra fish production are given below.

Firstly, it is necessary to implement the establishment of tra fish planned zones in Can Tho City. These zones have been planned but not yet established. They would pave the way for concentrated wastewater treatment systems to be set up, which will help reduce COD treatment costs.

Secondly, it is essential to set emission standards for tra fishpond wastewater released into public water bodies. This will spur the application of efficient technologies to reduce organic matter concentration in the wastewater to acceptable levels. Fish farmers will only think of waste treatment if standards are set by state law and enforced by capable agencies. At present, environmental protection fees apply only to the shrimp industry, not to the tra fish industry.

Thirdly, due to the high initial investment costs of the treatment technologies, farmers are unwilling to implement them without support. Thus, local governments should set up an environmental fund to provide long-term loans with preferential interest rates to fish farmers to enable them to build treatment systems. In addition, governments should think of establishing state-financed pilot waste treatment systems capable of treating input and output water to demonstrate the usefulness of these systems to farmers.

Finally, the COD treatment costs estimated in this study are high compared to the environmental fees specified by the state in Decree 67/2003/NĐ-CP—this will not encourage individual farmers to apply technical options. Therefore, state agencies should compare current environmental fees with the COD treatment costs of different technologies and adjust them if they are too low.

References

Can Tho Statistical Yearbook (2003) Can Tho Province Statistics Department, Can Tho City, Vietnam

Can Tho Statistical Yearbook (2004) Can Tho Province Statistics Department, Can Tho City, Vietnam

Field BC, Olewiler ND (2005) Environmental economics (Updated second Canadian edition), McGraw-Hill Ryerson Limited, Canada

Ky QV (2004) Research report on tra and basa fishpond water treatment options in Thotnot District. Can Tho City Environmental Monitoring Station, Can Tho City (In Vietnamese)

Part III
River Basin Management

Chapter 9
Application of an Educated Trade-Off Analysis Framework for the Ma Oya River Basin Development Project, Sri Lanka

Bhadranie Thoradeniya

Abstract This study explores economic aspects of a proposed development project by the National Water Supply and Drainage Board in Sri Lanka to mitigate water shortages in the Ma Oya river basin. The project entails constructing a balancing reservoir to store excess flows during the rainy season and releasing water from the reservoir during dry weather periods. A stakeholder-based approach, using a so-called Education Trade-Off or ETO framework, is adopted for the study. Two scenarios are evaluated: first, a base case representing a continuation of the current situation and, second, a scenario involving construction of the reservoir. Economic values are estimated for water supplied to domestic and commercial consumers according to tariffs applied, while environmental economic values are obtained in terms of effects on recreation, tourism, industrial use, agriculture, and wastewater discharges. Engaging stakeholders in the process allowed them to learn about the technical and economic feasibility of the proposed project, as well as reducing potential conflicts in resource use.

Keywords Sri Lanka • Reservoir • Costs and benefits • Environmental impacts, economic value

Introduction

Background

Ma Oya is one of the most water-stressed rivers in the wet zone of Sri Lanka. Excessive sand mining has threatened the river's bathymetry as well as the downstream ecology. It is also beset with reduced water level, lower groundwater table in the associated catchment, and diminished dry weather flows.

B. Thoradeniya (✉)
Division of Civil Engineering Technology, Institute of Technology, University of Moratuwa, Moratuwa, Sri Lanka
e-mail: b.thoradeniya@gmail.com

D. James, H.A. Francisco (eds.), *Cost-Benefit Studies of Natural Resource Management in Southeast Asia*, DOI 10.1007/978-981-287-393-4_9

Uncoordinated and uncontrolled use by multiple stakeholders was expected to diminish more rapidly the river's valuable resources within the next few decades. Hence, to mitigate expected shortages in the water supply and sanitation sectors because of increasing demands, the National Water Supply and Drainage Board proposed a development project. This project consisted of a multipurpose balancing reservoir at Yatimahana in the upper portion of the catchment and four low-flow weirs. The reservoir would store excess river flows during rainy seasons and then release the required flows, under control, during the dry weather periods.

Potential Impacts of Proposed Development

Implementation of the balancing reservoir could have adverse impacts on other sectors using the water in the Ma Oya river valley. For example, because of regulated water flows, the groundwater table of the river valley could remain low for a longer period, leading to production losses in the agriculture sector. Because of increased concentrations of pollutants in the river, water quality would deteriorate. Hence, the growth of industries, which require water as an input to production or through which to discharge pollutants into the river, would be limited.

Limitations of Traditional Project Evaluation

The general practice adopted in development projects is to analyse the technical and economic feasibility of a proposal. The social and environmental impacts are usually analysed only in the project area and mainly focused on the people affected by the project. A feasibility study for the proposed project was prepared by Sweco Grøner (2004). In as much as the potential social and environmental impacts were not adequately considered in the proposed Yatimahana project, conflicts in decision-making among stakeholders and possible opposition to the project were anticipated.

Stakeholders are usually unaware of the technical, economic, and environmental dimensions of their resource use. Hence, faced with scarce resources, they often decide on how to prioritise the uses of these resources emotionally rather than rationally. This research addresses the need for a methodology to analyse and value the conflicting resource uses by stakeholders in situations with scarce resources when decision-making is highly sensitive.

Research Objectives

The study aimed to develop, apply, and validate an Educated Trade-Off (ETO) framework relating to the proposed project at Yatimahana in the Ma Oya river basin, to enable stakeholders to make informed decisions about the project's development and other resource uses. It is based on an existing framework already constructed by Thoradeniya (2009). The ETO framework can be used for resource development projects or for resource management scenarios more generally. In resource management, the framework can be expected to help in decision-making regarding the priorities for resource use, while in resource development it is expected to assist in deciding whether to accept or to reject a project. The more specific objectives of this study were to:

- Identify the dominant stakeholders and impacts of the proposed development project
- Value the major social and environmental impacts of the Ma Oya river basin project to various stakeholders of the Ma Oya river and its valley
- Compare the net benefits of the current scenario and the proposed development project
- Use this knowledge to educate different stakeholders for decision-making
- Validate the ETO framework as a tool for stakeholder education in decision-making.

Methodology

Specification of Scenarios

Two scenarios were examined for the present study:

Base Case Scenario: Without Balancing Reservoir This scenario involves continuation of the current situation. Impacts would be the result of the projected demand for water from the water supply and sanitation sector together with the demands for irrigation, industry, and other sectors. The baseline case is not static. Climate change, population growth, development of the area, and other similar factors would have impacts over time. Lack of data precluded detailed quantitative analysis, so the study investigated this scenario qualitatively.

Development Scenario: With Balancing Reservoir This scenario specifies the proposed schedule of the balancing reservoir's operation. The reservoir is intended to store the buffer requirement of the water supply and sanitation sectors and thus would operate during prolonged periods of low flows. Such low flows would result in water deficiencies in other user sectors, in turn creating various social and environmental costs.

In the present study, to achieve consistency with the report prepared in 2004, monetary values were represented in terms of the currency value in 2004. Benefits and costs were estimated for a 20-year period.

Steps in Developing the ETO Framework

Development of the ETO framework for the purpose of evaluating the two scenarios involved five steps, described below.

Step 1 – Identify the multiple stakeholders and the natural resource uses/issues that need stakeholder participation for decision-making.

A stakeholder involvement process was designed and carried out to identify multiple stakeholders and resource uses/issues and collect data for the baseline situation. This process involved 'consultation' rather than 'participation'. Key players were lower-level administrative divisions known as Grama Niladhari Divisions (GND).

The Divisional Secretariat of the National Water Supply and Drainage Board was identified as the administrative unit that could initiate the field studies, namely, rapid assessment of resource uses. Relevant data was collected from the Secretariat, including identification of lower-level administrative divisions along both banks of the river.

Step 2 – Bound the technical requirements of natural resource uses/issues.

In this step, engineering information knowledge was used to estimate the critical bounds of the technical requirements for different resource uses. The minimum volume to satisfy the requirements for irrigation, human consumption, and environment defined the critical bound. For other river basin uses such as sand mining, clay mining, and discharging pollutants, the critical bound was defined as the maximum volume to ensure that the effects of such activities would not exceed the assimilative capacity of the river basin.

Step 3 – Estimate the economic value of the critical bounds of the technical requirements for natural resource uses/issues.

The starting point of the economic analysis of resource uses in the river basin was a financial analysis of those resource uses, with estimates of the receipts (benefits) and payments (costs) relevant to the investors or owners of the resource or development project. For the development project, and for the purpose of this study, the 'economic' values were defined in terms of the direct financial costs and benefits accruing to different resource users in the river basin. More generally, the term 'economic' refers to the efficiency of resource use for society as a whole (Jenkins and Harberger 1992).

Step 4 – Estimate the environmental value of the critical bounds of the technical requirements for natural resource uses/issues.

Implementation of the new project or prioritisation of the natural resources to different uses was anticipated to cause an imbalance in the existing system and result in social and environmental impacts on other users. Valuation of environmental and social impacts was carried out in this step. In the context of this study, environmental costs and benefits comprised those that were either 'not quantified' or 'underestimated' in the original economic analysis as part of the normal engineering practice, thereby comprising externalities of the development project.

All environmental benefits and costs of impacts were valued in base-year prices and the net present value of environmental effects was calculated using the usual formula and procedures for deriving NPV (Ranasinghe 1994; Pearce and Warford 1993). Direct and indirect valuation methods were used to evaluate the environmental impacts.

Step 5 – Estimate the net value of the critical bound of the technical requirement of natural resource uses/issues by combining the economic and environmental values.

Knowing the combined economic value of all uses of the river basin at the critical bounds of the technical requirements was expected to help stakeholders make decisions on which issues/uses to prioritise in any year with less water. For example, in the river basin, the top priority might be for the minimum amount of water required for consumption or irrigation, followed by the minimum requirement for power generation and balance for irrigation in a dry year. The farmers could be compensated for any loss of livelihood if the minimum requirements were not met. In a good year when there is surplus water, irrigation might be allocated much as it could absorb, or the water could be used even to recharge the aquifer (Seckler 1996).

Stakeholder Involvement and Resource Uses

Approach to Stakeholder Involvement

The selected sample consisted of one to five persons from each GND representing all resource use sectors. The final sample covered 427 stakeholders representing 145 GNDs under the administration of 16 District Secretariat Divisions (DSDs) on both banks of the river. The sample comprised 362 males and 62 females. A questionnaire was developed as the main survey tool.

Table 9.1 Resource use sectors in Ma Oya river basin

1. Drinking water	2. Bathing/washing	3. Water supply
4. Industrial use	5. Hydropower generation	6. Sand mining
7. Clay mining	8. Dumping garbage/waste	9. Tourism industry
10. Inland fishery	11. Recreation	12. Animal rearing
13. Landowners/farmers	14. Dug-well users	15. Environment

Resource Use Sectors

The collected data were used to identify key resource use sectors at each GND. Even minor uses that could be important for future decision-making were included; hence, the resource uses identified by even a single respondent were considered. In all, 15 resource use sectors were identified in the 16 DSDs, shown in Table 9.1.

Existing Resource Uses and Issues

Resource uses vary along a river. Hence, the user sectors which were already threatened by other use sectors or were in a conflict situation with other uses had to be identified at each division. The respondents were requested to indicate whether their resource uses have had any conflicts or whether they expected such a situation in the future. Responses were obtained on a three-point scale (Yes/No/No idea) for the present situation and for the future, expressed as percentages of responses for each division. On average, 89 % of the respondents indicated that their resource uses involved conflict situations at present and 91 % expected a similar situation in the future.

The sectors for which negative social and environmental impacts were indicated were the sand industry (259 respondents), the waste disposal sector (72 respondents), and the clay industry (51 respondents). A cross-impact matrix was constructed indicating the ways in which resource uses by each sector affected uses by other sectors, shown in Table 9.2. The stakeholders also identified major issues pertaining to different resource use sectors (see Table 9.3) and expressed their views on the reasons for the negative social and environmental impacts of such uses as well as possible remedial measures (Table 9.4). Some stakeholders were identified as being particularly vulnerable.

Bathing/washing and sand mining were the two direct uses of river resources occurring in all the DSDs, while hydropower featured least often. The majority (89 %) of the respondents from all DSDs agreed that their resource uses were threatened or in a conflict situation at the time of the study and most of them (91 %) did not expect the situation to be reversed in the near future. In fact, they thought that the situation would deteriorate under the assumption that there would be no change in the river system as in the Base Case Scenario. Other resource use sectors

Table 9.2 Resource uses impacting other sectors, identified by respondents

DSD	Total	Resource use sectors														
		1	2	3	4	5	6	7	8	9	10	11	12	13	14	15
Negombo	**09**	–	–	–	–	–	02	–	05	01	–	–	–	01	–	–
Katana	**18**	–	–	–	01	–	10	03	01	–	–	–	–	–	–	–
Divulapitiya	**64**	01	–	–	01	–	42	17	09	–	–	–	–	–	–	–
Meerigama	**26**	–	–	–	–	–	16	02	–	–	–	–	–	–	–	–
Warakapola	**23**	–	–	–	–	–	14	–	04	–	–	–	–	–	–	–
Galigamuwa	**05**	–	–	–	01	–	04	01	01	–	–	–	–	–	–	–
Kegalle	**03**	–	–	–	–	–	01	–	–	–	–	–	–	–	–	–
Wennappuwa	**17**	01	–	01	–	–	06	01	08	–	–	–	–	–	–	–
Dankotuwa	**27**	01	–	–	–	–	19	14	04	–	–	–	–	–	–	–
Pannala	**42**	–	–	–	01	–	25	12	04	–	–	–	–	–	–	–
Narammala	**08**	–	–	–	01	–	03	01	02	–	–	–	–	–	–	–
Alawwa	**42**	–	–	–	–	–	25	–	02	–	–	–	–	–	–	–
Polgahawela	**19**	01	–	–	–	–	14	–	–	01	–	–	–	–	–	–
Rambukkana	**29**	–	–	–	–	–	21	–	2	01	–	01	01	–	–	–
Mawanella	**23**	–	–	01	–	–	17	–	10	–	–	–	–	–	–	–
Aranayaka	**72**	01	01	04	–	03	40	–	20	–	01	–	05	–	–	–
Total	**427**	05	01	06	05	03	259	51	72	03	01	01	06	–	–	–

Table 9.3 Issues in resource uses by stakeholders

User sector	Main issues	Frequency
Sand mining	Drowning risks because of slippery and deep river bottom, poor accessibility for bathing	224
	Erosion of banks, hence uprooting bamboo and other trees, resulting to the collapse of roads and railways on banks	244
	Dried up dug wells, reduced water levels, salty well water	241
	Loss of crop production	167
	Adverse environment impacts, rising temperature, air pollution, dryness, etc.	16
	Distancing of families in the two river banks	10
Clay mining	Damages to village roads by transport of sand, etc.	36
	Drug addiction/theft by mining labourers	11
	Loss of habitable land	12
	Risk of diseases, poor health because of dust while transporting, non-filling of clay pits	61
Waste disposal	Fish kills	12
	Health hazards (bad smell, etc.), stagnant pools	64
	Unsuitable water for drinking, bathing/washing	73

Table 9.4 Respondents' views re development/restoration of river's environment

Sector/issues	View	Frequency
Washing/bathing	Construction/rehabilitation of river access points	79
Waste disposal	Banning of waste dumping into the river	29
	Education of the people	13
Sand mining	Limiting depths of mining/stopping of sand mining, need for proper management	143
	Effective and fair intervention of government officials	113
Land erosion bank slides	Construction of rock embankments, protection of banks	104
	Construction of bunds across the river	71
Environment	Long-term river protection, programming, education of people, specially of politicians	23
	Protection of natural swamps and depressions	24
	Reintroduction of river reservations	57

facing conflict situations are listed below. The numbers of GNDs affected are indicated in parentheses:

Dug-well users (111), landowners/farmers (100), bathing/washing (87), environment sector (49), sand mining (24), clay mining (13), water supply (14), drinking water (direct) (06), inland fishery (04), animal rearing (11), recreation (03), and dumping garbage/waste (2)

Almost all of the respondents (94 %) from each DSD said that they knew the reasons for these conflict or threatening situations. Most of them (91 %) attributed the conflicts to over-exploitation by other users. A majority (86 %) from each DSD likewise viewed the resource uses by other sectors as causing negative and social impacts, but only 11 % accepted that their own resource uses led to such negative impacts. The sectors that often caused negative social and environmental impacts were the sand industry (61 %), followed by the waste disposal sector (17 %) and the clay industry (12 %). The negative effects of extensive mining by the sand industry were most acute during dry weather low-flow periods.

The above analysis of the present resource use sectors and stakeholders helped to identify the sectors that would be significantly impacted by the proposed reservoir. In an average year under normal conditions, the river experiences 6–8 weeks of low-flow condition (DHI 1999). The proposed reservoir could result in longer low-flow periods, depleting groundwater and increasing impacts on the indirect resource use sectors, namely, the dug-well domestic water supply and rainfed agriculture sectors. Other sectors that would be impacted by the proposed reservoir because of the changes in water availability of the river included the recreation sector, industrial sectors, and tourism. These five impacted sectors were not included in the project's original economic analysis (Sweco Grøner 2004) but were considered in applying the ETO framework.

Technical Aspects of Water Resource Uses

General

The major variables differentiating the two scenarios, 'with' and 'without' the reservoir project, were the river water flow and the water level. A time series of daily water flows for a 20-year period (from 2008 to 2027) at the proposed location of the reservoir was already established by the project proponents, simulating reservoir operation based on modelled inflows and outflows of water and predetermined operating rules.

Critical Bounds of Direct Uses

As noted, the main objective of Scenario 2 was to meet the projected drinking water deficit. However, the project, designed as a multipurpose reservoir, was expected also to generate hydropower, supply water for three irrigation schemes, and meet forecast water deficits in the industrial sector. The project proponents identified water supply, hydropower, industries, and irrigation as the sectors of greatest importance for the river system. In developing the ETO framework, the critical bounds of the technical requirements for water supply, hydropower, industries, irrigation, and environment sectors were assumed to be the same as those set by the project proponents (Sweco Grøner 2004).

Various activities and sectors were not included or quantified in the original project report but were incorporated in the ETO framework. For the ecological sector, a previous study had estimated the flow requirement for preserving the ecology as 3 m^3/s (DHI 1999). This was considered an overestimate by the project proponents because of a seawater barrier now in place in the estuary but was nevertheless adopted in the present study.

A major use of the river is for waste disposal from cities, private dwellings, hospitals, and industries. There are both point and nonpoint waste disposal sources along the river. Surveys conducted for waste disposal showed that most of the wastewater treatment plants did not function properly and the wastewater discharged into the river was not tested. However, the samples of water tested at water supply intakes occasionally satisfied the standards for drinking water with conventional treatments. During the lean period, the 'with project' scenario would improve the river flows until the reservoir storage was released; thereafter, the 'without project' situation would prevail. As such, no adverse impacts associated with waste disposal were expected because of the proposed project.

Both river banks downstream from the proposed reservoir are either densely populated or used for rainfed agriculture; thus, biodiversity was not considered significant in the segment of the river covered by the study. No inland fisheries occurred at a commercial scale anywhere along the river.

Economic Value of Project

Incremental Costs and Benefits of the Project

Economic values were estimated in terms of the financial costs and benefits of the project due to an increased water supply and hydropower generation.

The project construction costs comprised Rs. 1,352 million in the first year and Rs. 2,028 million in the second year, with a one-time project rehabilitation cost of Rs. 33 million in the 10th year. An annual fixed operation and maintenance cost of Rs. 9.4 million was assumed for the 20-year period of project operation. Sweco Grøner (2004) stated that the residual values of assets were considered in the original economic appraisal. These had an economic life beyond the assessment period, although their values become almost negligible as a result of discounting.

The annual income for water supply was based on the assumption by the project proponents that 80 % of the water would be sold for domestic purposes with a tariff of Rs. 2.90 per m^3 and the balance of 20% would be sold for commercial purposes at a tariff of Rs. 42.00 per m^3. Constant tariffs were applied throughout the project period. Financial values were converted to economic values using conversion factors suggested for Sri Lanka by Curry and Lucking (1992).

Apart from the incomes derived from the water supply and hydropower sectors, other economic benefits considered in the ETO framework comprised a one-time increase in land value of Rs. 120 million, plus annual incomes of Rs. 100 million in the irrigation sector and Rs. 130 million in the industry sector, in line with estimates contained in the project proposal. Adjustments were made to avoid double counting of benefits. Benefits accruing to the irrigation sector were neglected in the economic analysis as the three main irrigation schemes were abandoned at the time of the study.

Acceptable Rate of Return

The minimum acceptable rate of return (MARR) for water supply projects was assumed to be 10 %, reflecting the opportunity cost of capital. The analysis of benefits and costs in the original project report yielded an EIRR of 8.28 %. Three additional cases (Cases 2, 3, and 4) were analysed for different scenarios. For Case 2, removal of land value increases reduced the EIRR from 8.29 to 7.95%. For Case 3, the benefits accruing from the irrigation sector were removed, resulting in the EIRR decreasing to 4.95 %. For Case 4, the planning period was extended from 20 to 40 years, producing an EIRR of 7.29 % again without the irrigation benefits.

Valuation of Social and Environmental Impacts

General

An important step in constructing the ETO framework required the valuation of social and environmental impacts because of changes in resource uses. Such impacts for both scenarios would depend on the changes in river water flow and water levels during the project's lifespan.

Under Scenario 1, the changes of river water flow directly depended on the rainfall pattern, proposed upstream diversions both to and from the river, and catchment development work. Available literature did not show any concrete proposals for other catchment development or river diversions to be implemented during the project life. There were also no studies to predict rainfall changes. Therefore, it was assumed that the water flow in the river would not be significantly impacted during the project period under Scenario 1.

The river water levels were already being impacted from sand mining industry, and this had caused some social and environmental issues. Since the river had been mined to the bedrock during the last 15 years, further mining of annual sand deposits would not significantly change the water level. Therefore, no significant social and environmental impacts were assumed under Scenario 1, as both water flow and water level could not be expected to undergo significant changes within the project life.

The major impacts associated with changes in river water flow and river water level under Scenario 2 were identified in the initial data analysis conducted when developing the ETO framework. The direct resource use sectors that could be impacted were the recreation, tourism, and industries sectors, while the indirect resource use sectors included rainfed agriculture and dug wells.

The indirect use sectors depended on groundwater levels whereas the direct users depended on river flow (discharge) or volume. The groundwater level in most of the river bank areas depends on the river water level. Significant changes expected with the implementation of the reservoir were reductions in flood height and prolonged periods of low water flows. These would result in low groundwater levels for longer periods. As a result, both indirect water use sectors, rainfed agriculture, and dug wells could face water scarcities.

Recreation Sector

A major recreation site enjoyed by the local people is a waterfall (Bo Ella) located upstream of the proposed dam. This waterfall is an open-access recreational site. Most visit the site on the independence day of Sri Lanka, 4th of February, as a tradition. The waterfall at the proposed dam site was expected to be inundated as a consequence of the reservoir, and the recreation site was expected to be lost. To estimate the value of this recreational site, visitors were surveyed on 4 February 2009 to collect data required for application of the travel cost method (TCM).

The total sample consisted of 70 people who were visiting the site for recreation. Data such as their age, annual income, and mode of transport were collected. The gender ratio was 57 males to 13 females. Their annual income ranged from zero to Rs 600,000. The visitors had used seven modes of transport and some of them had used more than one mode of transport. A majority (73 %) used their own vehicles while 26 % hired vehicles.

Three methods were tried to estimate the site's value: two variations of the travel cost models and a simple approximation. A zonal travel cost model with six zones was tried, but the data collected did not yield an inverse demand curve for trips to the site, so this was rejected as an appropriate valuation method. An individual travel cost model was also tried. Seven factors potentially influencing the number of visits were considered: (a) gender, (b) age, (c) work category, (d) distance travelled, (e) monthly income, (f) total travel cost, and (g) other expenses incurred at the site. None of the above factors showed any significant influence on the number of trips made by the individuals. The final method applied was a simple approximation, using the average value of total expenditure on travel and other expenses and the value of travel time, incurred by the visitors and the number of visitations per year. The resulting estimate was Rs. 2.59 million per year.

Tourism Sector

The river flows through a prominent tourism location in Sri Lanka called Pinnawala, where an elephant orphanage is found. Housed along the river banks, Pinnawala was initially conceived to provide shelter for baby elephants orphaned in the wild. The captive breeding programme by wildlife authorities resulted in more than 75 elephants at the orphanage; hence, they could be naturally herded. This world-famous tourist attraction depends on the Ma Oya river for the water needed by the elephants to cool their body system. The herd uses the river two times a day. The surrounding area is likewise being developed for tourism.

Variation of river water flows resulted in two types of negative impacts. First, the very low flows could result in insufficient water for elephant bathing. Prolonged lack of water would lead to the need for another water supply; otherwise, the orphanage would have to be closed. This loss could be estimated using the 'alternative project' or the 'travel cost method' to assess the value of the orphanage.

Second, the high flows (floods) did not permit the elephants to use the river for bathing because the small elephants could be washed away. In this instance, the tourism industries along the route taken by the elephants from the orphanage to the river would be affected, but not the orphanage itself. Thus, the specific days when the elephants could not use the river because of floods would spell economic loss in the area.

The proposed project (Scenario 2) would not create flows lower than those in Scenario 1 during the dry periods. There would be no incremental loss in Scenario 2 when the extreme low flow conditions prevailed. Instead, the proposed reservoir would actually be beneficial to the tourism sector. The mitigation of floods during high-flow periods would also benefit the tourism industry in the area. The additional net income earned by business establishments involved in tourism was estimated at Rs. 2.59 million per year.

Industry Sector

Industries located in river banks depend on two types of use from a river system: the extraction of raw water needed in manufacturing activity and the discharging of industrial wastes to the river system. The cost of raw water extraction is affected by both the quantity and quality of water used. In discharging wastes, the key factor is the assimilative capacity of the river. In the dry weather situations, the waste disposal rate could exceed the assimilative capacity.

A survey covering 20 industries along the river banks showed that most of them belonged to the four industrial estates that have been established along river banks, while others were individual enterprises, as shown in Table 9.5. Of the four industrial estates, the Nurani Industrial Estate is located about one km away from the river. The industries within Nurani Industrial Estate use the river only for waste

Table 9.5 Water demands of industries along river banks

Industrial estate	Respondents	No. of industries	Demand up to 2025 (m3/day)	Project allowance (m3/day)		
				2005	2025	Average
Thulhiriya	01[a]	05	8,500	4,500	18,000	11,250
Meerigama	01[a]	05	4,500	2,500	6,000	4,250
Makandura	12	15	Not known	6,000	13,000	9,500
Nurani	03	04	–	–	–	–
Individual	03	About 10	Not known	–	–	–
Total	20					25,000

Note: [a]At 'Thulhiriya' and 'Meerigama,' all the industries within the estate were supplied with water by one intake. Waste is also treated by one plant under one authority

disposal. Data were collected on the expansion of the industries requiring additional water supply or waste discharge facilities within the project life of the proposed reservoir. Six industries at Makandura and the Thulhiriya and Meerigama industrial parks have plans for expansion during this period. All such demands were well within the provisions proposed by the proposed reservoir project.

Prolonged periods of low water flows combined with reduced flood peaks can affect the industry sector differentially. For water intake, eight industries used tube wells and six used dug wells as their sources of water. One industry depended on an outside source: a tube well on a neighbouring compound. Five respondents, including the two from Thulhiriya and Meerigama, drew water directly from the river. Seven respondents agreed that the river flow, water level, and water quality influenced their water intakes. The benefits expected from reduced water input treatment costs by the industry sector were estimated at Rs. 0.45 million per annum.

To estimate the impacts of low flows on costs of discharging wastes to the river, the method of 'response cost' could be used to this sector because of the proposed reservoir project. However, the environmental authority has already raised the standards of permissible waste water releases to the open surface of water bodies, and this is already being implemented in the environmental clearance certification of industries. As such, the proposed project would not require the industries to enhance their treatment plant capacities and incur any additional costs.

Dug-Well Sector

Prolonged low flows causing low groundwater table would result in longer periods of water shortages for the users of dug wells to meet daily domestic needs. Most of the dug wells run completely dry or almost dry during the natural dry weather periods. This has resulted to extra costs for fetching water and poor sanitation conditions. The method of 'cost of productivity loss' was proposed as a possible valuation technique. Participants created a chart for the crops grown in the area during different seasons. Questions were posed to the group to facilitate discussions. This was followed by a household survey to collect individual data from 146 sample households. Of the 146 households, 126 (86 %) owned a dug well and others used common dug wells and the river for their domestic water requirements.

During the past 15 years, the river bed has deepened by about 10–12 m because of sand mining. This has resulted in the drying up of dug wells near the river. Thus, the dug wells were excavated repeatedly. Many of the respondents (84 %) were aware of the depth of their wells. The average depth of a well in the area surveyed was 11 m whereas the normal depth of a dug well is usually between 4 and 6 m. Some wells were around 20 m deep, and some people used tube wells which reached up to 25 m. A review of available information and existing studies indicated that impacts on the dug-well sector could be marginally positive when compared with Scenario 1, although this could not be further evaluated due to a lack of data.

Rainfed Agriculture Sector

The prolonged low flows resulting in low groundwater table would reduce crop production. The environmental valuation technique used to value the impacts to this sector was the 'cost of productivity losses' method. The household questionnaire survey for the dug-well sector was extended to collect the required data. The major crop in the area was coconut, grown by 65 % of the surveyed households. Banana was grown by 25 %. Other crops grown commercially were pineapple, paddy rice, pepper, betel, rubber, flowers, and yams (Manioc).

The river water levels were found to have a definite impact on the groundwater table of most of the Ma Oya lower reach areas, through the observation of water levels of the dug wells along the river banks: low water levels of the river corresponded with the low water levels of the dug wells, and high water levels corresponded with the high water levels. Thus, the magnitude of the impact of the proposed reservoir on rainfed agriculture depended on the possible variation of groundwater table impacted by the river water levels, especially during the dry months.

A study was made using available data for river flow and water level. The data and computations for predicted changes in river flow showed that in 88 % of the time, the variations would be less than 10 cumecs. During the dry season, the flows were expected to increase as a result of flow regulation at the reservoir. However, observations on the relationships between river flow and water level measured at the gauging stations of Giriulla and Badalgama showed that for a flow variation of 10 cumecs, the water level variation was only a fraction of a metre.

Coconut cultivators were a key group of stakeholders using river resources in the lower reach. Their major concern was the impact of the proposed balancing reservoir project on soil moisture retention in the river bank areas, which in turn would impact coconut production. Research has shown that soil moisture availability is positively correlated with coconut production. The cultivators were encouraged to practise mulching to maintain soil moisture within the root zone. The roots of a coconut plant are generally around 2.0 m deep. When the groundwater table is present around this depth, the required soil moisture within the root zone is maintained. Given that a coconut tree has a root depth of 2.0 m, Scenario 2 would not have a significant negative impact on coconut cultivation. Other commercial crops have much smaller root depths but similarly would not be negatively impacted by the proposed reservoir operations. Hence, it was concluded that the flow variation because of the reservoir would not impact rainfed agriculture of the river banks.

Summary of Social and Environmental Values

For all the relevant social and environmental sectors, the total annual net benefit of the project was estimated to be Rs. 1.84 million. The net present value of the total social and environmental benefits *EnvioPV* for different discount rates are given in Table 9.6. The estimate ranges from Rs. 12.9 million to Rs. 23.7 million depending on the discount rate applied.

Both the project feasibility report and the present study considered only the financial value of water supply based on expected water prices per unit. However, this may considerably understate the value of scarce water to dependent communities, as in the absence of the reservoir project, the next best supply of water would be considerably more expensive to develop.

Combined Economic and Environmental Values

Combined Economic Value

Table 9.7 displays the results obtained for the combined economic value (*CombNPV*) of the project, comprising the summation of the economic value of the project estimated by following normal engineering practice (*EcoNPV*) and the economic costs (*EC*) estimated for the social and environmental impacts of the project. Three cases have been evaluated. In Table 9.7, all values represent the incremental changes between Scenarios 1 and 2 for cases 1, 3, and 4. Case 2 is ignored because of its insignificant difference between Scenarios 1 and 2. The discount rate is 10 %. It can be seen that none of the cases has a positive combined NPV or a BCR exceeding 1, implying that from the perspective of social economic welfare, the project should be rejected. Of the three cases in Table 9.7, Case 1 is the least undesirable.

Table 9.6 Net present value of social and environmental benefits

	10 % (20 year)	8.29 % (20 year)	4.95 % (20 year)	7.29 % (40 year)
EnvioPV (Rs.)	12,946,246	17,695,097	23,027,993	23,727,467

Table 9.7 Combined economic value for cases 1, 3, and 4

Case	EcoNPV (i=10%)	B/C[a]	EIRR[a]	EC (i=10%)	CombNPV[c]	B/C[b]	EIRR[b]
Case 1	−412,627	0.871	8.29	+12,946	−399,681	0.875	8.34
Case 3	−1,116,227	0.651	4.95	+12,946	−1,103,281	0.665	5.02
Case 4	−825,166	0.742	7.29	+14,871	−810,295	0.747	7.34

Note: [a]Values estimated under normal engineering practice
[b]Values estimated under the application of 'Educated Trade-Off' framework
[c]Combined direct, social and environmental economic values

Stakeholder Education and Framework Validation

Stakeholder Workshops

Stakeholders were consulted to obtain their views on the proposed project; to educate them on technical, economic, and environmental (including social) aspects of the project; and to ascertain the impacts of what they learned. The sample included 218 stakeholders representing 10 different water uses. Eleven stakeholder workshops were conducted at the Divisional Secretariat offices situated along the river banks and one city centre located by the river bank. The spatial distribution of workshops allowed the participation of stakeholders representing proposed project areas and the entire downstream reach of the river.

The two major tools used for stakeholder consultations were a set of presentations for introduction to and education about the proposed development and a set of questionnaires for 'pre-' and 'post'-education sessions. The distribution of respondents among different use sectors is shown in Table 9.8.

Awareness and Expectation of Impacts

Of the 218 respondents, a majority (67 %) were unaware of the proposed reservoir project. Irrespective of awareness, a majority (67 %) also thought that they would experience positive impacts from the project compared with their present situation, while less than half (44 %) expected negative impacts.

Of the 71 who were aware of the project prior to stakeholder workshops, almost two-thirds (70 %) expected positive impacts, while only 49 % expected negative impacts. This meant that 19 % expected both positive and negative impacts from the project. Of the 145 stakeholders who heard about the project for the first time,

Table 9.8 Frequency and percentages of stakeholders from different use sectors

Sector	Frequency	Percentage
Officials of NWSDB	13	5.96
Users of pipe-borne water	9	4.13
Tourism sector	8	3.67
Recreation sector	22	10.09
Industrial sector	6	2.75
Dug-well users	26	11.93
Paddy field owners/farmers	28	12.84
Coconut landowners	25	11.47
Environmental organisations	25	11.47
Government administrative mechanism	56	25.69
Total	218	100.00

65 % and 43 % expected positive and negative impacts, respectively, from the project. Chi-square analyses showed that awareness of the project had no significant influence on the stakeholders' expectations.

The numbers of stakeholders with different expectations regarding impacts and the reasons for them were determined as follows. Reasons for positive impacts were good quality water supply and/or expansion of water supply 62 (43 %) and extra benefits such as electricity for the area 34 (23 %). Reasons for expecting negative impacts were reduced flow to downstream areas 30 (25 %), relocation 22 (19 %), and negative impacts for dug-well users 16 (14 %).

Willingness Before and After Stakeholder Education

The level of willingness to support the development project was measured using responses obtained on a scale of 11 points from +5 to −5 including zero. A score of +5 indicated full willingness to accept the project, 0 indicated indifference. and −5 indicated full opposition to the project. The summary of the responses received at 'pre-' and 'post'-stakeholder education sessions is compared in Table 9.9 where the 11 scale points are grouped into three categories: −5 to −1 as unwilling, 0 as neutral, and +1 to +5 as willing.

The changes that took place in the three groups were as follows:

- Among the stakeholders who were 'unwilling' prior to education, 50 % were convinced and were 'willing' for the project after the education session.
- Among the stakeholders who were 'neutral' prior to education, 86 % were convinced and were 'willing' for the project after the education session.
- Among the stakeholders who were 'willing' prior to education, only 3 % were negatively convinced and became either 'unwilling' or 'neutral' for the project after the education session.

Table 9.9 Summary of stakeholder responses between 'pre-' and 'post'-education

Post Pre	Unwilling	Neutral	Willing	Total
Unwilling	16 (42.1 %)	3 (7.9 %)	20 (50.0 %)	39
Neutral	1 (7.1 %)	1 (7.1 %)	12 (85.7 %)	14
Willing	5 (3.0 %)	1 (0.6 %)	159 (96.4 %)	165
Total	22	5	191	218

Hypothesis Testing

To validate the ETO framework as a tool that could significantly impact the decision-making of stakeholders, the following hypothesis was tested using the binomial distribution.

H_0 Stakeholder involvement in decision-making on development projects in a river basin remains the same or becomes less with improved access to information through the ETO framework.

H_1 Stakeholder involvement in decision-making on development projects in a river basin is rationally impacted with improved access to information through the ETO framework. For the purpose of the study, the term 'rationally impacted' was defined as having the majority (more than 50 %) of stakeholders being impacted.

Influence of Technical, Economic, and Environmental Aspects

Further analysis was conducted to determine the factors that have influenced stakeholders to change their views about the proposed project. The factors were categorised as technical, economic, and environmental (including social). The most influential technical factors were catchment data, reservoir data, and rainfall data, while the least influential was the location of the river and cities. The only economic factor used, namely, the net benefit of the project, had a considerable impact on stakeholders.

Among the social and economic factors, understanding of the indirect positive impact of the low-flow weirs on the dug wells in the vicinity caused the major change among the stakeholders. This was followed by their understanding that the project would not adversely affect the downstream dug wells and that there were no adverse effects on crops such as paddy rice and coconut.

Usefulness of ETO Framework

Stakeholder workshops were conducted to test the broader hypothesis that the ETO framework was a useful tool in educating stakeholders for decision-making with regard to natural resource uses. It was hypothesised that the framework was a useful tool if more than 50 % of stakeholders changed their perception of the proposed Yatimahana reservoir project. The hypothesis testing found that 78 % of the sample accepted that the education changed their views about the project, leading to rejection of the null hypothesis at the 95 % significance level. By accepting the alternative hypothesis, this confirmed validation of the ETO framework.

Two of the three main reasons for disliking the project were associated with the fear of reduced flow to downstream areas, especially the dug-well sector and the agriculture sector without irrigation facilities. In the post-education session, what the stakeholders learned, especially regarding the rainfall and catchment data of the proposed reservoir, backed their confidence to change their views about the project.

Conclusions and Recommendations

The study successfully applied the ETO framework to the Ma Oya river basin, enabling the stakeholders to make informed decisions regarding the proposed Yatimahana development project and other resource uses. It has also validated the ETO framework as a tool for stakeholders' education.

The methodology facilitated the identification of dominant stakeholders and the impacts of the development project, spatially distributed over 100 km. The economic analysis of the technical requirements and economic feasibility of Scenario 2 produced an estimated IRR of 8.29 % for the base case over the assumed 20-year project life.

An important feature of the framework is that it provided economic values for the social and environmental impacts of three key sectors, which had not been considered in the project's original feasibility study. For the recreation sector, inundation of a waterfall would produce a loss of Rs. 1.2 million per annum. For the tourism sector, controlled water release at the proposed reservoir, increasing the expected number of tourism days in Pinnawala, was estimated to result in an incremental income of Rs. 2.59 million per annum. In the industry sector, flood mitigation would save water treatment costs of Rs.0.45 m per annum.

Data analyses in the fourth step established that the 'dug-well' and 'rainfed agriculture' sectors along the river bank areas would not be impacted by the proposed development project. It was observed that the value of water was considerably understated if only the financial benefits were considered. Without the reservoir project, developing the next best supply of water would be significantly more expensive.

The incremental net benefit for Scenario 2 was found to marginally increase when the social and environment impacts were added to the initial economic estimates of costs and benefits. Irrespective of the outcome, application of the ETO framework educated the stakeholders in technical, economic, and environmental aspects to enhance the rationality of their decision-making. Statistically, it was demonstrated that the ETO framework was a useful tool in educating the majority of the stakeholders in making decisions about natural resource uses at 95 % level of significance.

To support ongoing work, it is recommended that a network of interested stakeholders be formed in the GNDs along the river using lower-level government officials as a catalyst. Networking would help to reveal stakeholders' concerns as well as identifying emerging issues at an early stage so that remedial actions could be implemented. The benefits gained by avoiding the social and environmental costs incurred by users of the water supply scheme should be emphasised. Immediate steps should also be taken to collect reliable technical data on water flow, water levels, and water quality.

References

Curry S, Lucking R (1992) Shadow prices for Sri Lanka: users' guide. National Planning Department, Colombo

DHI (Delft Hydraulics Institute) (1999) Western River Basin Sector Project TA 3030 –SRI, Working document A, Ma Oya River Basin. Detailed Basin Assessment, Delft Hydraulics Institute

Jenkins GP, Harberger AC (1992) Manual for cost-benefit analysis of investment decisions. Harvard Institute for International Development, Cambridge, MA

Pearce DW, Warford JJ (1993) World without end: economics, environment and sustainable development. Oxford University Press, New York

Ranasinghe M (1994) Extended benefit cost analysis: quantifying some environmental impacts in a hydropower project. Project Appraisal 9(4):243–251

Seckler D (1996) The new era of water resources management: research report 1. International Irrigation Management Institute (IIMI), Colombo

Sweco Grøner (2004) Improvement of water availability for water supply schemes from Maha Oya, final report. Volume I – Main Report. National Water Supply and Drainage Board, Ratmalana, Sri Lanka

Thoradeniya WBM (2009) Stakeholder Involvement in the Decision Making for Development Projects Using 'Educated Trade-offs'. PhD dissertation, Department of Civil Engineering, University of Moratuwa, Sri Lanka

Chapter 10
Policy Options for Sustainable River Sand Mining in Sri Lanka

L.H.P. Gunaratne

Abstract This study investigates policy options for sustainable river sand mining in Sri Lanka that minimize environmental degradation while meeting the requirements of the construction industry and local people. Four separate economic analyses are undertaken: comparison of the annual costs and benefits of selected mining sites, analysis of miners' views and preferences using choice modeling, evaluation of expert opinion using multi-criteria analysis, and a comparison of alternative sources of river sand. The comparison of the costs and benefits at sample mining sites reveals that the social cost of river sand mining exceeds the private costs. The results of the discrete choice experiment with the miners indicate they believe that the negative effects of sand mining can be partly mitigated by directing government funding to an environmental trust fund (ETF) with some level of comanagement. Strict rules, regulations, and awareness programs, as suggested by the media and environment groups, were not found to be productive.

Alternative policies for sustainable sand mining in three major rivers are evaluated ecologically, economically, socially, and technically using multi-criteria analysis. For the Ma Oya River, where there are more than 70 mining sites, the restriction of mining at vulnerable sites was found to be the best management alternative, followed by the establishment of an ETF. For the Mahaweli and Deduru Oya rivers, community-based management was found to be the best option. The best way to decrease pressure on rivers for sand extraction would be to use offshore sand, but price was a potential impediment in relying on this option.

Keywords Sri Lanka • Sand mining • Cost and benefit • Choice modeling • Multi-criteria analysis

L.H.P. Gunaratne (✉)
Department of Agricultural Economics and Business Management, Faculty of Agriculture, University of Peradeniya, Peradeniya, Sri Lanka
e-mail: lhpguna2010@gmail.com

D. James, H.A. Francisco (eds.), *Cost-Benefit Studies of Natural Resource Management in Southeast Asia*, DOI 10.1007/978-981-287-393-4_10

Introduction

River Sand Mining in Sri Lanka

Sri Lanka is an island with a land area of 65,525 km^2, endowed with 103 distinct rivers and streams and 94 coastal basins. Most of these rivers originate in the central highlands and travel to all corners of the country. Except for the longest river, the Mahaweli, which stretches for 335 km, all of the other rivers are less than 160 km in length. These rivers have always been an integral part of life for the people of Sri Lanka as the location of various resources, including water, and as a means of development for cities and villages throughout the country's 25 centuries of agricultural civilization.

River sand is an essential raw material in the construction industry, and mining furnishes the rural population with work as miners, transporters, and vendors. Sand is a resource that can be freely accessed, and mining activities have expanded dramatically – in some river basins by 2.5 times more than the number of original sites. As reported by Ingall (2006), this expansion is related to the boom in the construction industry that followed the Asian Tsunami of 2004. The tsunami changed the appearance of rivers and gave rise to a substantial number of environmental problems.

Unregulated and unmonitored sand mining has taken place without a clear regulatory framework, and this has aggravated environmental problems. In addition, sand prices have increased several times, thanks to a ban imposed on traditional sources of supply. Elevated sand prices have driven the exploitation of sand from fragile and remote places, causing more damage to the road infrastructure due to heavy loads carried on weak rural roads. This situation has been further complicated by politics. No tangible benefit can be derived from current institutional activities unless there is a mechanism that meets both demand for sand and the need for conservation goals. In addition, the needs of the rural poor, who benefit from river sand mining, need to be considered.

Indiscriminate river sand mining has resulted in a large number of environmental and social problems in Sri Lanka. Environmental problems include on-site effects such as the erosion of riverbanks, lowering of water tables, intrusion of salt water, damage to riverine vegetation, loss of habitat of the aquatic population, increase in mosquito-related health problems, and damage to bridges and structures. Off-site effects include the impairment of rural roads and increased coastal erosion. These environmental issues have affected most of the rivers and streams of Sri Lanka.

The strategies adopted so far have not been based on sound economic analysis and so have led to policy failures. Moreover, ad hoc and inconsistent policies have contributed to a short supply of sand, further degradation of the environment, and increased social upheaval. There is an urgent need to identify appropriate policy guidelines that guarantee environmental protection with minimum regulatory costs and high levels of public cooperation.

Institutional Context

Sand is the property of the state and a permit is required to mine and transport it (Mines and Minerals Act No. 33 (1992) of Sri Lanka). Artisanal sand mining was the norm until the introduction of the current Mines and Minerals Act (which replaced the former Mines and Minerals Law No. 4 of 1973) established the Geological Survey and Mines Bureau (GSMB), which regulates the exploration for and mining of quantities of sand deposits. Expressions of interest for mining tenders are issued by the Divisional Secretary because the custodian of rivers and state land is the government. Including the Mines and Minerals Act No. 33 of 1992, the National Environmental Act of 1980, the Coast Conservation Act of 1981, and other relevant legislation, regulations, and policy statements reflect Sri Lanka's constitutional, international, and national obligations. These policies support other national policies such as the National Environmental Policy, which was adopted by the cabinet of Sri Lanka on 15 May 2003. The Ministry of Environment and Natural Resource Management prepared a draft policy on sand as a resource for the construction industry and opened it for public comment.

Research Objectives

The main objective of this study is to identify appropriate policy options that minimize environmental degradation from river sand mining while meeting the requirements of stakeholders, including the construction industry and local people who depend on sand mining. The specific objectives are to:

- Present an overview of the present situation related to river sand mining, including the general causes of environmental degradation and the institutional and policy set-up.
- Compare the private profitability and the social cost of river sand mining.
- Analyze the preferences and trade-offs of various options for miners.
- Evaluate the policy alternatives under social, economic, ecological, and technical criteria, based on stakeholder views.
- Assess suitable alternatives for river sand mining that reduce the pressures on rivers while also sustaining the construction industry.

Overview of Sand Mining and the Sand Market

Demand for Sand

Sand is used for building construction, road surfacing, and landfilling around bedding pipes and telecommunication lines. The construction industry remains the main consumer of river sand, accounting for more than 95 % of demand. The total national demand for sand is difficult to estimate because of data limitations. Studies (Katupotha 1998; Dias et al. 1999) that have attempted to estimate sand demand have been found to be incomplete. However, the best estimate, based on reports by the Geological Survey and Mines Bureau of Sri Lanka, is that the demand has been increasing from 11.01 m^3 million in 2007 to 14.05 m^3 million in 2011.

The gap between licensed sand mining output and the estimated need for sand is high, indicating extensive illegal sand mining activity. Unsustainable sand mining occurs at two levels: one is illegal extraction over the quota by licensed miners with the help of police officers and local government officers; the other is mining from banned or vulnerable sites. In addition to environmental and other social costs, these practices deprive the government of income.

The shortage of sand has pushed up sand prices more than fourfold during the last decade, exerting pressure on vulnerable and remote sites. However, the royalty payment to the GSMB and the income that goes to local government is fairly low. The economic feasibility of other sources of sand is masked by the low price of river sand which includes only the direct costs such as license fees, wages, equipment, and transport. Environmental costs as well as other social costs, such as loss of livelihood and the damage to rural road networks, are not reflected in market prices.

Environmental Impacts of River Sand Mining

Environmental problems related to sand mining are reported from most parts of Sri Lanka. The problem is more severe in the districts of Kurunegala and Puttalam in Northwestern Province, as revealed in reports from various NGO environment groups such as *Lanka Jalani*, the Japan Water Forum, Associated Development Research Consultants, the NGO forum of the ADB, the Network of Women Water Professionals in Sri Lanka, and the *Navoda* Farmers' Organization.

Major negative environmental impacts include the lowering of riverbeds, the reduction in capacity of water intake, water pollution, saline water intrusion, riverbank and coastal erosion, soil quality deterioration, environmental changes, damage to riparian vegetation, and in-stream habitat and biota. Other detrimental impacts of sand mining are the alteration of flow patterns due to the modification of riverbeds, overloading of suspended sediment, damage to the channel beds due to the use of heavy equipment, permanent flooding conditions or the presence of pooled water, disturbance of the natural hydrology of the riparian zone because of infrequent elevated flow levels, the exclusion of large woody debris from the riparian zone, and reduced vegetative bank cover.

As reported by (Kamaladasa 2008), the amount of sand being extracted from the main rivers along the west coast of Sri Lanka is far in excess of the supply of sand from the upper parts of the catchment, causing the lowering of riverbed levels, affecting the drinking and agricultural water supply, and lowering the water table near rivers, in turn lowering water levels in wells. The drying up of irrigation channels has reduced the flooding of paddy fields with nutrient-laden water and increased the unsteadiness of the banks of rivers.

Coastlines are enriched by river sand entering the sea, and inland sand mining has hindered sediment flow, leading to coastal erosion. A study by Nanayakkara (1999) clearly identified the relationship between too much sand mining in the Ma Oya River and coastal erosion north of the Ma Oya River mouth. This study estimated that coastal erosion caused by the reduced sand supply from the Ma Oya, Kelani, and Kalu rivers was equivalent to an average land loss of about 5–6 ha per year.

Other losses caused by present sand mining practices include loss or damage to houses, churches, graveyards, and archaeological sites, loss of livelihood for coastal fishermen, and changes to natural estuarine vegetation.

Benefits and Costs of River Sand Mining

Conceptual Approach

A sand miner is concerned with his private costs when making decisions but not the costs borne by others, i.e., social costs. Social costs include both private costs and any external costs to society arising from sand extraction. When external costs arise, because environmental costs are not paid, market failures and economic inefficiencies exist at local or national level.

Damage costs of sand mining in Sri Lanka have been estimated by various researchers (Byrne and Nanayakkara 2002; Siripala 2002). However, river sand mining also has some positive impacts. The main benefit of river sand mining is the work it offers to sections of the community that have few other options. Rural people who live below the poverty line are relatively well paid for involvement in sand mining during certain months of the year.

Field Data Collection

The field surveys to collect data relevant to the resource use, income, private costs, and subsequent social costs were based on a structured questionnaire and were conducted at three sites: Site 1, Doluwa (Central Province); Site 2, Weragantota (Central/Uva Province); and Site 3, Alawwa (Northwestern Province). The first two sites are located along the Mahaweli River while the third site is adjacent to the Ma Oya River.

The data collected included the amount of damage to riverbanks; the amount of damage to wetlands; the amount of damage to sensitive areas; the number of trees removed; the amount of soil erosion; the loss of productivity on adjoining land; the amount of damage to structures such as bridges, water pumping stations, etc.; the number of licenses issued/the number of miners involved per site; the amount of sand removed per day; the number of people employed (in all categories) by origin (local or outsiders); satisfaction with the mining activities and perceptions of the mining activities; the direct production costs borne (wages, license fees) by the miner; the price of sand at the site; other transaction costs; and damage to roads and the number of complaints received by the local authority. This data was used to estimate monetary values using direct market-based approaches.

Private Costs and Benefits

Only one period of costs and benefits was compared without any discounting. The total production cost of site 1 was estimated to be LKR 2,163,893 per month, and the gross mining income was LKR 2,600,000 per month. Since there is a positive difference between income and expenditure (LKR 436,107 per month), sand mining is privately profitable at this site. Similarly at site 2, the total income received by selling sand was LKR 116,736,480 (=210 cubes per month × 12 months × 37 miners × LKR 1,252 per cube), and the total direct cost of sand mining was LKR 84,741,470, so the total annual income over expenditure was LKR 31,995,010. The income and expenditure data related to the third site revealed that there were positive private profits of LKR 13,605,120–38,810,880 (140 cubes × 24 miners × 12 months × LKR 1,300 per cube 38,810,880) [115 Sri Lanka Rupees = 1 USD as of July 2009, one cube = 2.8 cubic meters].

Social Costs and Benefits

Social cost comprises the summation of private cost and external cost due to the environmental and non-environmental effects of sand mining. Of all the potential environmental damage that can be caused by river sand mining, the following effects were prominent in the study area: riverbank erosion, productivity loss in paddy fields, and the non-environmental social costs of damage to rural roads.

It was found that the opportunity cost of the sand site was zero for sites 1 and 3. The survey revealed that the average opportunity cost of site 2 per mining location was approximately LKR 27,500 per year. There were no conceivable social benefits from the sand pits, i.e., no increase in fish stocks or restoration of water during dry periods. It was assumed that all environmental damage was attributable to river sand mining during the permit period.

No major loss to biodiversity was reported from the study areas. Though there were complaints related to health, these were not considered at the time of data collection, as people did not appear to be concerned with this aspect of sand mining. Other potential negative externalities such as saltwater intrusion, damage to infrastructure and houses, potential effects on cities downstream, loss of bathing water, and decline in inland fisheries were not prominent and hence were not quantified or monetized during the study.

Productivity Loss

Loss of income due to lowering of the water table was computed. The estimated yield loss, the extent of affected crops, and the loss of income were computed from information gathered from the survey and the information obtained from the Agrarian Services Center. At site 1 the total farm income loss was estimated at LKR 311,190. Similar computations were made for the other two sites, and the estimated losses amounted to LKR 596,000 for site 2 and LKR 186,000 for site 3.

Damage to Roads

The destruction of roads appeared to be the main non-environmental social cost of river sand mining in the study area. The valuation methods that were applied were based on the repair/replacement cost approach. For sites 1 and 3, the charge imposed by local authorities to cover the cost of damage to roads is LKR 100 for each cube of sand. For site 2, the charge imposed by the Road Development Authority is LKR 8,292 per month per miner. The total annual replacement cost was accordingly estimated at LKR 3,681,648 per year.

Riverbank Erosion

It was assumed that if sand mining needs to be continued, the riverbank should be protected by fixing a retaining wall or similar structure. The wall would cover only the sand site at the bank, and the residual effects, e.g., damage to the opposite side of the bank during the rainy season due to the presence of a dam on one side, were not taken into account. The length and height of the sand site was measured and

Table 10.1 External cost of sand mining (LKR) per year

Cost items	Site 1	Site 2	Site 3
Net income loss from crops	311,190	596,000	186,000
Damage to the riverbanks (adjusted per year)	1,900,000	4,300,000	1,321,000
Damage to roads	1,300,000	3,681,648	336,000
Opportunity cost of the land		1,017,500	
Total	3,511,190	9,595,148	1,843,000

Table 10.2 Summary of private and social costs and benefits (LKR) per year

Cost/benefit	Site 1	Site 2	Site 3
Private cost	25,966,716	84,741,470	38,810,880
Private benefit	31,200,000	116,736,480	52,416,000
Net private benefit	5,233,284	31,995,010	13,605,120
External cost	3,511,190	9,595,148	1,843,000
Social cost	29,477,906	94,336,618	40,653,880
Net social benefit	1,722,094	22,399,862	11,762,120

multiplied by per unit cost of restoration. The cost of construction of the retaining wall was estimated based on the specifications such as the size of the foundation, width and height, and the concrete mixture provided by the Irrigation Engineer, Irrigation Management Department, Kandy. The estimated social costs of sand mining in the study area are shown in Table 10.1.

The major social benefits of river sand mining are the provision of employment opportunities to local people, flood control, and the supply of raw materials to the construction industry. These are in addition to the private benefits of miners. Due to the vagueness of this information, only the net private profits were considered as a lower bound value of social profitability.

Summary of Costs and Benefits

The social costs and net private benefits were computed for each of the study sites, and a summary is presented in Table 10.2. In all cases, social costs exceeded private costs. Although the net social profits were nonnegative at all the study sites, the results should be interpreted with caution. The results must be considered in context, as none of these sites experiences other major environmental impacts such as saltwater intrusion or damage to infrastructure.

Preferences of Sand Miners

Significance of Miners' Preferences

Information regarding trade-offs in the behavior of resource users (i.e., sand miners) and broader social indicators of well-being is important in resolving potential conflict between private and public interests regarding the use of natural resources and the environment. Without considering miners' preferences and perceptions, realistic solutions cannot be sought. It should be recognized, however, that miners may not provide values that are widely shared or accepted by other members of society.

Analytical Framework

To assess the preferences of miners regarding trade-offs, a discrete choice experiment (DCE) was undertaken. The study was conducted at the sand mining sites of the Ma Oya River in the Polgahawela, Alawwa, and Narammala Secretariat Divisions in Kurunegala District. Information about all the license holders presently engaged in sand mining in the Ma Oya River was collected from the regional GSMB office at Kurunegala. The survey dealt with a random sample of 150 miners and mine workers.

DCE is a stated preference technique, in which respondents are asked to evaluate hypothetical scenarios, in contrast to revealed preference techniques that rely on modeling actual behavior. Discrete choice experiments were originally developed in the market research and transport literature, but they have now become familiar in environmental economics (Adamowicz et al. 1994, 1998; Louviere et al. 2000). DCEs are closely linked to two economic theories known as Lancaster's characteristics theory of value (Lancaster 1966) and the random utility theory (Thurston 1927; McFadden 1974). In choice experiments, respondents are presented with a series of alternative profiles of environmental goods or policies and asked to choose their most preferred choice (Bateman et al. 2002). These profiles are set out in terms of the attributes (characteristics) of these goods and policies (Adamowicz et al. 1994). Usually a monetary value is included as one attribute. Two components are required to implement a DCE: (a) a statistical design to create the hypothetical scenarios and (b) a statistical method to analyze the responses (Louviere et al. 2000). The widely used statistical method of analysis is the multinomial logit model, which is based on the behavioral assumptions of random utility theory (McFadden 1974). Fractional factorial designs can be used to reduce the number of profiles, thereby reducing the cognitive burden faced by the respondents in the choice experiment (Carson et al. 1994).

Table 10.3 Attributes and levels used in the choice experiment

Attributes	Level
Monitoring and regulation	Strict regulation, i.e., identify the vulnerable sites and police them
	Contribute to environmental trust fund and continue mining
	Status quo, i.e., no contribution to trust fund but continue mining
Awareness programs for resource users	By environmental organizations
	By the government, i.e., GSMB and local government environment officers
	Status quo, i.e., no awareness programs
Willingness to pay	Reduce by 50 % the amount of sand allowed to be mined with the current royalty
	Increase the royalty by 50 % and increase how much sand can be mined
	Status quo, i.e., no change in amount or royalty
Community-based management	Apply only for small sites
	Comanagement with local government for all sites
	Status quo, i.e., no community-based management

Survey Instrument

The survey instrument for the DCE was developed through focus group discussions and a mailed survey to environmental NGOs. The attributes and levels chosen for the DCE are summarized in Table 10.3.

The scenarios created by combining the four attribute and three levels resulted in 81 combinations of possible management actions. Two profiles were combined into one choice set and respondents were asked to choose one of these two. The data were collected in personal interviews where each respondent was shown only four choice sets, out of nine possible choice sets. Each choice set was printed on a separate colored sheet and laminated. The data was entered into MS EXCEL and the statistical analysis was undertaken using STATA v.10.

Results and Discussion

10.4 presents the MNL parameter coefficients, their standard errors, and p-values for each attribute for the entire sample. The respondents assigned the highest priority to continuing mining (with the existing royalty or an increased royalty) followed by the establishment of an environmental trust fund. According to Table 10.4, regarding the regulation and monitoring attribute, respondents accepted that the present regulations are either weak or inefficient. Also, they perceived that strict regulations result in bureaucratic inefficiency. They want to continue mining,

Table 10.4 Estimates of the multinomial logit model

Attributes	Level	Coefficient	SE	P value
Monitoring and regulation	Strict rules, regulations, and monitoring	0.1286	0.4856	0.945
	Establishment of ETF	0.6284	0.1910	0.001
	Status quo	−0.7570	0.5218	0.149
Awareness programs for resource users	Environmental organizations	0.1023	0.1669	0.540
	GSMB and environmental officers	−1.0401	0.2793	0.000
	No awareness programs	0.9378	0.3254	0.004
Willingness to pay	Increase the royalty by 50 % and increase allowable amount	1.4622	0.3680	0.000
	Reduce by 50 % allowable amount with current royalty	0.1211	0.1427	0.396
	No change in royalty or amount	−1.5833	0.6946	0.000
Community-based management	Fully managed by CBM	−1.2471	0.4728	0.008
	Comanagement with local government	0.4694	0.1873	0.049
	No CBM	0.8777	0.8402	0.297

Log likelihood = −157.51

and environmental damage could be corrected via their contribution to an environmental trust fund maintained by the GSMB, but rehabilitation should be done with the collaboration of local government. The surprising finding is that "no awareness" is statistically significant in the analysis, implying that local people do not perceive any tangible benefits from such activities. This is contrary to most arguments suggesting that providing persuasive instruments is effective in managing this issue. Another striking finding is that the respondents rejected the awareness programs conducted by environmental organizations.

It was found that the miners were willing to contribute more to royalties if they could mine more sand. This implies that they have accepted that the present royalty should be increased but that they should also benefit. With respect to community-based management, the respondents were not in favor of being fully managed by community-based organizations (CBOs). Instead they preferred the involvement of CBOs with some control by local government, i.e., comanagement.

The part-worth utilities presented are consistent with the field observations made by the researchers. The field survey showed that respondents placed the highest priority on continuing the present levels of mining even if it means increasing the royalty payment. They were not concerned about sustainability due to a lack of other livelihood opportunities.

The results of the discrete choice experiment suggest that the negative effects associated with sand mining can be partly mitigated by allocating government revenue to an environmental trust fund with some level of comanagement. Strict rules and regulation as well as the awareness programs, as suggested by the media and environment groups, may not be so effective.

Evaluation of Policy Alternatives

Holistic Assessment of Policy Options

Effective management of the environmental issues related to sand extraction is workable only by linking the activities, interests, and perspectives of all the groups concerned, including civil society, industry, nongovernment organizations, and government agencies. Current and potential policy alternatives for river sand mining were accordingly evaluated holistically using multi-criteria analysis (MCA), incorporating the assessments of policy options by relevant stakeholders.

Analytical Framework

MCA is a multi-objective decision support technique that uses a set of predetermined criteria to evaluate policy options. Such models can play an important role in environmental management. They have been used in studies of conservation requirements (Cochrane and Zeleny 1973; Voogd 1983; Nijkamp and Van den Bergh 1990; Ferdinando et al. 2000), marine reserves and mangrove forest (Gilbert and Janssen 1998; Janssen and Padilla 1999), forestry (Kangas et al. 2001), and energy system management (Munasinghe 1993).

The simplest MCA model is a weighted summation model. For each management option, an "effects table" is first constructed, with scores indicating the outcome or performance of each option with respect to each criterion. The data are then normalized for each criterion. Each normalized value is multiplied by its corresponding relative criterion importance weight, and a total score for each option is calculated by summing its weighted values. A ranking of options can then be derived. The evaluation procedure can produce a complete ranking (A > B > C > D), the best alternative (A > (B,C,D)), a set of acceptable alternatives ((A,B,C) > D) or an incomplete ranking of alternatives (A > (B,C,D) or (A,B) > (C, D)). Data used in the model can be quantitative, qualitative, and/or monetized.

Specification of the MCA Model

For the present study, DEFINITE, a multi-objective decision support system was used (Janssen et al. 2001). The system supports the whole decision process from problem definition to report generation. In DEFINITE, the focus is on methods to support the choice phase. The results are sensitive to the evaluation method, uncertainties in scores, weights, and prices used.

The data to populate the model were gathered in two different ways. First, a survey was mailed to environmental NGOs, to establish the set of management alternatives. Second, these were then scored by field engineers attached to GSMB using expert judgment, expressing the relative performance of the four options in terms of an effects table.

Management Options

The management alternatives incorporated in the model are described below.

Complete Ban As a consequence of environmental concerns raised by NGOs and the media, sand mining has been completely banned in certain rivers in Sri Lanka by the Supreme Court, thanks to public interest litigation. Some institutions, such as National Water Supply and the Drainage Board, believe that there should be stricter regulation of some rivers, with a complete ban on sand mining needed considering its impact on water intake levels.

Restrict Access to Vulnerable Sites This is based on technical classification of sites. At present all the sites along the Ma Oya River have been technically evaluated by the GSMB. This approach could be extended to mining sites at other rivers, and then mining in potentially vulnerable sites would not be allowed.

Do Nothing Currently, although the mining licenses are issued by the GSMB and a certain level of monitoring is conducted by local government, illegal mining takes place. Therefore, the "do nothing" option represents a continuation of present practices.

Environmental Trust Fund With a reduction in sand supply caused by the restriction of access to vulnerable sites, the construction industry would face a serious sand shortage which would further escalate prices. The current supply cannot be sustained if damage to the environment is not dealt with. An environmental trust fund could be used to rehabilitate the environment. Moreover, a trust fund could finance other environmental protection and natural resource management programs and promote the use of innovative solutions. Also, it could be used to support education and training programs related to the protection of the environment.

Community-Based Sand Mining The draft Act on sand mining prepared by the Ministry of Environment and Natural Resources has identified this option. Permits would be provided to registered community-based organizations whose membership comprises existing river sand extractors. These community-based organizations would be assigned single extraction rights for specified extraction zones in those rivers for which harvest limits are introduced. Each mining zone would be supported with a single landing site and would be awarded to the community-based organization for a period of five years, within which an annual extractable quota would be assigned.

Evaluation Criteria

Criteria in the model were defined in terms of four policy objectives: ecological, social, economic and technical.

Under the social criterion, a number of effects were considered. They included the effects on livelihood, health, loss of residences, cattle rearing, bathing water, and pollution of drinking water. The economic criteria included the cost of repairing infrastructure, mine owners' income, income from crop production, income of sand vendors, and impacts on inland fisheries. The main environmental criteria considered were saltwater intrusion, loss of biodiversity, decline in groundwater levels, imbalance in river flows, and soil quality. The technical criteria included the use of mining equipment and technology and scale of mining.

Criteria weights were chosen for the social, economic, environmental, and technical criteria, provided by experts and by stakeholders. In all cases, the respondents placed equal importance on the social, economic and environmental criteria but relatively less on the technical criterion.

Results

An effects table was compiled, as described above, for each of the three sites. By way of illustration, the effects table for Site 1 is shown in Table 10.5. The other two effects tables have not been reproduced here, for reasons of brevity. The effects tables were used as direct inputs to the analysis using DEFINITE, which yielded several graphical outputs. Bar charts so obtained were used to interpret the results. An overall ranking was produced using a combination of expert and stakeholder weights. Sensitivity analysis was also carried out.

Ranking of Management Alternatives

For the Ma Oya River, a high variation in the performance of management options under the economic, environmental, and technical criteria was observed. "Do nothing" appeared to fare best under the economic criterion, while it was the least preferred option in terms of the social, environmental, and technical criteria. As expected, a complete ban was the best option under the environmental criterion. Overall, it was found that continuing the current process with the restriction of mining at vulnerable sites was the best alternative, followed by the establishment of an environmental trust fund.

In the evaluation of options for the Mahaweli River, a complete ban was not included. This is because the Mahaweli is the longest river in Sri Lanka, and it flows

Table 10.5 Effects table for the Ma Oya River

Evaluation criteria	Complete ban	Restrict access to vulnerable sites	Make allowance for environmental trust fund	Do nothing
Social				
Effect on livelihood	– –	–	+	+ +
Effect on health	+ +	+ +	+	– –
Loss of residences	+ +	+	+	–
Water for cattle rearing	+	+	+	–
Loss of bathing water	+ +	+ +	+	–
Pollution of drinking water	+ + +	+	+	–
Economic				
Cost of repairing infrastructure	+ + +	+ + +	+	–
Effect on crop production	+ + +	+ +	0	–
Income of mine owners	– – –	– –	–	+ +
Income of laborers	– – –	– –	–	+ +
Income of sand transporters	– – –	– –	–	+ +
Income of building material suppliers (alternatives to sand)	+ + +	+ +	0	–
Income of building material suppliers (except alternatives to sand)	– –	–	0	+ +
Cost of riverbank conservation	+ +	+	+	–
Effect on inland fisheries	+ +	+	0	–
Price of sand	– –	–	0	+ +
Environmental				
Saltwater intrusion	+	+	0	–
Loss of biodiversity	+ +	+	+	–
Loss of riverine vegetation	+ +	+	+	– –
Decline in groundwater levels	+ + +	+ +	0	–
Imbalance in the natural flow system of rivers	+ +	+	+	–
Quality of soil	+ + +	+ +	0	–
Technical				
Use of technology	+	+	– –	–
Scale of mining	+ + +	+ +	– –	–

across a steep enough and big enough land area for there to be no major environmental problems other than overextraction, compared to other rivers.

For this reason, community-based sand mining was incorporated in the analysis. This particular management option performed well under all the criteria except for economic and was found to be the best option overall. Identification of vulnerable sites and the restriction of mining at these sites appeared to be environmentally more attractive.

For the Deduru Oya River, there was a high variation in performance between the management alternatives for sand mining sites. Currently sand mining is banned in this river by the government, mainly due to environmental concerns. A complete ban is not economically attractive, so it was not accorded a better social preference, mainly due to the loss of livelihood that it would entail. Although the analysis ranked community-based sand mining as the best alternative, there was not a substantial variation among the management options. The multi-criteria analysis was clearly capable of accommodating all the dimensions of the key issues (social, economic, environmental, and technical) and evaluating all the possible management options. Scores were obtained from experts and from a group of people who had environmental concerns and were aware of the issues. Therefore, the results of the analysis can be treated as a holistic evaluation of potential policies for each of the rivers studied. Analysis found that there is a wide variation among the alternatives so a single criteria evaluation may fail at field level. Sensitivity analysis did not result in any major changes in the preceding findings.

Alternative Sources of Sand

Potential Alternatives

If the demand for sand is partly absorbed by other means, subsequent environmental issues could be mitigated. Four major alternatives to river sand have been identified in Sri Lanka: offshore sand, land-based sand, dune sand, and quarry dust. Table 10.6 shows the availability these alternatives (Dias et al. 1999).

Of these alternatives, land-based sand has already been used as an alternative to river sand in landfilling for road construction. A considerable amount of land with sandy soil is currently cut for property development, so there is potential to use fine and aggregate land-based sand. However, land-based sand cannot be considered as a successful alternative to river sand for a number of reasons. For instance, yield depends on the type of soil and the washing method. Also, the nature of the aggregate is usually coarse, so land-based sand cannot be used for plastering. In addition, the environmental impact of using land-based sand has not been assessed, and there is the possibility that huge pits would be left after digging, creating environmental hazards similar to that of clay mining. However, studies carried out at the University of Moratuwa, Sri Lanka, revealed that land-based sand could be introduced on a project scale, where the unit cost depends on the scale of operation (Nanayakkara 1999). The cost computation based on a proposed project that included the cutting and washing of soil found that the price would range between LKR 1,230–1,500 per cubic meter.

Dune sand is available in Sri Lanka along the northwest coast between Chilaw and Kalpitiya, across Mannar Island and the Pooneryn Peninsula, along the northeast coast between Pulmoddai and Point Pedro and on the southeast coast from east

Table 10.6 Alternatives to sand mining (Dias et al. 1999)

Type of sand	Annual availability (approx.)
Offshore sand	31.5 million cu.m
Land-based sand	9.6 million cu.m
Dune sand	0.3 million cu.m.
Quarry dust	9.96 million cu.m

of Ambalantota to Trincomalee (Nanayakkara 1999). According to a study by NARESA (1991), an area of 7,500 ha of dune sand is available on the island. Although dune sand was found to be an appropriate substitute for river sand, the wider application of dune sand cannot be recommended due to its role in storm protection, as experienced during the Tsunami of 2004. Also, the quantity of dune sand available is insignificant in relation to the demand for sand.

Quarry dust is the only non-sand option being considered as a realistic alternative to river sand. Studies have shown that quarry dust in Sri Lanka is usually obtained as a by-product of aggregate production and is widely used for the production of cement blocks. Siriwardana (1992) showed that quarry dust could be successfully used in making concrete. Although no specific operations are needed other than crushing, quarry sand is not available at present in significant amounts. The computation, based on a proposed project, revealed that the end price ranges between LKR 2,280 and 3,000 per cubic meter, depending upon bulk density (Nanayakkara 1999).

Offshore sand also has potential as an alternative to river sand. Large sand reserves in the sea, north of Colombo, have been identified and could supply sand for the foreseeable future. Although offshore sand mining is a new technology for Sri Lanka, it has been practiced for decades in many other countries such as India and the United Kingdom. However, this type of sand is less popular with the construction industry due to the presence of salt and shells. Previous investigations of the technology required have shown that the salt content can be altered, leaving the presence of shells as the only slightly negative factor for the construction industry.

Three major processes are involved with offshore sand mining as far as cost component identification is concerned. They are investigation and monitoring costs, the dredging cost, and the cost at the discharge point. The analysis used secondary data available at the Sri Lanka Land Reclamation and Development Corporation (SLLRDC) and the University of Moratuwa, Sri Lanka.

Offshore Sand Mining Process

The economics of offshore sand mining depend on the scale of the dredging operation in the sand mining locality. As the economy of scale is increased, the cost of the operation eventually declines; thus large-scale operations can reduce the unit cost of sand supplied at the stockpile. Trailing suction hopper dredgers

(TSHDs) have dredging capacities that range from several hundred cubic meters of sand per day to more than Sri Lanka's annual requirement of sand in a few weeks. However, economical rates can only be achieved when the usage time of the machine is maximized. Therefore, round-the-clock operation of the dredger is usually preferred. In this analysis, it was assumed that a TSHD with a relatively small capacity would be operated in Sri Lanka. Moreover, 2-week shifts have been assumed in order to increase the usage time. Since there is evidence that unprocessed, sea-dredged, and well-drained fine aggregates with coarse aggregates have been successfully used in concrete preparation, this analysis assumes that the offshore sand, after being pumped into discharging bays, would be allowed to drain for sufficient time and then moved to the stockpiles. It is also assumed that the drying of the sand is mainly due to draining out the seawater and not due to evaporation. However, previous experiences also show that there are cases where processing aggregates have been carried out before usage.

A project on offshore sand dredging should follow a proper seabed investigation. This investigation should include a bathymetric survey and seabed sampling to understand the topography and material characteristics of the ocean floor and to identify possible borrow areas. An EIA would also have to be carried out. Costs of offshore sand mining have been estimated using records available from the Sri Lanka Land Reclamation and Development Corporation. The cost per cube is 3,447.86 LKR, and the selling price is 3,800 LKR per cube after the inclusion of a profit margin.

The current tender prices of offshore sand are shown in Table 10.7. The price in Colombo without the transport cost is in the range of LKR 3,900–4,900 per cube. With a reasonable (depending on the distance) transport cost, the offshore sand price at a construction site is in the range of LKR 5,900 and 7,900 per cube. The price of the sand at the river sites varies from about LKR 1,200 to 2,100 per cube during the dry season, while the price at the construction site with the transportation cost ranges between LKR 6,000 and 7,000 per cube depending upon the season. Since the difference is marginal, offshore sand appears to be the best alternative to river sand. Currently offshore sand is mainly available in the suburbs of Colombo (Western Province) so a reasonable fraction of the total sand requirement could be replaced by offshore sand. (Currently 40 % of the sand demand is created by the construction boom in Western Province.) Increased use of offshore sand would reduce the pressures on the Kelani and Ma Oya rivers, which are the main suppliers of sand to Western Province.

Stakeholder discussions revealed that there are a number of limitations to offshore sand as an alternative to river sand. These include: the need for a large initial investment, so traditional industries may be reluctant to enter into the business; the traditional thinking of construction companies that believe that river sand is the best kind of sand; the possibility of the presence of chlorides and shells that may affect workability, permeability and the durability of buildings; and high transportation costs to the interior from the shore. Moreover, the offshore sand price is sensitive to the scale of operation so there needs to be an assured demand before the business can be undertaken by firms.

Table 10.7 Offshore sand tender prices without transportation (per cube) (Source: Paper advertisement by Sri Lanka Land Reclamation and Development Corporation, August 2009)

Places	Prices
Wattala and Ja-Ela (Muthurajawela)	3,900.00
Kelaniya, Kiribathgoda	4,600.00
Colombo and suburban	4,900.00
Average price	4,466.67

This implies a need for policies and other awareness activities to promote the use of offshore sand. If the government declares that all large development projects (i.e., highway construction, airports, harbors, etc.) must use offshore sand, then the industry can be initially sustained. Also, the government could use regulation to persuade all large private sector construction projects to move from using river sand to using offshore sand. Of the total sand demand, 88 % comes from home building and other construction which is usually advised by architects and engineers. Therefore, effective awareness and promotional programs that have the support of existing organizations such as the Institute for Construction Training and Development (ICTAD), the Chamber of Construction, and the Sri Lanka Institute of Engineers could be launched to familiarize the marketplace with offshore sand. Moreover, if necessary, the implementation of a subsidiary scheme at the beginning might be helpful to move constructors toward offshore sand. Once offshore sand has become widely accepted, these campaigns and the reduced price level can be terminated. Offshore sand can then be competitively supplied to the market.

Potential Negative Externalities Associated with Offshore Sand Mining

A number of potential environmental impacts have been cited. One key impact is increased turbidity which has a negative effect on the marine ecosystem, fauna, and benthic colonies, partly due to the overflow from dredging canals. This could be minimized by filling hoppers to safe levels and monitoring the hoses to prevent leaks and restricting dredging operations during the inter monsoon and northern monsoon period, i.e., when the sea is calm. A large number of fish species are present in the pelagic waters of the existing environment for offshore sand mining. These commercially important fish species are commonly found in western coastal waters. However, on the west coast, where offshore sand mining would be initiated, none of the phytoplankton or zooplankton species is endemic, threatened, or endangered. There could be some biological impacts due to stockpiling, but this can be ignored because a huge area is still available as a conservation zone and an urban buffer zone.

A number of registered fisherman associations and cooperatives exist in the area. Marine and lagoon fishing and the fish processing industry provide a major income for a substantial segment of the population within this area. The major social impacts of offshore sand mining are damage to fishing nets and fishing craft, a reduction in the fish population, obstructions to laying nets, and damage to prawn grounds. These impacts could be mitigated by completing dredging operations in the shortest possible time, and providing fishermen with information about the dredging period and the area of mining activities. There would be no substantial difficulty due to salinity or the turbidity of surface or groundwater because it is not used for drinking and commercial purposes. The transportation of the sand could avoid the main periods of traffic in order to prevent congestion on roads.

Offshore sand may contain approximately 12 % shell content. At the dredging stage, sand should be monitored to remove particles larger than 10 mm in size. To reach the acceptable chlorine content (0.075 %), the sand stockpile could be exposed to monsoon rains and would not need supplementary washing.

National Environmental Regulation No. 01 (1993) indicates that all offshore mining and mineral extraction processes need to be environmentally screened. There are two methods of screening: an Initial Environmental Examination (IEE) or an Environmental Impact Assessment (EIA). As part of this procedure, the Project Proponent (PP) needs to submit preliminary information to a Project Approving Agency (PAA) such as the Central Environmental Authority (CEA). After submitting information to environmental scoping, the PP should be given Terms of Reference (TOR) for an Environmental Impact Assessment Report (EIAR). In addition, approval from the GSMB is needed for an exploration license for offshore mining and an industrial mining license is needed for the extraction of offshore sand. Approval from the Coastal Conservation Department is required in order to lay pipelines through a defined coastal zone, and also approval from the Marine Pollution Protection Agency is necessary for the removal of ship-generated waste. The authorities allow dredging at selected sites located 20 km offshore at a water depth of over 15 m.

Conclusions and Recommendations

River sand is an open access resource, and conventional attempts to convert this into private property by quotas and tariffs have not been effective, mainly due to institutional failures related to monitoring and regulation. In Sri Lanka, a third of the supply that caters for the current demand of sand is met by illegal means, thus creating environmental problems. Based on the findings of the study, the following recommendations can be made.

Market prices of river sand do not reflect any of the social costs. Therefore, it is suggested that the GSMB should take an active role in assessing sand availability, i.e., the preparation of a sand management plan for all potential sand sites. With the support of the law enforcement authorities, mining of sand from vulnerable sites should be completely banned and monitored.

The royalty payments and land rent collected by GSMB and local government have remained at very low levels. However, during the last few years, the price of sand has increased at least fourfold – this increase has not been passed on to the government. The increasing costs of environmental rehabilitation due to the overexploitation of sand resources have been borne by government institutions. In analyzing the preferences, it was found that miners are willing to contribute more to the royalty given to the government in order to continue sand mining. The results of the discrete choice experiment further suggest that the negative effects associated with sand mining could be partly mitigated by increasing government revenue toward an environmental trust fund, with some level of comanagement. Strict rules and regulation as well as awareness programs, as suggested by the media and environmental groups, may not be productive, according to the perceptions of miners.

The evaluation of management alternatives based on expert views gave consistent results. For the Ma Oya River, where there are more than 70 sand mining sites, the restriction of access to vulnerable sites was found to be the best option, followed by the establishment of an environmental trust fund. Community-based management was found to be the best option for the Mahaweli and Deduru Oya rivers.

Offshore sand was found to be the best alternative to river sand. In Sri Lanka, the construction industry in Western Province is responsible for 40 % of the total sand demand. If this river sand were to be replaced by a reasonable amount of offshore sand, then the immediate pressures on rivers could be mitigated. However, there is less preference for offshore sand and its price is slightly high. Action is needed to move the construction industry toward the use of offshore sand, which would absorb part of the demand for river sand, especially in Western Province.

As shown in the analysis, at present the price of offshore sand is slightly higher than that of river sand. Moreover, there is less preference for offshore sand due to the possibility of the presence of shells and chlorides. Prices could be brought down by expanding the size of offshore sand operations because offshore dredging is sensitive to economy of scale. This could be achieved by the creation of demand for offshore sand by increasing awareness of the benefits of offshore sand with engineers, architects, and other craftsmen. The existing Mines and Minerals Act No. 33 (1992) could be amended so that the use of offshore sand is compulsory for large construction projects and landfilling.

A river sand tax could be introduced in order to encourage the construction industry to move toward the use of offshore sand. This tax should make offshore sand cheaper than river sand. It could be easily computed using the Bill of Quantities (BOQs) prepared by architects and engineers. The payments could be collected at the BOQ approval stage by the relevant authority, i.e., municipal councils.

References

Adamowicz WL, Louviere J, Williams M (1994) Combining stated and revealed preference methods for valuing environmental amenities. J Environ Econ Manag 26:271–292

Adamowicz W, Boxall P, Williams M, Louviere J (1998) Stated preference approaches for measuring passive use values: choice experiments and contingent valuation. Am J Agric Econ 80:64–75

Bateman IJ, Carson RT, Day B, Hanemann M, Hanleys N, Hett T, Jones-Lee M, Loomes G, Mourato S, Ozdemiroglu E, Pearce D, Sugden R, Swanson J (2002) Economic valuation with stated preference techniques: a manual. Edward Elgar, Cheltenham

Byrne G, Nanayakkara A (2002) Alternatives for River Sand, Sri Lanka. Unpublished Final Report. Coastal Resources Management Project, Component B: Institutional Strengthening (ADB TA No. 3477-SRI)

Carson RT, Louviere JJ, Anderson DA, Arabie P, Bunch DS, Hensher DA, Johnson RM, Kuhfeld WF, Steinberg D, Swait J, Timmermans H, Wiley JB (1994) Experimental analysis of choice. Mark Lett 5(4):351–368

Cochrane LJ, Zeleny M (1973) Multiple criteria decision making. University of South Carolina Press, Columbia

Dias WPS, Perera AADAJ, Nanayakkara SMA, Sahayan SJM, Sathyaprasad IMS (1999) Interim sand study, draft final report, Department of Civil Engineering, University of Moratuwa

Ferdinando V, Tunesi L, Agardy T (2000) Zoning marine protected areas through spatial multiple-criteria analysis: the case of Asinara Island, National Marine Reserve of Italy. Conserv Biol 16:515–526

Gilbert AJ, Janssen R (1998) Use of environmental functions to communicate the values of a mangrove ecosystem under different management regimes. Ecol Econ 25(3):323–346

Ingall R (2006) The trails and tribulations of illegal sand mining, Sunday Times, June. http://www.sundaytimes.lk/060611/ft/Theillegalsand.html. Retrieved January 2008

Janssen R, Padilla JE (1999) Preservation or conservation valuation and evaluation of a mangrove forest in the Philippines. Environ Resour Econ 14:297–331

Janssen R, van Herwijnen M, Beinat E (2001) DEFINITE for windows: a system to support decisions on a finite set of alternatives. Institute for Environmental Studies (IVM), Vrjie Universiteit, Amsterdam

Kamaladasa B (2008) Issues and challenges in river management due to excessive sand mining, Irrigation Department of Sri Lanka, Colombo. http://www.riversymposium.com/index.php?element=KAMALADASA. Retrieved March 2009

Kangas J, Kangas A, Leskinen P, Pykalainen J (2001) MCDM methods in strategic planning of forestry on State-owned lands in Finland: applications and experiences. J Multicrit Decis Anal 10:257–271

Katupotha KNJ (1998) Survey on economic-environmental linkages between the demand and supply of sand and coral in the west and southwest coasts of Sri Lanka, National Sand Survey, Environment Action 1 Project. Ministry of Forestry and Environment, Colombo

Lancaster K (1966) A new approach to consumer theory. J Polit Econ 74:132–157

Louviere J, Hensher D, Swait J (2000) Stated choice methods: analysis and application. Cambridge University Press, Cambridge

McFadden D (1974) Conditional logit analysis of qualitative choice behaviour. In: Zarembka P (ed) Frontiers in econometrics. Academic, New York

Munasinghe M (1993) Environmental economics and biodiversity management in developing countries. Ambio 22(2–3):126–135

Nanayakkara SMA (1999) Alternatives for River Sand (ABSE) (International Association for Bridge and Structural Engineering) report, 80, pp 120–125

NARESA (1991) Natural resources of Sri Lanka: conditions and trends. NARESA, Colombo, Sri Lanka

Nijkamp P, Van den Bergh JCJM (1990) Sustainable economic development in regional systems, Research Memorandum, Economische Faculteit, Vrije Universiteit, Amsterdam
Siripala N (2002) Financial and economic feasibility of sand mining, TA No 3477-SRI. Coastal Resource Management Project, Colombo, Sri Lanka
Siriwardana KADSW (1992) Role of quarry dust in concrete. University of Moratuwa, Moratuwa, Sri Lanka
Thurston LL (1927) A Law of comparative judgment. Psychol Rev 3:273–286
Voogd H (1983) Multicriteria evaluation for urban and regional planning. Pion, London

Part IV
Protected Areas

Chapter 11
A Cost-Benefit Analysis of Ecotourism: A Case Study of Xe Pian National Protected Area, Champasak Province, Lao PDR

Malabou Baylatry, Sivannakone Malivarn, and Philayvanh Viravouth

Abstract The tourism sector in Lao PDR is growing rapidly, especially ecotourism. The Government of Lao PDR is attempting to implement ecotourism within protected areas across the country as a means of protecting natural resources, reducing poverty, and encouraging local sustainable development. This study contains a cost-benefit analysis of a project designed to promote ecotourism in the Xe Pian National Protected Area (NPA). Two scenarios are evaluated: first, the status quo, in which no additional investments to promote ecotourism are undertaken, and, second, the ecotourism project for Xe Pian NPA. The latter scenario involves major investments to encourage community participation in sustainable tourism, agriculture, and forest management and protection.

A net present value of \$ 1.8 million was obtained for the status quo scenario using a 10 % discount rate and 10-year planning horizon. The ecotourism scenario results in an NPV of \$ 6.7 million and a benefit-cost ratio of 2.84. The results clearly indicate that the ecotourism project can generate greater net benefits for society than the status quo. The study identifies three main issues that need to be addressed in order to promote ecotourism in Xe Pian: weak enforcement of conservation regulations, limited funding resulting in ineffective management of National Protected Areas, and the need to adopt a community-based tourism approach. Further more detailed recommendations are offered by the authors for policy planners in the Government of Laos, as well as the project's administrative bodies.

Keywords Lao PDR • Cost-benefit analysis • Ecotourism • Sensitivity analysis • Economic value

M. Baylatry (✉) • S. Malivarn • P. Viravouth
Greater Mekong Subregion Division, The Lao National Mekong Committee Secretariat, Ministry of Natural Resources and Environment, Khunbulom Road, P.O. Box 7864, Vientiane Capital, Lao PDR
e-mail: mbaylatry@gmail.com; m_baylatry@yahoo.com; sivannakone@gmail.com; vphilayvanh@yahoo.com

D. James, H.A. Francisco (eds.), *Cost-Benefit Studies of Natural Resource Management in Southeast Asia*, DOI 10.1007/978-981-287-393-4_11

Background

Ecotourism and Economic Development

Tourism is one of the largest and rapidly expanding global economic activities, contributing to high rates of economic growth and development in host countries, significant foreign currency inflows, infrastructure development, and the introduction of new management and educational experiences that support relevant sectors of the economy and promote the social and economic development of a country as a whole (Mirbabayev and Shagazatova 2006).

Tourism can be subdivided into two broad categories: conventional mass tourism which will continue to dominate the industry and alternative tourism. Alternative forms of tourism typically are small scale, have low density and attempt to attract a special segment of society, namely, tourists with above-average incomes and higher education. Alternative forms of tourism can be defined as cultural, educational, scientific, adventure, and agritourism or green tourism (Mieczkowski 1995). Ecotourism's main characteristic is that it is nature oriented and nature based, but not always exclusively conducted in a wilderness surrounding. Thus, ecotourism may overlap with other forms of alternative tourism. As such, it can play an important role in achieving the general objective of ecologically sustainable economic development.

Rich biodiversity and unique natural areas within developing countries actively encourage tourism by domestic and international visitors. At the same time, it is recognized that tourism can also produce serious adverse environmental impacts. Social scientists and decision-makers accordingly are seeking alternatives to conventional mass tourism and the dramatic expansion of highly organized package tours. Community-based ecotourism is one of the most environmentally compatible alternatives for the economic use of land resources in developing countries, with the hope of minimizing negative ecological and sociocultural impacts of visitors at recreational locations.

Tourism in Lao PDR

After Lao PDR opened the country to international tourists in the early 1990s, the country has welcomed more than two million visitors. The number of visitors to Lao PDR increased continuously from 1996 with an average growth rate of 20.53 % per year. Tourist arrivals to Laos continued to increase in 2009, exceeding two million and generating a total revenue of US $268 m, despite political instability in some countries in the region and the world economic turmoil that affected the whole region. The Lao National Tourism Administration predicted that total revenue from tourism in Lao PDR would be almost $400 million in 2011, generated by more than 2.6 million visitors.

Ecotourism has become an important economic activity in Lao PDR, and the country is now widely recognized as an ecotourism destination. Culture- and nature-based tourism as well as ecotourism has accounted for more than half of the total revenue of the entire Lao tourism industry. The Government considers ecotourism to be an effective means of generating income for local people and raising awareness of environmental conservation, encouraging local production, and protecting multiethnic culture and traditions (Harrison and Schipani 2007). In 2008, Lao PDR was named by the New York Times as the number one vacation destination. The National Tourism Authority of Lao PDR has stated that "Laos will become a world renowned destination specializing in forms of sustainable tourism that, through partnership and cooperation, benefit natural and cultural heritage conservation, local socio-economic development, and spread knowledge of Laos' unique cultural heritage around the world."

Ecotourism in Lao PDR has generated significant financial benefits and has offered relevant stakeholders, particularly local people and land use planners, a broader perspective on an alternative livelihood that is already taking place on forest land.

Xe Pian NPA

In 1993, the Government of Lao PDR declared 18 areas throughout the country as "National Protected Areas" (NPAs). This declaration was set out in Prime Minister's Decree No. 164/PM on *Stating National Forestry Reservation over the Country* (29 October 1993). The goals of the NPA system are (1) protection of forests, wildlife, and water, (2) maintenance of natural abundance and environmental stability, and (3) protection of natural beauty for leisure and research.

Xe Pian NPA was one of the areas declared under Prime Ministerial Decree No. 164/PM. It is located in Champasak and Attapeu provinces in southern Laos, with a total area of 240,000 ha. Much of the land is gently rolling, covered mainly by semievergreen forest. The northern part of the park borders with Kele and Phon Ngam villages and this area contains several large marshy wetlands. The southern boundary borders with Cambodia. The eastern boundary runs between Champasak and Attapeu provinces. The western edge joins with Houay Mesang village and Khmer state line.

The characteristics of Xe Pian NPA make it highly suited for the development of ecotourism. The area is endowed with 39 mammals, 44 reptiles, 21 amphibians, and 176 fish species. Biodiversity values of Xe Pian place it in the top third of Lao NPAs and in the top ten in Southeast Asia.

Impacts of Human Activity in Xe Pian NPA

Population has grown rapidly inside and outside Xe Pian NPA. There are 69 villages within Xe Pian NPA and 48 villages located around the park's boundary with a total population in excess of 20,000 people. Enormous pressure has been placed on Xe Pian NPA (National Statistic Centre Laos 2009). Local people belonging to Brau, Su, and Lao ethnic groups, especially the poor, have occupied Xe Pian NPA for at least 200 years, practicing a subsistence lifestyle based on paddy farming and shifting cultivation; fishing in small wetlands; hunting; and collecting forest resources for food, medicine, and building materials for their daily consumption and/or for commercial purposes.

The general disturbance of habitats and wildlife by human activities, as well as the construction and upgrading of road on the north side of the NPA, has led to unsustainable resource use. Inadequate management capacity and human resources have led to deforestation in the area, unsustainable resource use, and overharvesting of non-timber forest products (NTFPs). In 1993, the IUCN reported a large elephant population and also found evidence of many muntjac, wild pig, sambar, mouse deer, otter, gray langur, gaur, banteng, tiger, dhole, civets, sun bear, and black bear in Xe Pian NPA. Gibbons were reported living in high densities in semievergreen forest. However, in 2005/2006, according to a survey of Xe Pian NPA conducted by World Wildlife Fund (WWF), dramatic declines have occurred in areas that previously had rich wildlife populations. Pangolins were reported to be very common in local markets. No elephants were found even at salt licks, indicating a very low population. No tracks or signs of tigers or leopards were found throughout the survey. Gaur and sambar were at very low densities, although muntjac and wild pig appeared to be common.

To protect Xe Pian NPA, fences and guards are needed to control rural people in the reserve area. From 1998 to 2002, no funding was available to set up a monitoring and management system. It is estimated that the monitoring process in Xe Pian NPA requires 600 man-days a year, or US$ 3 per hectare of forest habitat, and no funds exist to provide this level of support (Poulsen and Luanglath 2005).

Prospects for Ecotourism in Xe Pian NPA

Xe Pian people as well as the government clearly need to take action to reduce adverse impacts on the environment and encourage a pattern of development that is consistent with the general policies of the Government of Lao PDR. According to the government's policies, "development" in NPAs does not refer to large projects such as mining and dams but rather to improvement in the livelihoods of local people. Ecotourism development in NPAs, if carefully planned and managed, and introduced with the participation of and benefits to key local stakeholders, can contribute both to the national economy and to NPA management.

Aims of the Study

Research Objectives

The main objective of the study has been to conduct an economic analysis of ecotourism in Xe Pian NPA, applying the methods of environmental economics to provide sound information to Lao policy-makers regarding the benefits of sustainable ecotourism to local communities. By conducting a CBA of possible state investment in ecotourism for five villages in Pathoumphone district (Kiat Ngong, Pha Pho, Pha Lai, Ta Ong, and Ngong Ping), it is hoped that the study will help to fill important gaps in knowledge and lead to improved policy formulation and decision-making. More specific objectives of the study are as follows:

- To assess the current situation and trends of ecotourism in the study area
- To identify existing policies, action plan, programs, projects, legislation, and measures in place supporting ecotourism development
- To identify and classify possible negative socioeconomic and environmental impacts from ecotourism development
- To analyze the costs and benefits of promoting ecotourism activities in National Protected Areas and identify measures to reduce undesirable impacts

Research Questions

The authors of study have investigated the general question or whether a 10-year ecotourism project in Xe Pian NPA would result in private gains but also whether negative externalities associated with adverse environmental effects might outweigh the collective benefits of Xe Pian communities. The major research questions accordingly are as follows:

- What is the current situation and what are the trends regarding ecotourism in Xe Pian NPA?
- What are the existing policies, action plans, programs, projects, and legislation relating to Xe Pian NPA to support ecotourism development?
- What are the expected adverse socioeconomic and environmental impacts resulting from ecotourism development?
- What is the distribution of costs and benefits to local communities and local government managing the park?

- What measures might be introduced to reduce potential negative impacts of ecotourism?

Methodology

Framework for Analysis

Cost-benefit analysis (CBA) was selected as the appropriate framework through which to conduct the economic analysis. Two scenarios dealing with local communities were assessed: a business-as-usual (BAU) scenario and a second scenario in which an ecotourism project is introduced in Xe Pian NPA. The following assumptions were adopted:

- The planning period assumed for the ecotourism project is 10 years. This presumes that a 10-year period is long enough to highlight all of the consequences of changes in land use to promote community-based ecotourism, including evidence that the project may yield significant benefits to local communities and improve their financial sustainability. Local stakeholders and communities living in and around Xe Pian NPA should have an opportunity to gain a clearer understanding of ecotourism policy and regulatory mechanisms during that time. In addition, they may be more actively involved in the planning and management of the protected area, especially in relation to ecotourism.
- The estimated total amount of funding available to invest in the project is $1.3 million, comprising $1.2 million for improving ecotourism facilities and $85,000 to cover training expenses during the project.
- US dollars are used to monetize all values in the calculations. The exchange rate is one US dollar per 8,027 kip (2011).
- To calculate the economic net present value (NPV) of the project, a discount rate of 10 % is used, representing the opportunity cost of capital invested.

Steps in the CBA

Nine steps were taken in carrying out the CBA:

Step 1: Define the Referent Groups

The two main referent groups are (1) the local people in Xe Pian NPA and (2) the local government. These two groups are the main ones involved in implementing the project.

Step 2: Select the Portfolio of Project Options

There are two scenarios. *Scenario 1* is a base-case scenario in which there is no intervention by public authorities other than that under normal conditions. *Scenario 2* is the ecotourism scenario in which extra conditions and measures are imposed for environmental protection and conservation of natural resources. Obtaining data for each scenario was possible as there are two different kinds of households within the study area: households practicing conventional agriculture and exploitation of the forest and other households already engaged in ecotourism activities. The revenue from ecotourism in Xe Pian NPA is assumed to be generated by an entrance fee and license or permit fee.

Step 3: Identify Potential Impacts of the Project

Impacts refer to all inputs and outputs. Inputs are usually costs such as those for improving facilities, training, and so on. Outputs are mainly benefits, but sometimes it can comprise costs, such as those associated with adverse environmental impacts. The positives and negatives to villagers and the park authority for each of the scenarios vary.

Under the base-case scenario, Xe Pian NPA is assumed to be managed inadequately, due to limited budgetary resources. The majority of protected areas in Laos, especially Xe Pian, are suffering from a shortage of funding, leading to the degradation of natural resources in protected areas. Kyophilavong (2009) has estimated the cost for management and protection of Xe Pian NPA as approximately \$148,800 per year. Moreover, the park management is required to address the cost of deforestation, loss of forest existence value, cost of CO_2 emissions, and soil erosion without receiving any benefit. For local communities, costs include a loss of forest products such as timber and NTFPs, as well as the cost of conducting agriculture, fishing, and forestry-related activities. Xe Pian communities are almost all subsistence rice farmers who depend heavily on natural resources in the protected area to secure a balanced diet and building materials. The most important products for the local population are malva nut fruits from the malva nut tree (*Scaphium macropodum*).

Under the ecotourism project, local villagers have substantial control over, and involvement in, its development and management. A major proportion of the benefits remain within the community. Additionally, the project has the potential to create an alternative source of income to Xe Pian NPA, such as ecotourism revenue, CO_2 credits, and reserve forest value.

Step 4: Quantify the Outputs and Impacts

For the base-case scenario, the following estimates or assumptions were made regarding the quantification of outputs and impacts. At the site selected for the study, the deforestation rate is 1.12 % per year, with forest clearing amounting to 18 ha per year. CO_2 emissions are estimated at 1,984 t per year. The loss in forest value is $385 per ha per year. Prevention of soil erosion involves an expenditure of $644 per ha per year. By contrast, the ecotourism project will assist all stakeholders to increase their capacity to implement conservation practices and better handle the potential positive and negative impacts of tourism. Accordingly, the project is expected to minimize external and internal pressures on the forest.

Step 5: Monetize the Outputs and Impacts

In this step, all impacts are valued in monetary units, so that costs and benefits can be compared. Generally, many of the environmental goods and services associated with the protected area are not traded in the market, so prices do not exist to use as reference values. For the purpose of the study, various indirect valuation methods have been adopted in order to derive the monetary values used in the calculations. In particular, the benefit transfer approach was applied to estimate the cost of soil erosion, carbon sequestration, and the existence value of the forest.

Soil erosion is a serious environmental problem in mountainous region of Laos. It is caused by wind, water, human activity, and animals, and it represents a serious threat to sustainable agriculture. Some areas in the region experience high levels of soil erosion because of high rainfall. There is an acute need to undertake actions to restore vegetation cover, including building gabions and other soil erosion prevention strategies. Erosion leads to a decline in soil fertility and lower land productivity in affected areas. In the study, the cost of soil erosion was estimated at $644 per ha per year.

To estimate the value of carbon stocks, it was assumed that carbon stocks per ha of forest would be 112 t/C/ha (Brown and Pearce 1994). A conservative estimate of $5 t/C was used in the study, implying that conversion of forest due to expanding agriculture would involve a loss of stored carbon value amounting to $558 per ha. Based on a study by Rosales et al. (2005), the existence value of the forest was estimated at $385 per ha per year.

Step 6: Calculate NPV, BCR, and IRR

The NPV, BCR, and IRR of the ecotourism project are calculated using the standard formulae for these evaluation criteria. All values for the project are compared with those of the base-case scenario.

Step 7: Describe the Distribution of Costs and Benefits

Under the base-case scenario, villagers would be the "winners" in the short run. The general community, however, would be the "losers" due to resource degradation and the costs associated with adverse externalities that are generated. Ecotourism has been hailed as a win-win proposition for conservation, the host country, the traveler, and the industry. The main achievement of the project would be an improvement in the well-being of residents within the five villages covered by the study, resulting from gains to the local economies and reduced pressure on forest resources. An improvement in animal habitat can also be expected, as villagers often report poaching, logging, and hunting to the authorities. In this way, the task faced by local government is made easier and local government benefits as a result.

Step 8: Perform Sensitivity Analysis

Sensitivity analysis is applied to assess uncertainty surrounding the assumptions relating to predictions of variables and their future values in the economic evaluation (Folmer and Gabel 1998). Ecosystems typically are complex, so it can be difficult to predict changes in physical impacts as well as responses and the adaptive behavior of social and economic agents. In the study, different assumptions are tested in terms of their effects on NPV.

Step 9: Make Recommendations

Documenting the main findings of the study and offering policy recommendations is the last step in conducting the CBA. Scenarios are ranked in terms of their net benefits, and the scenario is recommended on economic grounds that has the greatest net benefit. Other recommendations are also made relating to management of the protected area.

Data Collection

Primary Data Collection

Primary data were collected from three separate groups. The first two groups were chosen as target groups to address quantitative aspects of the study. The third group was consulted using in-depth interviews to collect more general information relating to the study site.

Group 1 comprises five target villages: Kiat Ngong, Pha Pho, Pha Lai, Ta Ong, and Ngong Ping. These villages are located in the soft adventure zone. Face-to-face interviews, involving a sample of 135 households, were conducted with local residents to determine who might be involved in ecotourism activities. This group potentially represents the gainers from the ecotourism project. Information provided by this group assisted in evaluating the direct costs and benefits of the ecotourism project.

Group 2 comprises farmers not involved in tourism, living in the five villages. Face-to-face interviews were also used in this task. This group was selected to predict the circumstances without the project. The research team randomly surveyed 135 households.

Group 3 includes key informants whose working mandates involve management and administrative responsibilities with Xe Pian NPA communities. Among the people who provided information for the analysis of this study through in-depth interviews were the Xe Pian NPA Management Unit, who has managed Xe Pian Park, and village headmen of the five target villages.

Secondary Data Collection

Secondary data was collected from various sources and in-depth interviews with relevant bodies and government agencies. They included Lao National Tourism Administration, Champasak Provincial Tourism Department, Pathoumphone District Tourism Office, Laos World Wildlife Fund, Xe Pian National Protected Area Management Unit, Pathoumphone District Agricultural and Forestry Office, Champasak Provincial Water Resources and Environment Office, and Department of Planning and Investment, Champasak Province.

Description of Study Area

Study Site

The study area is in Pathoumphone district in Champasak Province. The villages of Kiat Ngong, Pha Pho, Pha Lai, Ngong Ping, and Ta Ong were chosen for the study as these represent the nearest locations from Pakse township and the usual departure points for most foreign trekkers and ecotourists. The site can be reached by buses, vans, and trucks. Kiat Ngong is about 1 h travel from Pakse via rough road. From Kiat Ngong to Pha Pho, it takes about 30 min, and reaching Pha Lai takes an extra hour. From Pha Lai to Ngong Ping and Ta Ong, the access is a quite difficult trail, and it takes at least an hour to reach each place by its local transport such as boat (available in the rainy season), bicycle, motorbike, tractor, or walking.

Villages in Study Area

Kiat Ngong Village

Kiat Ngong is one of 21 villages located inside Xe Pian National Park. It is located approximately 70 m above sea level, occupying an area of 7,392 ha. The landscape is diverse, with forested mountains, upland plateaus, and lowland plains along the Mekong River. Its territory borders the 18th Km road to the north, Xe Pian NPA to the south, Kale and Pha Pho village to the east, and Ta Tooh village to the west. There are 215 households in the village, with a total population of 1,007. Based on the village statistics for 2000, the settlement covers an area of 90 ha. The area under rural production covers 197 ha. Forest comprises 5,893 ha. Kiat Ngong is well known for its elephant rides to Phou Asa mountain and boat trips at Kiat Ngong wetland for bird watching.

Pha Pho Village

Pha Pho village was established in 1310 on lowland in Xe Pian NPA. The total area comprises 597 ha of which settlement covers 90 ha, paddy fields 209 ha, and the rest is forest. Its population numbers 1,628. There are presently 339 households in the area. Of these, 171 have an electricity service, 23 lack electricity, and 86 obtain electricity from generators. The site is famous for elephant walks and trekking to Pha Lai village. In addition, Pha Pho is a focal point or coordinated tours involving three other villages: Pha Lai, Ngong Ping, and Ta Ong. Pha Pho provides package tours for visitors to experience the natural beauty associated with the three villages.

Pha Lai Village

Pha Lai village was established in 1510 before the declaration of the protected area. The total area available to the village is 3,432 ha. Settlement covers an area of 7 ha, rural production 145 ha, and forest 2,849 ha. The total population is 1,734, with 309 families and 268 houses. The richest households number is 14, rich households 27, better-off households 208, and poor households 19. Pha Lai has no electricity service, so 244 houses lack electricity and only 24 houses use generators to obtain electricity.

Ngong Ping Village

Ngong Ping village has a total area of 10,043 ha including 28.20 ha of paddy fields, 4.09 ha of construction area, 137.89 ha of agriculture, 4.68 ha of citrus plantation, 5 ha of graveyard, 175.78 ha of production forest area, and 9,687.36 ha of protected area. Ngong Ping village borders with Pha Lai village to the north, Phak Ka and Phon Gnam villages to the west, Ta Ong and Ta Wang village to the east, and Xe Pian NPA to the south.

Ta Ong Village

Ta Ong is located in a remote area of Xe Pian NPA. It is 61 km from Pakse township. There is only one dirt trail to access the village, but an alternative is trekking. Ta Ong borders Ngong Ping and Ta Wang at the northern part, and Kong district to the south. The total area comprises 24,912.95 ha including 23,993 ha of protected area, 750 ha of production forest, 150 ha of agriculture, 35 ha of paddy fields, and 1.34 ha of settlement. The majority of the village's inhabitants belong to the Brao ethnic group, while the Lao group comprises 20 %.

Demographic Characteristics

The survey results revealed that the average age of respondents was 45.9 years. There was no significant difference among the household groups. One in two of the respondents had finished primary school, while a quarter had no opportunity to attend school due to war and living in a rural area. Only few respondents had finished secondary or high school.

Most of the villagers surveyed were male and the majority surveyed were farmers. In order to earn sufficient income, local people engage in extra activities such as running a grocery store, collecting NTFPs, and raising livestock. The average sources of income comprised cereals (32 %), ecotourism (14 %), natural products (12 %), livestock grazing (11 %), vegetables (8 %), nonfarm employment (10 %), and other activities (13 %). Xe Pian villagers have larger families and own more land than average. The average family size is 5.8 persons per household. On average, a household owns 1.2 ha to produce rice and cash crops. The average income per household was estimated at $544 per year. Generally, both ecotourism and non-ecotourism groups have similar household economic characteristics. The ecotourism group owned more fixed assets than the non-ecotourism group, including televisions, tractors, mobile phones, bicycles, motorbikes, and a car or van. Both groups have similar monthly expenditures: $40 per person for the ecotourism group and $34 per person for the non-ecotourism group.

Land Use

Land use at the sites surveyed usually included areas of rice production, other cash crops, teak plantation, citrus, and forest. For rice production, a typical estimate of gross revenue is $569,411 per year, from an area of 605 ha. Production costs, excluding the value of family labor, are $224,205 equivalent to $370 per ha. Only a small percentage of rice farmers use chemical fertilizers.

Other cash crops that people grow include coffee, cabbage, chili, and long bean. Gross revenue from other cash crops is $128,473 per year and total costs are $18,750 per year. Chemical fertilizers are not used by villagers for these other crops.

Forest Use

Forests are a source of construction material for households, particularly timber. Most households collect 2–4 m^3 to repair their houses. In the study area, the total volume of wood collected amounted to 120 m^3 with an equivalent market value of $2,990. Net value after deducting labor cost, transportation, and marketing costs is estimated at $2,000 equivalent to $66 per ha per year.

One of the main use values for the forest for almost all the members of the community is malva nuts, which villagers sell to Vietnamese and Chinese merchants. A large amount of effort is used in their collection. The estimated total quantity collected is 2,250 kg per year, with a market value of $11,144 per year. Net returns, after deducting collection and marketing costs, are $9,144 per year.

Water for Domestic Consumption

In the study area, there is no water supply for domestic consumption. Water use in Xe Pian NPA comes primarily from streams, ponds, boreholes, wells, and rainwater. Supply is not available all year round and the months of shortage are usually between March and May. Water supply for rice production is from rainfall and wells. Most of the people who live in Kiat Ngong and Pha Pho villages are better-off households. Their water source is mainly from borehole and well. Other inhabitants draw water from a nearby stream.

Forest Management Regulations

Ministerial Decree 164/PM contains a number of general legal restrictions on activities that can be conducted in the protected area. The area is also managed under relevant national and provincial laws. They include Decree No.185/CCM on the Prohibition of Wildlife Trade (21 October 1986); Decree No.118/CCM of the Council of Ministers on the Management and Protection of Wild Animals, Fisheries, Hunting and Fishing (5 October 1989); Instructions for the Implementation of

Decree No. 118; Decree No.169/PM of the Prime Minister on the Management and Use of Forests and Forest Land (3 November 1993); Decree No.1074/MAF Introducing Wildlife Conservation (11 September 1996); and The Forestry Law No. 01-96 (11 October 1996).

National conservation policies relating to Xe Pian NPA are formulated by the Department of Forestry and disseminated to provinces and NPAs by the Division of Forest Resources Conservation. The main management agency is the Xe Pian Protected Area Management Unit which operates within the Agriculture and Forestry Office (PAFO) for Champasak Province, with extensions situated in the relevant District Agriculture and Forestry Offices (DAFO).

Base-Case Scenario

General Assumptions

The base-case scenario focuses on the conditions that are expected to prevail over the next 10 years in the absence of the ecotourism project and is constructed from information provided by villagers who do not engage in tourism activities. The local government maintains its established responsibilities and performs its usual functions in relation to logging problems, poaching, forest fires, and so forth. Villagers are predicted to rely increasingly on natural resources in the area, making their behavior difficult to control because fewer people will be inclined to cooperate with forest conservation objectives. Benefits are assumed to continue for villagers over the 10-year period as they engage in rice production and other cultivation, overfishing, overcutting of timber, and over-harvesting of NTFPs.

The Ecotourism Project

Ecotourism in Champasak Province

The tourism sector in Champasak Province has grown strongly, expanding each year. One of the main reasons is that the Government of Lao PDR has targeted tourism as a high priority in provincial development policy. Moreover, Champasak has good facilities and infrastructure to support tourism, including good telecommunications systems, water supply, electricity, and roads.

Champasak Province represents the third largest tourism market in Lao PDR after Luang Prabang and Vientiane, attracting 10 % of the country's total visitors. A great opportunity exists for Xe Pian NPA to become a major tourist attraction in the region, with proper development of its natural and cultural attractions.

Ecotourism in Xe Pian NPA

Xe Pian NPA is one of the most important tourist destinations within Champasak Province. It will continue to attract greater numbers of tourists as ecotourism grows within the whole country. With proper development, it should be possible to achieve conservation objectives through comanagement involving local communities.

The ecotourism demand in Xe Pian NPA is international or non-regional with a small percentage of domestic tourists. International or non-regional tourists visiting Xe Pian NPA travel in packaged tour groups or independently. The four major nationalities that represent this market are all European: France (50 %), Belgium (7 %), Switzerland (4 %), and Spain (3 %). Domestic tourists account for only 1 %. The average length of stay in Xe Pian NPA is one day. Most tourists do not stay overnight. The average amount of money spent by each tourist while in local villages is $18.

Among the five target villages selected for the study, Kiat Ngong has been the most successful in implementing ecotourism. It has already benefited from a project introduced by the World Wildlife Fund (WWF) Laos in 1997. Broad participation in tourism by villagers is excellent, the distribution of benefits from tourism is widespread, and the activity contributes significantly to the alleviation of poverty in Xe Pian NPA. Basic infrastructure for ecotourism in Xe Pian NPA is established. Training to support ecotourism has been undertaken. Mechanisms for generating revenues from ecotourism for conservation and village development are also in place and have already begun to provide benefits. The marketing of ecotourism and private sector tours is improving. Awareness regarding ecotourism in Xe Pian NPA and the need to protect its wildlife, forests, wetlands, and rivers among officials, villagers, the private sector, and visitors is growing. For all of these reasons, the WWF project has been successful.

Description of Ecotourism Development Project

In general, the ecotourism development option aims to take advantage of the core attractions in Xe Pian NPA and social development based on culture and heritage. The study area will be characterized by tightly controlled, medium-density facilities aimed at the ecotourism market. The hub for development will be small-scale facilities in the five selected villages catering to visitors seeking nature-based attractions. Integrity of the natural attractions will be maintained, and a range of activities for visitors will be provided.

Upgrades of facilities include road access, public transport, medical care, public toilets, and other improvements. The park gains from an entrance fee from both domestic and international visitors of $1 and $4, respectively. The park authority will receive a 15 % license fee from ecotourism activities. It is assumed that the growth rate for international visitors is 20 % per year, while for local visitors it is 5 %.

For the construction of lodges, 10 ha of paddy field would be resumed to build eco-accommodation for tourists. The main targets for this scenario are small groups and individual ecotourists interested in nature and the cultural attractions of Xe Pian NPA. Wildlife viewing opportunities would form an important part of the overall product offering in addition to well-managed natural settings for recreation and accommodation. High-quality interpretive services form an important part of this scenario. In spite of the relatively lower numbers of visitors, with high-quality infrastructure and facilities, visitors would be encouraged to stay longer.

Additional features of the scenario are that strict limits will be applied on the conversion of forest to agriculture, thus reducing the pressure on forest and controlling deforestation. Timber extraction and the collection of NTFPs will be prohibited. These arrangements would initially be introduced at a pilot site. Villagers would be hired as patrol officers to monitor and control conservation activities.

Costs and Benefits of Base-Case Scenario

Benefits to Local Communities

Benefits accruing to local communities are derived principally from their agricultural activities including rice production, other cropping, fishing, and collecting timber and NTFPs. It is assumed that the value of their production remains constant throughout the 10-year period, but in order to maintain it, villagers need to clear and cultivate more forested land, as well as incurring increasing costs of production.

Costs to Local Communities

In the base case, local communities are assumed to face rising costs of production as the quality of the resource base declines and productivity decreases. Differential rates of increase in production costs per hectare are assumed: 10 % per year during the first 5 years, then 20 % per year for the sixth and subsequent years. The costs of fishing and collecting timber and NTFPs are assumed to increase 10 % each year due to a rising real wage rate as well as decreasing productivity of the resource.

Costs to Local Government

The Xe Pian Management Unit is obliged to meet the costs of maintaining a similar level of forest-related benefits and prevent ongoing losses in the existence value of forests and ecological functions. These costs were valued in terms of preventive costs. In the base case, the deforestation rate is 1.12 % per year, and 18 ha of forest is cut down every year by villagers as they expand their agriculture land. Monetary value of the relevant loss of forest is included in the cost estimate. The decrease in carbon storage was valued as a foregone benefit, using the benefit transfer technique. The value is based on an estimate of the changes in carbon stocks as closed secondary forest is converted to permanent agriculture. The cost of preventing soil erosion is monetized in terms of the investment required to prevent soil loss. The lowest-cost technology of planting vetiver strips is used in the calculation. The unit cost per hectare is multiplied by the forest area that would be converted.

Results

The results under the assumptions adopted in the base case are shown in Table 11.1. A 10 % discount has been applied. The present value of benefits is $8.153 m and the present value of total costs is $6.398 m. NPV in the base case is thus $1.755 m.

Costs and Benefits of Ecotourism Scenario

Benefits to Local Communities

The ecotourism project provides direct economic benefits by creating new job opportunities in the ecotourism industry, including acting as trekking guides, conducting elephant rides, elephant queue organizer, tourist receptionist, boatmen, homestay managers, cooking units, camping, bird watching, and traditional massage. Indirect economic benefits are received by locally run shops and handcraft stores. Such benefits were verified in the survey of the five villages covered by the study.

The research team discovered also that ecotourism enhances social cohesion among Xe Pian residents, encouraging local residents to work together, share common issues, and help each other to overcome difficulties. Ecotourism provides an opportunity for locals to interact with tourists and witness their positive behavior such as refraining from littering and leaving plastic bags on the ground. Local people are induced to adopt similar good behavior themselves. The improvements in construction facilities within the community have played a crucial role in creating social happiness and enjoyment. The ecotourism program has supported

Table 11.1 Base-case scenario

Year	Present value	1	2	3	4	5	6	7	8	9	10
Benefits											
Rice	5,832,423	949,200	949,200	949,200	949,200	949,200	949,200	949,200	949,200	949,200	949,200
Teak	167,439	27,250	27,250	27,250	27,250	27,250	27,250	27,250	27,250	27,250	27,250
Other cultivation	789,409	128,473	128,473	128,473	128,473	128,473	128,473	128,473	128,473	128,473	128,473
Fish	693,225	112,819	112,819	112,819	112,819	112,819	112,819	112,819	112,819	112,819	112,819
NTFPs	652,860	106,250	106,250	106,250	106,250	106,250	106,250	106,250	106,250	106,250	106,250
Timber	18,372	2,990	2,990	2,990	2,990	2,990	2,990	2,990	2,990	2,990	2,990
Ecotourism project	0	0	0	0	0	0	0	0	0	0	0
Value of forest ($/year)	0	0	0	0	0	0	0	0	0	0	0
Carbon storage value	0	0	0	0	0	0	0	0	0	0	0
Residual value – ecotourism project	0	0	0	0	0	0	0	0	0	0	0
Total Benefits	*8,153,729*	*1,326,982*	*1,326,982*	*1,326,982*	*1,326,982*	*1,326,982*	*1,326,982*	*1,326,982*	*1,326,982*	*1,326,982*	*1,326,982*
Costs											
Capital costs											
Ecotourism project	0	0	0	0	0	0	0	0	0	0	0
O&M Costs											
Rice	5,643,058	406,800	488,160	585,792	702,950	843,540	1,012,249	1,214,698	1,457,638	1,749,166	2,098,999
Teak	87,809	14,812	14,672	14,532	14,392	14,252	14,112	13,972	13,832	13,692	13,552
Other cultivation	260,097	18,750	22,500	27,000	32,400	38,880	46,656	55,987	67,185	80,622	96,746
Fish	77,682	5,600	6,720	8,064	9,677	11,612	13,935	16,722	20,066	24,079	28,895
NTFPs	117,911	8,500	10,200	12,240	14,688	17,626	21,151	25,381	30,457	36,548	43,858
Timber	38,835	1,800	2,340	3,042	3,955	5,141	6,683	8,688	11,295	14,683	19,088
Soil erosion	70,114	11,411	11,411	11,411	11,411	11,411	11,411	11,411	11,411	11,411	11,411

Value of forest ($/year)	41,916	6,822	6,822	6,822	6,822	6,822	6,822	6,822	6,822	6,822	6,822
Carbon storage value	60,968	9,922	9,922	9,922	9,922	9,922	9,922	9,922	9,922	9,922	9,922
Ecotourism project	0	0	0	0	0	0	0	0	0	0	0
Total costs	*6,398,389*	*484,417*	*572,747*	*678,825*	*806,216*	*959,206*	*1,142,940*	*1,363,603*	*1,628,627*	*1,946,944*	*2,329,292*
Net benefits	*1,755,340*	*842,565*	*754,235*	*648,157*	*520,765*	*367,776*	*184,042*	*−36,621*	*−301,645*	*−619,962*	*−1,002,310*
Incremental benefits	**0**	0	0	0	0	0	0	0	0	0	0
Incremental costs	**0**	0	0	0	0	0	0	0	0	0	0
Incremental net benefits	**0**	**0**	**0**	**0**	**0**	**0**	**0**	**0**	**0**	**0**	**0**
Net present value	$0.00										
Benefit-cost ratio	0										
Internal rate of return	0										

locals in maintaining cultural practices and sites, with the aim of attracting tourists to the area to visit them. Local people have strongly supported these efforts, evidenced by the ways they have chosen to maintain the traditional styles of their houses and in their dress.

Benefits to Local Government

The benefits of the park to local authorities are measured in terms of revenue from entrance fees to the park paid by tourists and the reduction in management costs to protect the forest, preserve biodiversity, reduce soil erosion, and maintain carbon stocks.

Costs to Local Communities

Costs incurred by local communities include the opportunity cost of paddy field area required to build an ecolodge, forgone income from the harvesting of timber and NTFPs, and increasing costs to train people to practice agriculture more efficiently and compensate for the lower rate of deforestation otherwise occurring in the base case. Local communities would also be obliged to incur part of the capital and operating costs of the ecotourism project.

Costs to Local Government

The costs to be met by local government are mainly operation and management costs associated with controlling potential adverse environmental impacts of ecotourism, including salaries for park staff, patrolling costs, printing entrance tickets, conducting monthly stakeholder meetings, monitoring and protecting biodiversity and forest condition, preventing soil erosion, and encouraging awareness of conservation values by local people. If too many tourists visit the site, they may compete with local people, causing resentment, social inequality, and a loss of traditional culture.

Results

Results for the ecotourism scenario are presented in Table 11.2. The present value of benefits is estimated at $11.883 m, and the present value of costs at $$5.087 m. The present value of net benefits is thus $6.796 m. Compared with the NPV of the base case, the ecotourism scenario has an incremental NPV of $5.041 m. This

Table 11.2 Ecotourism scenario

Year	Present value	1	2	3	4	5	6	7	8	9	10
Benefits											
Rice	5,561,287	935,200	935,200	935,200	935,200	935,200	907,144	879,930	853,532	827,926	786,530
Teak	85,600	0	16,350	16,350	16,350	16,350	16,350	16,350	16,350	16,350	16,350
Other cultivation	764,856	128,473	128,473	128,473	128,473	128,473	124,618	120,880	117,254	113,736	110,324
Fish	693,225	112,819	112,819	112,819	112,819	112,819	112,819	112,819	112,819	112,819	112,819
NTFPs	0	0	0	0	0	0	0	0	0	0	0
Timber	0	0	0	0	0	0	0	0	0	0	0
Ecotourism project	4,354,979	314,507	376,686	452,023	542,427	650,913	781,095	937,314	1,124,777	1,349,733	1,619,679
Value of forest ($/year)	23,203	3,776	3,776	3,776	3,776	3,776	3,776	3,776	3,776	3,776	3,776
Carbon storage value	165,719	26,970	26,970	26,970	26,970	26,970	26,970	26,970	26,970	26,970	26,970
Residual value – ecotourism project	234,778	0	0	0	0	0	0	0	0	0	608,953
Total benefits	*11,883,648*	*1,521,746*	*1,600,274*	*1,675,611*	*1,766,015*	*1,874,501*	*1,972,773*	*2,098,039*	*2,255,478*	*2,451,310*	*3,285,401*
Costs											
Capital costs											
Ecotourism project	1,153,488	1,217,905	10,000	10,000	10,000	10,000	10,000	10,000	10,000	3,000	2,000
O&M costs											
Rice	2,462,742	400,800	400,800	400,800	400,800	400,800	400,800	400,800	400,800	400,800	400,800
Teak	87,809	14,812	14,672	14,532	14,392	14,252	14,112	13,972	13,832	13,692	13,552

(continued)

Table 11.2 (continued)

Year	Present value	1	2	3	4	5	6	7	8	9	10
Other cultivation	115,211	18,750	18,750	18,750	18,750	18,750	18,750	18,750	18,750	18,750	18,750
Fish	34,410	5,600	5,600	5,600	5,600	5,600	5,600	5,600	5,600	5,600	5,600
NTFPs	0	0	0	0	0	0	0	0	0	0	0
Timber	0	0	0	0	0	0	0	0	0	0	0
Ecotourism project	1,233,629	115,459	130,854	148,558	168,955	192,501	219,732	251,282	287,902	330,483	380,079
Total costs	*5,087,289*	*1,773,326*	*580,676*	*598,240*	*618,497*	*641,903*	*668,994*	*700,404*	*736,884*	*772,325*	*820,781*
Net benefits	*6,796,359*	*−251,581*	*1,019,598*	*1,077,371*	*1,147,518*	*1,232,598*	*1,303,779*	*1,397,636*	*1,518,594*	*1,678,985*	*2,464,619*
Incremental benefits	**3,729,919**	194,764	273,292	348,629	439,034	547,519	645,791	771,058	928,496	1,124,328	1,958,419
Incremental costs	**−1,311,100**	1,288,910	7,929	−80,585	−187,719	−317,302	−473,946	−663,199	−891,743	−1,174,620	−1,508,511
Incremental net benefits	**5,041,019**	**−1,094,146**	**265,363**	**429,214**	**626,753**	**864,822**	**1,119,737**	**1,434,257**	**1,820,239**	**2,298,948**	**3,466,930**
Net present value	**5,041,019**										
Benefit-cost ratio	−2.84										
Internal rate of return	56 %										

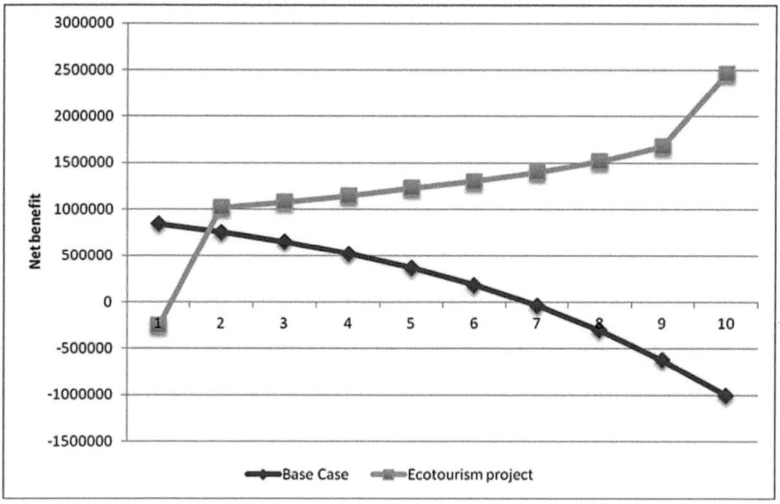

Fig. 11.1 Comparison of net benefits in base-case and ecotourism scenarios

provides convincing evidence that ecotourism is more desirable in economic terms, compared with a continuation of agricultural activities and exploitation of forest resources within Xe Pian NPA. A diagrammatic representation of net benefits for each of the scenarios is shown in Fig. 11.1.

Sensitivity Analysis

Sensitivity analysis was carried out in relation to the discount rate, the rate of deforestation, and the projected number of tourists visiting the area. Three discount rates were applied: 8 %, 10 %, and 12 %. A slower rate of deforestation, reducing it from 18 ha per year to 7 ha per year, was also tried. The growth rate for visits was reduced from 20 % per annum to 5 %. In all of these cases, however, the results revealed that the ecotourism project was economically superior to the base case. A similar exercise, excluding the environmental benefits, again showed the ecotourism option to be better than the base case, indicating that ecotourism would be financially more beneficial than the base case, as well as being more desirable in social economic terms.

Conclusions and Policy Recommendations

The main conclusion resulting from this CBA is that ecotourism in Xe Pian NPA can produce significant economic gains to local communities, compared with a continuation of traditional agricultural and forest-based production. This is consistent with the findings of studies in other countries (Mirbabayev and Shagazatova 2006). The economic analysis confirms that the Xe Pian ecotourism project would assist in preserving Champasak's environmental and cultural heritage. Under the ecotourism scenario, the Government of Lao PDR would empower the community to manage their own economic activities, with financial returns accruing to local people. The project still allows ongoing agriculture and fishing, but strict controls would be applied to conserve the natural forest. The advantage of protecting the forest is that more tourists will be attracted to Xe Pian NPA. The costs associated with the ecotourism project will increase each year due to a larger number of tourists, but ongoing maintenance of the area is not expected to cost more than $0.01 m per year. To manage Xe Pian NPA effectively, at least $0.72 m is needed per year instead of the $20,000 that is presently provided. Under the ecotourism project, the local community and park management would receive at least $0.4 m per year. In the tenth year, total revenue from the park could be expected to reach $1.6 m.

Ecotourism is not a complete solution for the conservation and protection of National Protected Areas, nor can it alone become the best economic means of raising local communities from poverty. Ecotourism must be well planned and monitored and based on wide participation of local communities to maximize economic benefits. As a result, cooperation between government agencies and local communities is one of the most important factors in implementing a successful ecotourism development strategy.

References

Brown J, Pearce DW (1994) The economic value of carbon storage in tropical forests. In: Weiss J (ed) The economics of project appraisal and the environment. Edward Elgar, Cheltenham, pp 102–123

Folmer H, Gabel HL (1998) Principles of environmental resources economics. Edward Elgar, New York

Harrison D, Schipani S (2007) Lao tourism and poverty alleviation: community-based tourism and the private sector. Curr Issues Tour 10(2&3):194–230

Kyophilavong P (2009) Evaluation of macroeconomic policy in Laos, Discussion paper (no. 171). Economic Research Center, Graduate School of Economics, Nagoya University, Nagoya

Mieczkowski Z (1995) Environmental issues of tourism and recreation. University Press of America, Lanham

Mirbabayev B, Shagazatova M (2006) The economic and social impact of tourism. National Graduate Institute for Policy Studies. Available at http://www.grips.ac.jp/alumni/UzbekistanTourism(2).pdf

National Statistic Centre Laos (2009) Statistical year book 2009. Lao Department of Statistic, Vientiane, Laos.

Poulsen MK, Luanglath K (2005) Projects come, projects go: lessons from participatory monitoring in Southern Laos. Biodivers Conserv 14:2591–2610

Rosales RM, Kallesoe M, Gerrard P, Muangchanh P, Phomtavong S, Khamsomphou S (2005) Balancing the returns to catchment management: the economic value of conserving natural forests in Sekong, Lao PDR. IUCN Water, Nature and Economics, Technical paper no. 5, IUCN — The World Conservation Union, Ecosystems and Livelihoods Group Asia, Colombo, Sri Lanka

Chapter 12
A Cost and Benefit Analysis of the Community Forestry Project in Chumkiri District, Kampot Province, Cambodia

Kalyan Hou, Sothunvathanak Meas, Maredi Im, Chanthy Ros, and Han Keam

Abstract Open-access forest schemes in Cambodia have contributed to a serious decline in timber and non-timber forest products collected from forests. Authorities have responded by introducing a major programme of Community Forestry (CF) in more than 237 CF areas covering 71,724 ha and involving 411,440 people. The authors of this study carry out a cost-benefit analysis to estimate the incremental net benefit of three Community Forestry management options, relative to the base case scenario, in Chumkiri district of Kampot Province to determine whether a community forest management scheme would be economically desirable. In the first management option, the Conservation Option, only 389.5 ha out of 992 ha of forest is used for timber and non-timber product collection. The present value of total costs over a 30-year time period at a 10 % discount rate is approximately USD 821,000. The option could generate benefits of up to USD 6.30 million measured in terms of present value. The second option, the Exploitation Option, involves the exploitation of forest products with CF management. The present value of total costs over a 30-year time period at a 10 % discount rate would be USD 1.16 million, and the present value of benefits would be about USD 5.01 million. The third option, which combines the Conservation and Exploitation Options, shows that the present value of total costs over a 30-year time period at a 10 % discount rate would be nearly USD 1 million, while the present value of benefits would be up to USD

K. Hou (✉) • M. Im
The Center for People and Forest (RECOFTC), Forestry Administration, Phnom Penh, Cambodia
e-mail: kalyan@recoftc.org; maredi.im@recoftc.org

S. Meas
The Learning Institute, Phnom Penh, Cambodia
e-mail: mvathanak@learninginstitute.org; mvathanak@yahoo.com

C. Ros • H. Keam
National Committee for Sub-national Democratic Development (NCDD), Kampong Thom Province, Cambodia
e-mail: keamhan@kpt.ncdd.gov.kh; chantyrous@yaoo.com; rourschanthy@kam.ncdd.gov.kh

D. James, H.A. Francisco (eds.), *Cost-Benefit Studies of Natural Resource Management in Southeast Asia*, DOI 10.1007/978-981-287-393-4_12

5.92 million. The analysis reveals that the Conservation Option is the most economically viable, followed by the Combined Option and lastly by the Exploitation Option. Sensitivity analysis indicates that the Conservation Option ranks first even under the assumptions of a higher price for timber and non-timber products, as well as under different time horizons and discount rates.

Keywords Cambodia • Forest management • Cost-benefit analysis • Conservation option • Exploitation option

Introduction

Background

Natural resources, especially forest resources, play a significant role in the economy of Cambodia as well as in the livelihood of rural Cambodians. Although most of the population is primarily dependent on agriculture, forest resources are highly valued for daily consumption and use. Forests provide resources for construction materials, human food, animal fodder, medicines and income generation and a place for pursuit of spiritual well-being.

The management of forests in Cambodia has gone through various stages over the last 50 years. Before 1975, legal forest harvesting took place under a system of 'collection permits' that had been established during French colonial times (McKenney and Prom 2002). After paying a certain fee for the permits, loggers then had a fixed time to fell trees. The loggers did not have to make management plans or define the harvesting area. From 1975 to 1979, there was no record of forest management programmes in the government. In the 1980s, the management task fell to the control of the Forestry Department. However, the pace of forest exploitation by powerful interest groups in the country increased starting in the late 1980s, which continued until before the elections in 1993. Following the elections, the government went back to the system of issuing permits and having private industrial forest concessions.

From 1973 to 1976, the forest of Cambodia covered approximately 12.7 million ha or 72 % of the country's total land area. Over the years, the forest area has been gradually lost. In 2002, the area covered by forests had decreased to 10.4 million ha or 57 % of the country's total land area. Large areas of once forest land have been degraded (McKenny and Prom 2004). Seeing the problem of forest destruction and degradation almost everywhere in the country, the government started to look for other approaches to forest management that called for more participation from local communities. With this in mind, Community Forestry (CF) began to receive considerable attention as a potential alternative (or complement) to forest concession management (McKenney and Prom 2002).

Broadly speaking, CF is an effort to support and empower communities to continue their traditional uses of forest resources and to encourage sustainable practices. CF plays an important role in the reformation of the forestry sector. It is also a strategy for poverty reduction of the Royal Government of Cambodia (Chea Sam Ang et al. 2005). As an alternative management system, CF enables the effective participation of local people and communities in managing forest resources by focusing on the forest resources on which they depend (CBNRM LI presentation 2005).

Legal instruments were enacted to guide and support the CF formulation and implementation. These include the Forestry Law, CF Sub-Decree and CF Prakas on CF guidelines. So far, over 237 CF areas have been established totalling 71,724 ha and involving 411,440 people (McKenney and Prom 2002). Most of these CF areas were established through projects with the support from local NGOs, international organizations (IOs) and government agencies.

Community Forestry in Chumkiri, Kampot Province

One example of these CF areas in Cambodia is the CF project in Chumkiri, Kampot Province, which is supported by the Community Forestry Research Project (CFRP). This CF area was developed in early 2001 and covers 992 ha of degraded forest connected to a population of 633 households from three villages. CF rules and regulations were formulated and were recognized at the provincial level. Five committee members were selected from the three villages to lead and implement the CF activities.

After more than 4 years of implementation, it is timely to assess whether or not CF had been a good approach for better forest management in the area in terms of its costs and benefits to the local community and to other key stakeholders. Hence, this research assessed the costs and benefits of the CF programme. It also determined the factors that contributed to the optimization of costs and benefits of the CF programme.

Research Objectives

The overall aim of the study was to assess the costs and benefits of the ongoing CF project in Chumkiri, to identify the factors influencing the outcomes of CF implementation and to compare these to a non-CF area. The specific objectives were as follows:

- To identify the main costs of both a CF- and a non-CF-managed forest in terms of project costs and costs to community members (e.g. time, money)
- To identify the main benefits of both a CF- and a non-CF-managed forest in terms of environmental benefits and biodiversity conservation

- To quantify the main benefits of both CF- and a non-CF-managed forest in terms of economic/tangible benefits and improvements in local livelihoods
- To compare the costs and benefits of the CF project with the costs and benefits in a non-CF area
- To develop policy recommendations and areas for future research

Methodology

Selection of Study Sites

The research was conducted in two communities: one where the primary focus was CF management and one without any CF management. The community under the CF initiative consisted of three villages: Prey Yav, Tbaeng Pok and Damnark Snuol villages in Srae Knong commune. The community without CF (non-CF) consisted of three villages: Thmei, Kandal and Damrei Koun villages in Chumpouvoan commune and Chumkiri District, Kampot Province. The selected communities under both CF and non-CF management had similar key biophysical and socio-economic characteristics. These similar characteristics included forest type, social structure and income level. In the non-CF community, the forest area was degraded, and people accessed forests outside their commune's forest area. This case was mirrored throughout the whole country.

Environmental Problems in Study Area

Environmental problems were threatening the livelihoods of local poor people in rural and remote areas. In fact, both target communes in Chumkiri District were constantly facing natural disasters (storms, floods, etc.), temperature changes, irregular rainfall, loss of NTFPs and timber, lack of water and soil erosion, which affected their livelihoods.

The results of the study indicated many environmental problems in the non-CF area. These included:

- Loss of forest products, both timber products and NTFPs. This has resulted in the loss of jobs or market in the forestry sector.
- Irregular weather. These included inadequate or irregular rain, strong winds and others.
- Silted soil erosion channels. These have caused some rice fields to no longer yield rice.
- Lack of reliable sources of water. The water here is for household consumption and agricultural use.

- Changing soil quality. The soil has become less fertile and farmers were starting to use chemical fertilizer to increase yields.
- Variable local production. Production relied on the weather and rate of soil erosion.
- Nearly extinct wildlife in the non-CF area. People reported never seeing wildlife because the latter's habitat had been lost through the clearing forests for/by human activities.

The results of the study in the CF area showed fewer problems, especially on forestry-related offences. These problems included:

- Lack of reliable sources of water. This water was for household consumption and agricultural uses.
- Drought. This has become a critical environmental problem in local communities in the study area. Drought has resulted to low agricultural production.
- Use of chemical fertilizers. Local people complained that some of the community members still used chemical fertilizer and pesticides to gain higher yields and to kill insects next to the CF area.

The well-managed habitat in the CF areas has steadily increased and enhanced wildlife.

Approach to Economic Analysis

The analysis undertaken comprised six steps: (1) evaluating the problem in the base case scenario (non-CF management of the forest), (2) assessing possible CF management options, (3) estimating the total annual costs and benefits of each option, (4) calculating the present values of net benefits, (5) comparing the net benefit values of the potential CF management options with the base case scenario, and (6) conducting a sensitivity analysis.

The study focused on both the direct and indirect use values of the forest. Direct values comprised NTFPs, timber forest products and wildlife collected from the CF and non-CF areas. The indirect values of the forest included watershed management, soil erosion and environmental values (ETAP 1998). The ETAP paper (ETAP 1998) cited a survey showing reduced soil erosion in the CF area and soil erosion problems in the non-CF area, but nothing about soil erosion values. This raised the larger question of how to include and utilize indirect values and present them in a way relevant to this study.

Data Collection

The study used both primary and secondary data (2000–2006) for both forest areas (CF and non-CF). Primary data were collected through household surveys, focus group discussions (FGDs) and direct community observation. Secondary data were gathered from existing reports, official records and research papers.

For this study, no information was obtained on the capture of wildlife in either area. Capturing wildlife is banned by the CF by-laws and regulations. The Cambodian forest law bans hunting, which includes trapping. Under the current forestry and wildlife law, hunting wildlife is prohibited; hence, illegal hunting in the study area, mostly in the non-CF area, is decreasing. However, some members of the community still hunt for their household's own consumption. The research team observed, for instance, that wildlife was being sold in the market. Nevertheless, information on hunted wildlife has become very difficult to ascertain. The research team found it difficult to get the required information from the respondents through predetermined/pretested interview questions.

Primary Data Collection

A final draft of the questionnaire was tested with both genders through household interviews within the target communities. The pretest included 15 randomly selected households (HH) in the CF and 20 randomly selected households in the non-CF area. The objective of the project was to test the 'wording' and flow of the draft questionnaire and to ensure that the data collected are sufficient to answer the research objectives. No major changes were made based on the pretest. The household data for the study were collected from a survey of 137 randomly selected households, with 69 households in the CF area and 68 households in the non-CF area. A household survey was used to collect primary data from the household units in the study areas. The households were asked about the utilization as well as the marketing (e.g. where these products were being sold) of timber and non-timber forest products. Focus group discussions were also held, involving 55 participants: 27 people from the non-CF area and 28 people from the CF area. The FGDs took place for one and half days in each site.

Secondary Data Collection

Collection of secondary data focused on several studies and reports produced from the study sites by university students, CFRP staff and others. These included Rapid Rural Appraisal on the Users' Right System of Natural Resources (2000), Forest Use and Product Flow in the Chumkiri District, Kampot Province (2001), and Economic Analysis on Community Forestry Establishment, Chumkiri District, Kampot Province, and others (2004). The findings of these reports were used as baseline data for the research. However, there was insufficient data from these sources to support the required analysis. Therefore, other data were gathered from interviews and reports from CIDSE, Forestry Administration and district officers in Chumkiri, Kampot Province.

Key Characteristics of Study Sites

Stakeholders in Study Area

Many stakeholders and institutions working in each commune were involved in the development activities of the communities in the study areas. They advised and helped strengthen the capacity of the communities. The main stakeholders for this study were documented below:

- Local community residents (six villages): These people were direct users of the forest resource in the study area.
- Community Forestry Committee: This committee was directly involved in the management of forest resources in the CF study area. This committee was elected by the CF community members. Their main role and responsibility is to make decisions on how to use and manage forest resources within the CF area.
- Businessmen/outsiders (entrepreneurs): They were involved in buying and selling forest products from the local communities in the study area. They purchased most of the forest products that were being collected legally and illegally within the areas and sold these for profit.
- Commune Council (CC) and village headmen: They made all decisions involving the village. They played an important role in assisting other developers and all government sectors in the development of the village and in conflict management.
- District authority: The government staff at a district level oversaw the communes.
- They had a similar role as the communes but at the district level.
- Development Partnership Agency (DPA): DPA was a local organization from the Cooperation International for Development and Solidarity (CIDSE) organization. The CIDSE supported the project activities of the CF programme in the study areas. DPA was a development organization and partner that provided and shared experiences and served as consultant in matters related to CF.
- Triage and other levels of the Forestry Administration: This is the government professional institution with a mandate over forests. It played a significant role in supporting the CF, in serving as consultant in forest management, in repressing illegal activities and in resolving problems and/or conflicts.
- Military and police: These institutions and officers intervened to help resolve conflicts.

- Other government departments such as the Department of Agriculture and the Department of Environment. These were involved in the consultation process, and they supported and/or conducted other necessary activities such as agriculture and environmental extension in these areas.

Noncommunity Forestry Communities

There are seven villages in Chumpouvoan commune, Chumkiri District. The commune has 1,608 households with a population of 7,823 people, including 4,129 females. The total land area of this commune is 3,977 ha, comprising 992 ha of forest land, 1,491 ha of rice fields, 153 ha of farmland and 426 ha of residential land. The commune centre is 2 km from the district centre. The forest area is gradually being degraded by human activities such as illegal harvesting of forest products, burning of the forest for farming and encroaching on the land to secure tenure. Deforestation has also led to soil erosion from the upland areas to lowland areas. The results of the FGD indicated that soil erosion was a problem in the non-CF area. The respondents believed that when the forest was cleared, raindrops directly contacted with and exposed the topsoil, hence leading to soil loss. They also said that erosion washed away the organic matter, microorganisms and other important nutrients needed for plant growth, which decreased crop production.

Of the seven villages, the study was conducted in three villages of the non-CF area, including Damrei Koun, Kandal and Thmei villages. This area was chosen for the non-CF site because it has a similar population density, forest size and condition, villagers' livelihood activities, security and accessibility as the CF site. Local people rely heavily on agricultural production, animal raising, sale of labour, small businesses and collection of forest products for sale and for household use.

Based on the interviews and observations of the villagers in this study area, this area was covered by primary forest that was rich in many wood species and biodiversity before 1998. At the time of the study, however, the forest had been degraded because of several factors. First, there were illegal activities on forest resources by local communities living in the area and by outsiders accessing the forest area without obtaining permission from authorities. Secondly, the high demand and high value of forest products in the market have encouraged entrepreneurs to collect these resources from local people and sell these to more central markets and abroad. Finally, there was no clear management system in place, especially on the rights of local resource users to manage the forest area.

Timber Forest Products

Villagers reported accessing the forest area far from their home or a distance of 10–50 km in areas such as Phnum Prahout and Taken Koh Slar. It took the local people

4–15 days per trip. They also reported spending 30 days per trip to gather forest resources. Respondents reported that the households collected logs (53 %), harvested sawn wood (24 %), harvested small poles (26 %) and collected big poles (21 %). Regarding the amount of timber collected, adjusted for comparing to a forest of equal size, people reported harvesting 888 m^3 of logs, about 685 m^3 of sawn wood, 1,806 m^3 of small poles and 832 m^3 of big poles. Based on market values, the total benefit of timber as a product was valued at nearly USD 206,000 per year.

Non-timber Forest Products

To access forest areas, people said they had to walk from 1 to 30 km and spend from about 0.5 to 5 days to collect NTFPs. The six locations are identified in the text: Toul Khlong, Phnum Prahout, Phnum Thom, Prey Lech, Toul Samrong and Taken Koh Slar. The survey showed that 87 % of the households collected fuel wood for their consumption, adding up to 2,269 m^3 per year. Exactly 69 % of the households collected bamboo culms for their consumption and sale, amounting to 203,130 culms per year. A total of 32 % of the households collected bamboo shoots for consumption and fewer amounts for sale, amounting to 4,024 kg per year. Meanwhile, 35 % of the households collected 480 kg of mushroom per year, while 13 % harvested 254 kg per year of wild vegetables.

Only 1 % of households collected honey from the non-CF area, equalling 10 l per year. A total of 7 % of households collected wild fruit from the forest, totalling 149 kg annually. A few households (3 %) collected 24 kg per year of traditional medicine. The total monetary value of non-timber products collected in this area was estimated to be USD 19,100. Generally, people in non-CF area collected NTFPs for sale, except for the wild fruits and traditional medicines used for household consumption. The value of NTFPs in the non-CF market was USD 36,958.

Perceptions of Non-use Values of Forestry Resources

In the non-CF area, a few respondents had received lessons in natural resources management regarding fire prevention, fire protection and reforestation. However, no activities showed application of these lessons in that area. Some villagers had also attended a few training sessions and study tours on agriculture, animal raising, health, sanitation and use of clean water.

Community Forestry Communities

There are six villages in Srae Khnong commune, Chumkiri district. The commune has 1,106 households with a total population of 5,633 people, including 3,031 females. The total land area of this commune is 2,873 ha. It consists of 1,080 ha of forest land, 1,396 ha of paddy field, 104 ha of farmland and 293 ha of residential land. This commune is 12 km away from the Chumkiri district centre.

Three villages in the CF area were selected for the study. These included Prey Yav, Tbaeng Pok and Damnak Snoul. The total forest area under CF covers 992 ha of forest land. The total population of the three villages is 3,085 including 1,629 females. This population lives in 633 households.

This CF area was named the 'Damnak Nak Ta Thmor Poun Community Forestry'. It was established in March 2001 by the Community Forestry Research Project (CFRP). When setting up the CF, the CFRP team conducted several studies and consulted with various stakeholders. The CF members and committee implemented a capacity-building process and found solutions to deal with conflicts and problems in their area.

For the CF study site, the CF forest is located approximately 1–3 km away from the village. The respondents reported spending about 0.5–2 days in travelling, but majority spent 0.5 day on their trips. However, some villagers reported accessing forest resources outside the CF area, about 10 km away, for which they travelled back and forth for 2 days. Most of them, however, accessed forest resources within the CF boundary.

The results of the FGD indicated that soil erosion was not yet a big problem. The respondents reported that in the past, sandy soil was always a problem for their farms near the mountain because the rain eroded sandy soil from the mountain top. As a result, their production of rice and other crops was very low. Sometimes, the sand made it difficult for them to cultivate rice and other crops such as corn, beans, watermelons and cucumbers. They believed that CF had helped reduce soil loss and erosion rates.

Timber Forest Products

Respondents reported that only 4 % of the households had collected logs, totalling 18 m^3 per year. Majority (65 %) of the households reported collecting small poles, totalling 223 m^3. A total of 22 % of the households reported collecting big poles, totalling 156 m^3. The total value of the timber products collected was USD 7,357 per year.

Non-timber Forest Products

The survey showed that 80 % of the households collected fuel wood for their consumption. This totalled 4,399 m^3 per year; so on the average, each household used 6.94 m^3 of fuel wood per year. Most of them (81 %) also gathered mushrooms totalling 4.229 kg per year. A majority (58 %) of the households collected 41,062 culms of bamboo per year. An almost equal number (59 %) collected bamboo shoots that totalled 18,105 kg per year. People dried bamboo shoots for use later in the year or reserved them for other purpose. Many of them (67 %) likewise harvested wild vegetables totalling to 14,843 kg per year. Almost half of the households (48 %) gathered 2,459 kg of wild fruit per year, while 22 % collected traditional medicines totalling 587 kg per year. In addition, 9 % of households collected vines, totalling 1,312 kg. Six per cent of households collected rattan, and this added up to 147 kg per year. NTFPs were collected for household use, sharing with neighbours or selling at a low price. This meant that there was a limited market for NTFPs in the CF area. The value of NTFPs was USD 23,448 per year.

Perceptions of Non-use Values of Forestry Resources

In the CF area, the people and community received lessons about natural resources management, and they applied these by establishing the CF. They learned and applied the idea of CF as a common pool resource; the concept of CF, forestry laws, regulations and by-laws of CF; the concept of forest conservation; and the principle of participation in other community development activities. They also learned how to use and manage natural resources more sustainably in CF and how to protect the forest from fire and illegal activities. Moreover, they learned about forest benefit sharing, reforestation, planning for forest benefits for the next generation and the forest's influence on local weather and flooding. The respondents reported that they gained knowledge and benefited from their participation in trainings, workshops, study tours and meetings. From meetings, they learned about animal husbandry, new models of agriculture, health issues, gender in society and CF.

Assessment of CF and Non-CF Management Options

Management Options

The study assessed a relatively narrow range of CF management schemes that could be applied in the current situation. The management options are described below.

Conservation Option The forest area is divided into a conservation zone and a multiple use zone. Under this management scheme, only the forest in the multiple use zone would be opened for exploitation. This zone is divided into five parts, and the forest products/resources are collected from each zone in a 5-year rotation.

Exploitation Options All forest zones are divided into five zones for exploitation in a rotation pattern every 5 years.

Combined Options This combines the features of the Conservation and Exploitation Options. The conservation zone represents only one third of the total forest area; the remaining area is fully exploited under a 5-year rotation. Before assessing the CF management options, the base case scenario was examined. This was to establish the initial situation regarding the livelihood of the population living under the non-CF management, namely, the present values of net returns. The estimation of people's incomes in the base case helped in evaluating the changes in their welfare after introducing the CF management scheme.

Base Case Scenario (Non-CF Management)

The base case scenario assumed that the local population and people outside the commune under an open-access scheme used the forest area of 992 ha located within the commune. The present value of net economic benefits in the non-CF management was calculated, taking a production time period of 30 years, at a 10 % discount rate. The major category of net benefits was net returns accruing to the population that was harvesting timber forest products and non-timber forest products (NTFP) as described in Table 12.1.

Cost of Forest Products Collection

The main assessment methods and data used to estimate the cost of harvesting forest products were separated into three components: capital investment costs, operation and maintenance costs and ongoing costs. The capital costs of the non-CF management scheme involved purchasing of materials used for timber and non-timber collection especially ox, oxcart, knife, axe, line, shoe cloth and mosquito net. These costs would be incurred only once, in year 1. It was assumed that the ox and oxcart would last for the whole project period of 30 years and that the knife, axe, line, shoe cloth and mosquito net would be replaced every 2 years.

The operational and maintenance costs involved the costs to cover the repair of the materials quoted above. The informal fee paid at the checkpoints was the cost involved in transporting forest products collected across the informal checkpoint kept by the military police or agent of the forest administration. The costs of timber and non-timber product collections were included in the costs of purchasing food for the duration of the journey (to gather materials). The cost of disease was the cost to cover medical expenses to treat any illness incurred during the journey in the forest. Under a non-CF scenario, forest productivity could be degraded. In addition,

Table 12.1 Timber and non-timber forest products in non-CF scheme (field survey 2007)

Timber forest products	Quantity collected	Price (USD/m3)	Non-timber forest products	Quantity collected	Price (USD per unit)
Log (m^3)	888	39.41	Fuel wood (m^3)	2,269	3.69
Sawn wood (m^3)	685	184.73	Bamboo (trees)	203,130	0.05
Small pole (m^3)	1,806	14.16	Bamboo shoot (kg)	4,024	0.12
Big pole (m^3)	832	22.41	Mushroom (kg)	480	0.12
Total (m^3) 4,210			Wild vegetable (kg)	254	0.10
Or 4.24 m3/ha/year			Wild fruit (kg)	149	0.10
			Traditional medicine (kg)	24	0.94

Table 12.2 Costs incurred in the non-CF management scheme

Items costs	Year 1	Year 2	Year 3	Year 30	PV
Capital investment	274,972	0	9,975	0	290,161
O&M (transportation and material)	7,005	7,005	7,005	7,005	66,033
Accessing forest fee or check point	31,720	31,720	31,720	31,720	299,025
Costs of timber product collection	10,120	10,626	11,158	41,656	152,272
Costs of NTFPs collection	5,189	5,448	5,721	21,358	78,074
Diseases	3,600	3,780	3,969	14,817	54,162
Total costs	**332,606**	**58,579**	**69,547**	**116,556**	**939,727**

Note: PV value was calculated over a 30-year period with a 10 % discount rate. All values are in USD

if environmental degradation was expected, then the costs of managing and harvesting might well increase over time. Consequently, the study assumed that there was a 5 % increase in the cost of illness and in timber and non-timber products collection.

Table 12.2 shows that the present value of the total costs of timber and non-timber (NTFPs) collection was USD 939,727. With 627 households involved in the non-CF management scheme, these costs would be about USD 1,499 per household.

Revenue Under the Non-CF Management Scheme

The revenue under the non-CF management scheme was divided into two categories: (1) the direct revenue from timber and non-timber collection and (2) the environmental benefit generated from carbon storage and biodiversity gain. The value of timber and non-timber products was equal to the volume of timber collected multiplied by its financial price as shown in 12..1. Therefore, under the open-access forest regime, it was expected that the volume of timber and non-timber product would decrease over the period of a year. Table 12.1 shows that the volume of timber harvested would include 4,210 m^3 of timber forest products and 2,269 m^3 of fuel wood. Over 992 ha of forest area, a harvest rate of 7.33 m^3 per ha and per year was implied. Consequently, the study adopted a 3 % reduction each year for the whole project period.

The study used existing secondary data sources for the value figures of biodiversity and carbon. The value of biodiversity was USD 300/ha (Bann 1997). The value of carbon was USD 5 per ton (Hansen and Top 2006). The carbon storage (values are shown in Table 12.6.) calculation used the following formula for forest in a stocked plot (Hansen and Top 2006): $y = 2.39*\text{Ln}(X)-7.79$.

This formula can be used when the above-ground biomass (AGB) is higher than 40 Mg/ha in a plot. However, when the forest in a stocked plot has an AGB less than 40 Mg/ha, the formula cannot be used. In the latter case, an approximation of AGB was made.

Table 12.3 shows that the present value of benefits from the collection of timber and NTFPs was about USD 1.69 million, equivalent to 34 % of total benefits. With 627 households involved in the non-CF management scheme, this benefit would be about USD 2,695 per household. The carbon storage and biodiversity benefits under the non-CF management scheme declined year after year, from USD 651,300 in the initial year to only USD 30,700 in year 30.

Conservation Option (CF Management)

The Conservation Option assumed that within the forest area of 992 ha located within the commune, only 602 ha would be kept for conservation. The remaining 390 ha would be used by the local population and people outside the commune under a payment fee system of the CF management scheme. The 390 ha of forest was divided into five core zones: 54 ha, 55 ha, 56 ha, 60 ha and 51 ha. These zones were used for a 5-year rotation scheme. With better management, a forest growth rate of 3 % was used in calculating the benefit gained. Estimates of the costs and benefits of this option were derived from the CF management of Damnak Neak Ta Thmar Poun community forest, which had similar characteristics to the non-CF area.

Table 12.4 highlights the timber and NTFPs that could be collected from the CF area under the Conservation Management Option. The price and quantity collected are also shown in Table 12.4. In the CF area, no sawn wood and honey were collected compared to the non-CF area in this study.

Table 12.3 Benefits in the non-CF management scheme

Benefit	Year 1	Year 2	Year 3	Year 4	Year 30	PV
Value of timber products	205,725	199,553	193,566	187,759	85,048	1,546,129
Value of NTFPs	19,100	18,527	17,971	17,432	7,896	143,544
Carbon storage	383,458	345,112	310,601	279,541	18,061	1,912,630
Biodiversity gain	267,840	241,056	216,950	195,255	12,616	1,335,947
Total benefits	**876,122**	**804,247**	**739,088**	**679,987**	**123,621**	**4,938,250**

Note: PV value was calculated over a 30-year period with a 10 % discount rate. All values are in USD

Table 12.4 Timber and non-timber forest products in the conservation option scenario (field survey in 2007)

Timber forest products	Quantity collected	Price (USD/m3)	Non-timber forest products	Quantity collected	Price (USD per unit)
Log (m^3)	18	39.41	Fuel wood (m^3)	4,399	3.69
Sawn wood (m^3)	0	184.73	Bamboo (trees)	41,062	0.05
Small pole (m^3)	223	14.16	Bamboo shoot (kg)	18,105	0.12
Big pole (m^3)	156	22.41	Rattan (kg)	147	0.49
Total (m^3)	**397**		Mushroom (kg)	4,229	0.12
Or 1.02 m3/ha/year					
			Honey (litre)	0	9.85
			Vine (kg)	1,312	0.07
			Wild vegetable (kg)	14,843	0.10
			Wild fruit (kg)	2,449	0.10
			Traditional medicine (kg)	587	0.94

Cost of Forest Products Collection

As discussed earlier, the analysis of costs can be divided into three components: capital investment costs, operation and maintenance costs and ongoing costs. The capital costs of the non-CF management scheme involved:

- Community forest establishment costs that accrued only in the first 6 years.
- The purchase of materials used for timber and non-timber collection, especially ox, oxcart, knife, axe, line, shoe cloth and mosquito net. These costs would be incurred only once, in year 1. The study assumed that the ox and oxcart would last for the whole project period of 30 years and that the knife, axe, line, shoe cloth and mosquito net would be replaced every 2 years.

The operational and maintenance costs involved the costs to repair the materials quoted above. The costs of timber and non-timber products collection included the costs of purchasing food during the journey and of incurring any medical expenses.

Under the CF option, the time spent out of the village involved in CF activity was reflected in the labour costs. The CF management contribution was the fee collected from villagers to cover the daily expenses of managing the CF community. The forest plantation costs covered the costs of reforestation after cutting.

Under a CF scenario, forest productivity could be improved. Also, if environmental degradation was reduced or turned around, then the costs of managing and harvesting might well decline over time. Consequently, the study assumed a 5 % reduction in the cost of collecting timber and non-timber products.

Table 12.5 shows that the present value of the costs of collecting timber and NTFPs was more than USD 820,678. With 633 households involved in the CF management scheme, these costs would be about USD 1,296 per household. This cost was lower than the costs incurred in the non-CF management scheme.

Revenue Under the Conservation Option

The revenue under the CF management scheme was divided into two categories: (1) direct revenue from timber and non-timber collection and (2) the environmental benefit generated from carbon storage and biodiversity gain. Revenue from timber and non-timber collection was generated for the villagers involved in the collection and from the CF management under the licence form. The revenue from the licensing was set at 50 % of the total financial value of timber and non-timber collection.

This study assumed the value of biodiversity to be USD 300/ha (Bann 1997). The value of carbon was USD 5 per ton (Hansen and Top 2006). The carbon storage values are shown in Table 12.6.

Table 12.6 shows that the present value of benefits from timber and NTFPs collection for both villagers and CF management was about USD 1.13 million, equivalent to only 18 % of the total benefits. With 633 households involved in the CF management scheme, the benefits generated from timber and non-timber products would be about USD 1,792 per household. Majority of the benefits could be attributed to the carbon storage and biodiversity gain from the CF management scheme. This benefit was approximately constant over 30 years of the project return period. Its present value was equal to USD 5.16 million.

Exploitation Option (CF Management)

The Exploitation Option assumed that all the forest area of 992 ha located within the commune would be used by the local population and by the people outside the commune under a payment fee system of the CF management scheme. The 992 ha

Table 12.5 Costs incurred in the conservation option

Costs	Year 1	Year 2	Year 3	Year 30	PV
CF establishment cost	9,465	9,465	9,465	0	41,221
Capital investment	414,029	0	11,552	0	422,931
O&M (transportation and materials)	5,867	5,867	5867	5867	55,308
Costs of timber product collection	604	574	545	136	3,976
Costs of NTFPs collection	5,915	5,620	5,339	1,336	38,950
Opportunity own labour costs	9,653	9,653	9,653	9,653	90,997
CF management contribution	1,185	1,185	1,185	1,185	11,172
Cost of forest plantation	16,200	16,500	16,800	15,300	156,121
Total costs	**462,918**	**48,863**	**60,405**	**33,478**	**820,678**

Note: PV value was calculated over a 30-year period with a 10 % discount rate. All values are in USD

Table 12.6 Benefits in the conservation option

Benefit	Year 1	Year 2	Year 3	Year 4	Year 30	PV
Value of timber products	7,357	7,578	7,805	8,039	17,338	90,483
Value of NTFPs	23,448	24,151	24,876	25,622	55,256	288,374
Carbon storage	354,102	339,090	325,807	334,351	363,514	3,218,830
Biodiversity gain	247,350	230850	214,050	196,050	196,950	1,944,938
Value of timber products for CF	74,886	76,273	77,659	83,206	70,727	721,685
Value of NTFPs for CF	2,759	2,841	2,927	3,014	6,501	33,926
Total benefits	**709,902**	**680,783**	**653,124**	**650,284**	**710,285**	**6,298,236**

Note: PV value was calculated over a 30-year period with a 10 % discount rate. All values are in USD

of forest was divided into five core zones of equal surface area for exploitation for a rotation period of 5 years.

Cost of Forest Products Collection

The methods and data used to estimate the costs of forest products collection were the same as under the Conservation Option. Under a CF scenario, if forest productivity could improve and if environmental degradation would be reduced, then the costs of managing and harvesting might well decline over time. Consequently, this study assumed a 5 % reduction in the cost of timber and non-timber products collection. Table 12.7 shows that the present value of the costs of timber and NTFPs collection was about USD 1.161,461 million. With 633 households involved in the non-CF management scheme, the costs would be more than USD 1,835 per household.

Table 12.7 Costs in the exploitation option

Costs	Year 1	Year 2	Year 3	Year 30	PV
CF establishment cost	9,465	9,465	9,465	0	41,221
Capital investment	414,029	0	11,552	0	422,931
O&M (transportation and materials)	5,867	5,867	5,867	5,867	55,308
Costs of timber product collection	604	574	545	136	3,976
Costs of NTFPs collection	5,915	5,620	5,339	1,336	38,950
Opportunity own labour costs	9,653	9,653	9,653	9,653	90,997
CF management contribution	1,185	1,185	1,185	1,185	11,172
Cost of forest plantation	52,350	52,650	52,950	51,450	496,904
Total costs	**499,068**	**85,013**	**96,555**	**69,628**	**1,161,461**

Note: PV value was calculated over a 30-year period with a 10 % discount rate. All values are in USD

Revenue Under the Exploitation Option

Table 12.8 shows that the present value of the benefits from timber and NTFPs collection for both villagers and CF management was about USD 2.71 million, equivalent to only 54 % of the total benefits. With 633 households involved in the CF management scheme, the benefit generated from timber and non-timber products would be more than USD 4,281 per household. The carbon storage and biodiversity gain from the Exploitation Option would decline year after year from USD 512,675 in the initial year to only USD 148,973 in year 30.

Combined Option (CF Management)

The Combined Option assumed that within the forest area of 992 ha located within the commune, only 300 ha would be kept for conservation purposes. The remaining 692 ha would be used by the local population and by the people outside the commune under a payment fee system of the CF management scheme. That forest area would be placed under a 5-year rotation plan.

Cost of Forest Products Collection

The methods and data used to estimate the costs of forest products collection were the same as under the Conservation Option. Table 12.9 shows that the present value of the costs of timber and NTFPs collection was about USD 0.99 million. With 627 households involved in the non-CF management scheme, the costs would be more than USD 1,610 per household.

Revenue Under the Combined Option

Table 12.10 below shows that the present value of the benefits from timber and NTFPs collection for both villagers and under the CF management was about USD 1.91 million, equivalent to about 32 % of the total benefits. With 633 households

Table 12.8 Benefits in the exploitation option

Benefit	Year 1	Year 2	Year 3	Year 4	Year 30	PV
Value of timber products	7,357	7,578	7,805	8,039	17,338	90,483
Value of NTFPs	23,448	24,151	24,876	25,622	55,256	288,374
Value of timber products for CF	241,993	243,380	244,765	250,310	237,836	2,296,986
Value of NTFPs for CF	2,759	2,841	2,927	3,014	6,501	33,926
Biodiversity gain	210,840	158,550	105,600	51,450	52,350	766,982
Carbon storage	301,835	232,890	160,734	164,950	96,623	1,536,328
Total benefits	**788,232**	**669,391**	**546,707**	**503,386**	**465,904**	**5,013,078**

Note: PV value was calculated over a 30-year period with a 10 % discount rate. All values are in USD

Table 12.9 Costs in the combined option

Costs	Year 1	Year 2	Year 3	Year 30	PV
CF establishment cost	9,465	9,465	9,465	0	41,221
Capital investment	414,029	0	11,552	0	422,931
O&M (transportation and materials)	5,867	5,867	5,867	5,867	55,308
Costs of timber product collection	604	574	545	136	3,976
Costs of NTFPs collection	5,915	5,620	5,339	1,336	38,950
Opportunity own labour costs	34,275	34,575	34,875	33,375	326,513
CF management contribution	9,653	9,653	9,653	9,653	90,997
Cost of forest plantation	1,185	1,185	1,185	1,185	11,172
Total costs	**480,993**	**66,938**	**78,480**	**51,553**	**991,069**

Note: PV value was calculated over a 30-year period with a 10 % discount rate. All values are in USD

Table 12.10 Benefits in the combined option

Benefit	Year 1	Year 2	Year 3	Year 4	Year 30	PV
Value of timber products	7,357	7,578	7,805	8,039	17,338	90,483
Value of NTFPs	23,448	24,151	24,876	25,622	55,256	288,374
Value of timber products for CF	158,439	159,827	161,212	166,756	154,283	1,509,335
Value of NTFPs for CF	2,759	2,759	1,379	2,759	2,759	23,428
Carbon storage	327,711	285,990	243,270	249,651	230,068	2,387,986
Biodiversity gain	228,915	194700	159,825	123,750	124,650	1,618,937
Total benefits	**748,629**	**675,005**	**598,368**	**576,577**	**584,354**	**5,918,543**

Note: PV value was calculated over a 30-year period with a 10 % discount rate. All values are in USD

involved in the CF management scheme, the benefit generated from timber and non-timber products would be more than USD 3,020 per household. The carbon storage and biodiversity gain from the Combined Option would decline year after year from USD 557,000 in the initial year to only USD 355,000 in year 30.

Results of CBA

The cost-benefit analysis of alternative options within the CF management scheme is a method that compares the net benefit of the three alternative options within the CF management scheme to the base case situation, namely, non-CF management of the forest area. Such comparisons are useful for ascertaining whether a new alternative is better than the status quo or not.

Incremental Net Benefits

In the cost-benefit analysis, net benefits were the difference between total economic benefits and total direct costs in each CF management option. If net benefits were positive with higher values, this meant that the option would provide more benefits directly to the villagers involved and to their community than the current non-CF management option.

The incremental net benefits were the difference between the net benefits of each option of CF management and the net benefits accruing in the base case (non-CF management). The present values of net costs and incremental net costs were estimated over a 30-year period, with a discount rate of 10 %. The higher the incremental net benefits, the more economically viable was the option. Positive incremental net benefits would mean that the alternative option would yield greater economic benefits to the villagers and to their community compared to the base case scenario.

Table 12.11 below shows that the Conservation Option and the Combined Option were the most preferred options based on economic efficiency grounds. The incremental net benefits of the two options were positive. Carbon storage and biodiversity gain represented a large share in the Conservation Option, while revenues for the community generated from timber and non-timber products were the main elements of benefit for each of the two options. The incremental net benefit of the Exploitation Option had a negative value of USD 0.15 million.

Table 12.11 Present value (PV) of incremental net benefit, ranked by most benefit (field survey in 2007)

	NPV net benefit (USD million)	Incremental net benefit (USD million)	Benefit cost ratio
Conservation option: CF management scheme	5.48	1.48	7.67
Combined option: CF management scheme	4.93	0.93	5.97
Base case (non-CF management scheme)	4.00	0.00	5.25
Exploitation option: CF management scheme	3.85	−0.15	4.32

Sensitivity Analysis

Sensitivity analysis was conducted to investigate the effects of varying key assumptions (costs, revenues and discount rate) on the present values of incremental net benefits. The analysis provided a measure of the degree to which these variables could deviate from their estimated values before the preferred options ceased to be economically viable. Three scenarios were tested.

Scenario 1: Change the Discount Rate

For all options, the discount rate was increased from 10 to 15 %. The calculations showed that the Conservation Option and Combined Option were still economically viable. The incremental net benefit of the Exploitation Option was negative.

Scenario 2 : Increase of 10 % in Price of Timber and Non-timber Products

For all options, the price of timber and non-timber products was assumed to increase by 10 %. The results indicated that the Conservation Option and Combined Option were still economically viable. The incremental net benefit of Exploitation Option was negative. The ranking was identical to that of Scenario 1.

Scenario 3: Project Duration Is 20 Years

If a project return period of 10 years shorter than the initial 30 years was assumed, the Conservation Option was the most economically viable, and it ranked highest of all alternative options. With only a 20-year return period, the Conservation Option and Combined Option were still economically viable. The incremental net benefit of the Exploitation Option was negative. The ranking was identical to Scenario 1.

The three sensitivity analysis options above illustrate that with each variation of the basic assumptions, the ranking of the options remained the same: the Conservation Option was always the most economically viable, with the Combined Option still being viable. The Exploitation Option, in all scenarios tested, was negative, thus proving that this option would bring very little benefit to the communities.

Conclusion

The principal objective of this study was to identify the best practical options for preserving the forest cover from degradation. Three CF management options were assessed, and their incremental net benefits were calculated. These options as well as the base case scenario (non-CF management) are described below:

Base Case (Non-CF Management) A total of 627 households were involved in the non-CF management scheme of the Chumpouvoan commune. This commune extracted annually from open access forest a total of 4,210 m^3 of timber and 2,269 m^3 of firewood or 6.53 m^3/ha/year. Over a 30-year period, at a 10 % discount rate, the PV of total benefits including the carbon sequestration value and biodiversity gain would be about USD 7,876 per household.

Conservation Option A total of 633 households were involved in the CF management scheme of the Chumpouvoan commune. The commune extracted annually from the CF a total of 397 m^3 of timber and 4,399 m^3 of firewood or 12.3 m^3/ha/year. The present value of revenue received by the CF management team from sales of forest products, over a 30-year period, at a 10 % discount rate, would be USD 0.76 million. Over a 30-year period, at a 10 % discount rate, the PV of total benefits including the carbon sequestration value and biodiversity gain would be about USD 9,950 per household.

Exploitation Options A total of 633 households were involved in the CF management scheme of the Chumpouvoan commune. The commune extracted from the CF a total of 397 m^3 of timber and 4,399 m^3 of firewood or 12.3 m^3/ha/year. The revenue received by the CF management team from sales of forest products, over a 30-year period, at a 10 % discount rate, would be USD 2.33 million. Over a 30-year period, at a 10 % discount rate, the PV of total benefits including the carbon sequestration value and biodiversity gain would be about USD 7,920 per household.

Combined Options A total of 633 households were involved in the CF management scheme of the Chumpouvoan commune. The commune extracted from the CF a total of 397 m^3 of timber and 4,399 m^3 of firewood or 12.3 m^3/ha/year. The revenue received by the CF management team from sales of forest products, over a 30-year period, at a 10 % discount rate, would be USD 1.53 million. Over a 30-year period, at a 10 % discount rate, the PV of total benefits including the carbon sequestration value and biodiversity gain would be about USD 9,350 per household.

In conclusion, the results of the sensitivity analysis suggest that the Conservation Option always ranked first, and the incremental net benefit was always a positive value even when the input variable was altered. This proved that the Conservation Option in a CF area offers advantages compared with that in a non-CF-managed area. Hence, Cambodia should strengthen plans for CF in the future as the benefits and costs were evident in this study.

Further research could be done on this subject to help implement the concept and management practices, rules and regulations of CF. Communities were shown to benefit from CF in the long term. Through the CF, they learned that it was vital to encourage biodiversity, to increase the community's wealth and to prevent the forest from being degraded. Most importantly, they learned that the key rule in sustainability is to preserve the forest cover for future generations.

References

Bann C (1997) An economic analysis of tropical forest land use options, Ratanakiri Province, Cambodia, EEPSEA research report series. Economy and Environment Program for South East Asia, Singapore

CBNRM Learning Institute (2005) The development of community-based natural resource management (CBNRM) in Cambodia, Selected papers on concepts and experiences. CBNRM Learning Institute, Phnom Penh, Cambodia

Chea Sam Ang, Sokh Heng, Prak Marina (2005) Report, Benefit Sharing in Community Forestry and Poverty Alleviation in Cambodia. Forest Administration, Phnom Penh

ETAP (1998) Environment: concepts and issues – a focus on Cambodia. UNDP/ETAP Reference Guide Book, Phnom Penh, Cambodia

Hansen KK, Top N (2006) National forest benefits and economic analysis of natural forest conversion in Cambodia. Forest Administration, Phnom Penh, Cambodia

McKenney B, Prom T (2002) Natural resources and rural livelihood in Cambodia: a baseline assessment, Working paper no. 23. Cambodia Development Resource Institute, Phnom Penh

McKenny B, Prom T (2004) Focusing on Cambodia's high value forests: livelihoods and management, Cambodia Development Resource Institute and Wildlife Conservation Society, Special report. Cambodia Development Resource Institute, Phnom Penh

Chapter 13
Economic Benefit of Management Options for a Suburban Forest (Kho Hong Hill) in South Thailand

Saowalak Roongtawanreongsri, Prakart Sawangchote, Sara Bumrungsri, and Chaisri Suksaroj

Abstract Kho Hong Hill (KHH) is located close to Hat Yai City in Songkhla Province, South Thailand. Almost half of its forested area has been deforested as a consequence of population increase, urbanization and the ongoing conversion of forest areas into rubber plantations. This study assesses the net economic benefits of three forest resource management strategies based on information derived from interviews and discussions with relevant KHH stakeholders. The three strategies are represented as three scenarios: Scenario I is the base case or the business-as-usual (BAU) scenario; Scenario II involves the establishment of protected areas to preserve the remaining forest areas in KHH; and Scenario III assumes the implementation of forest restoration and rehabilitation initiatives designed to increase the forest areas on KHH. Six ecosystem services are selected as the environmental variables for the study: (a) the provision of timber, (b) carbon dioxide sequestration, (c) oxygen generation, (d) water supply, (e) flood control and (f) biodiversity. Market valuation is used to estimate the values of provision of timber, CO_2 sequestration, O_2 generation, water supply and flood control, while the value transfer approach is used to value of the service derived from KHH biodiversity. The results show that under Scenario I, the annual benefits from the ecosystem services from KHH would become negative after 15 years, whereas positive net present values would be yielded under Scenarios II and III. The study recommends that rubber farmers should be encouraged to convert their rubber plantations back to

---

S. Roongtawanreongsri (✉)
Environmental Economics Research Unit, Faculty of Environmental Management, Prince of Songkla University, Hat Yai, Songkhla 90112, Thailand
e-mail: saowalakroong@hotmail.com

P. Sawangchote • S. Bumrungsri
Department of Biology, Faculty of Science, Prince of Songkla University, Hat Yai, Songkhla, Thailand
e-mail: prakart.s@psu.ac.th; sara_psu@hotmail.com

C. Suksaroj
Department of Civil Engineering, Faculty of Engineering, Prince of Songkla University, Hat Yai, Songkhla, Thailand
e-mail: suksaroj@gmail.com

D. James, H.A. Francisco (eds.), *Cost-Benefit Studies of Natural Resource Management in Southeast Asia*, DOI 10.1007/978-981-287-393-4_13

forest areas. Introducing a payment for ecosystem services (PES), mechanism is also recommended to induce rubber farmers to reforest. Further studies would be needed, however, to establish an effective PES system.

Keywords Thailand • Forest • Economic benefit • Ecosystem services • Management options

Introduction

Ecosystem Services

Ecosystem services are commonly defined as benefits that people obtain from ecosystems (USDA 2010). Ecosystem services can also be the processes through which natural ecosystems, and the plants, animals and microbes that live in those environments, sustain human life. Ecosystems provide valuable services to humans through their natural functions. As habitats become fragmented, the services that natural systems provide become less effective (Tawna 2010). Climate change, pollution, over-exploitation and land-use change are some of the drivers of ecosystem loss, as well as resource challenges associated with globalization and urbanization (USDA 2010).

The concept of ecosystem services encompasses the delivery, provision, production, protection or maintenance of a set of goods and services that people perceive to be important (Chee 2004). It includes both tangible and intangible goods and services. Forest ecosystems, in particular, play an important role in providing the goods and Earth's life-supporting system. Tangible goods and services from forest include timber, non-timber forest products (NTFPs), water supply, medicine, tourism, biomass and nutrition. Intangible services from forest are providing clean air and oxygen, regulating the atmospheric abundance of carbon dioxide (Murray 2000; Stinson and Freedman 2000), assimilation of waste, recreation and aesthetic benefits, educational opportunities and spiritual enrichment (Tong et al. 2007). Losing or degrading them can cause significant harm to the nation's economy and the welfare of humans (USDA 2007).

Kho Hong Hill

Kho Hong Hill (KHH) is situated in Hat Yai District and Thung Yai District, Songkhla Province in the south of Thailand. It is the last forest area closest to Hat Yai City, the well-known urbanized area on the east coast of lower south Thailand. It lies north–south and is 5.6 km in length. The highest point is 371 m above sea level. The total area is 1,212.42 ha. Of this, 699.7 ha (57.71 %) remains as primary

Table 13.1 Area of forest and other encroachment on KHH

Type	Area (rai)	Area (ha)	Percentage
Forest	4,373.14	699.70	57.71
Rubber	2,757.35	441.18	36.39
Land with no cover	342.29	54.77	4.52
Buildings	104.82	16.77	1.38
Total	7,577.60	1,212.42	100.00

and secondary forest. Thirty six per cent of the total area has been converted to rubber plantation (see Table 13.1).

The Hill provides ecosystem services to a population of 157,682 living in the Hat Yai Municipality and 67,892 living within 3 km from KHH (Hat Yai Municipality 2010), not yet mentioning other off-site benefits. Previous studies show that the forest is still fertile with primary forest area intact. The ecosystem support system of KHH is recognized by biologists in the Prince of Songkla University (PSU) in terms of oxygen provider, CO_2 sequestration and a source of water supply for the downstream community. Moreover, the forest is home to many wild animals and plants, some of which are endangered or rare species, i.e. slow loris and Malaysian giant ant which can be found only in fertile areas, and a temporary home for the plain-pouched hornbill. Biodiversity is one of the greatest benefits from this forest.

KHH is an ever green forest with a mixture between dry forest and moist forest. Because this type of forest comprises of both deciduous and non-deciduous plants, organic matter can be generated throughout the year. Nutrient cycling and energy flows within the forest are thus at a maximum.

Threats to Ecosystem Services Provided by KHH

Like other threatened forest areas around the world, almost 40 % of KHH forest area has been deforested because of the pressure from increasing population, income and consumption to make way for rubber plantations or the construction of housing and urbanization because of its closeness to the city. Recently, the problem has become more serious. Flooding has occurred in areas that have never been flooded before, soil erosion is increasing, and there is prolonged drought during the dry season and a loss of wildlife, to name just a few examples. Because many of these services from forests are usually viewed as free benefits to society, or 'public goods', lacking a formal market, these natural assets are traditionally absent from society's balance sheet, and their critical contributions are often overlooked in public, corporate and individual decision-making. When forests are undervalued, they are increasingly susceptible to development pressures and conversion. Recognizing forest ecosystems as natural assets with economic and social value can help promote conservation and more responsible decision-making (USDA 2010).

The preliminary survey conducted with people living within three km from the centre of the hill revealed that people do not fully appreciate the ecological value of the forest. Compared with the tangible benefit of rubber plantation, many of the benefits of KHH are perceived to be intangible and immeasurable, thus having no meaning to them. To make matters worse, institutional arrangements establishing who has authority over the forest are weak. People who use the land in KHH claim that they have a customary right as their ancestors took over areas of land on the hill before any legal issues arose regarding encroachment of the forest. Some people have encroached the forest illegally due to the weakness of law enforcement.

Initiatives to Protect Ecosystems of KHH

Academics from Prince of Songkla University have realized the importance of KHH and the seriousness of the problem of deforestation. They have initiated a social movement to conserve KHH under the project 'Help Conserve Kho Hong Hill'. Various activities aimed at restoring and conserving the forest have already been undertaken. An ultimate goal of the project is to use KHH to sustain human well-being for the people in Hat Yai City and nearby areas. Accordingly, a best management option is being sought to provide the right direction on how to conserve it. To reverse the loss and degradation of ecosystem services, economic and financial motivations must include a conservation objective, and the value of ecosystem services needs to be incorporated into any decision-making (USDA 2010).

Research Objectives

The general objective of the research is to determine the value of net economic benefit of different management options for KHH in order that the involved parties will be informed of the most economically efficient option. The more specific objectives are:

- To determine economic value of some ecosystem services of KHH, namely, timber, freshwater supply, CO_2 sequestration, flood protection, O_2 provider and biodiversity
- To examine three likely scenarios (management options) which are possible for KHH, including the status quo
- To conduct a cost-benefit analysis of different management options for KHH, taking into account those ecosystem service values in objective 1
- To compare management options to see which option yields the maximum benefit to society
- To use this information in preparing for policy analysis and implication for Kho Hong Municipality, Hat Yai Municipality and Thung Yai Tambon Authority Organization

Methodology

Incorporating Environmental Values in Policy Analysis

It is now widely accepted that in order to make sound decisions in favour of sustainable resource use, the total economic value of natural resources and environment should be included in policy analysis. As pointed out by Ranganathan et al. (2008), 'Decision- makers - including those whose goals and actions might not at first seem connected to ecosystems - need to examine the dependence and impacts of their goals on ecosystem services. Making decisions for policy implementation by taking ecosystem services into account can strengthen decisions.'

The key principle is that any producer surpluses and/or consumer surpluses that are predicted to occur over time under the 'without' or baseline case but are reduced or foregone under the 'with' alternative policy scenario case are considered an economic cost, while increased or new producer and consumer surpluses generated from the 'with' alternative policy case are considered an economic benefit. One means of doing this is to carry out an economic cost-benefit analysis (CBA). As is well recognized in the professional literature, provided the discounted incremental economic benefits of a proposed policy exceed the discounted incremental economic costs, then the proposed policy is deemed to provide a net benefit to the community and an improvement in the economic efficiency of resource management, relative to a base case scenario.

Categorizing Ecosystem Services

The concept of ecosystem services was comprehensively documented by the Millennium Ecosystem Assessment (MA), which set out between 2001 and 2005 to assess the consequences of ecosystem change for human well-being and to establish the scientific basis for actions needed to enhance the conservation and sustainable use of ecosystems and their contributions to human well-being. Categorization of ecosystem services has been attempted by numerous authors and organizations including Olewiler (2004), MA (2005a, b), Anielski and Wilson (2005), Ranganathan et al. (2008) and Stenger et al. (2009). A widely accepted categorization is the following (MA 2005a; Irwin and Ranganathan 2007; USDA 2010):

- *Provisioning services* or the provision of food, fresh water, fuel, fibre, genetic resources, natural medicine, pharmaceuticals and other goods
- *Regulating services* such as air quality regulation, climate regulation, water regulation, erosion regulation, water purification and water treatment, disease regulation, pest regulation as well as pollination and natural hazard regulation
- *Supporting services* such as soil formation, photosynthesis and nutrient cycling

- *Cultural services* such as educational, aesthetic, cultural heritage and spiritual values as well as recreation and tourism

Valuing Ecosystem Services

The total economic value of the environment comprises the direct use value plus the indirect use value plus the non-use value of the environment. The challenge of valuing the environment often lies within non-market valuation, particularly indirect economic benefits of environment and non-use value (Brown et al. 2007). Nonmarketed benefits are often large and sometimes more valuable than the marketed benefits (MA 2005b). Valuation is necessary if markets and institutions are to be established to promote the efficient and sustainable use of ecosystem services (Abel et al. 2003). Such valuation is also required to incorporate environmental values within a CBA. There are several valuation techniques to estimate ecosystem service values (Farber et al. 2002; Brown et al. 2007) and numerous studies on valuing ecosystem services. Selected examples involving the valuation of ecosystem services in general are Curtis (2004), Olewiler (2004), Anielski and Wilson (2005), Tianhong et al. (2008) and Tong et al. (2007).

Valuing Forest-Based Ecosystem Services

Examples also exist for valuing forest-related ecosystem services. They include a study estimating the annual economic value of certain ecosystem services (water conservation, soil conservation and gas regulation) by forest ecosystems in the Xingshan County in China, using simulation models and geographic information system (Guo et al. 2001); valuations by the Millennium Ecosystem Assessment (2005b); and a study of forest benefits such as timber, hunting, CO_2 fixation and non-use and recreation benefits by Moons (2002). Jim and Chen (2009) reviewed studies on the major ecosystem services provided by urban forests in China, including microclimatic amelioration (mainly evapotranspiration-cooling effects), carbon dioxide sequestration, oxygen generation, removal of gaseous and particulate pollutants, recreational and amenity. Kuosmanen and Kortelainen (2007) proposed a new approach for environmental valuation within an environmental CBA framework that is based on data envelopment analysis (DEA) and does not require any monetary estimation for environmental impacts using traditional revealed or stated preference methods.

Table 13.2 Valuation techniques for ecosystem service accounting of KHH

KHH ecosystem services	Valuation techniques application
Provisioning services	
Timber	Market approach
Freshwater supply	Market approach
Regulating services	
CO_2 sequestration	Market approach/replacement cost approach
Oxygen provider	Market approach/replacement cost approach
Flood control	Market approach
Supporting services	
Habitat/biodiversity	Value transfer

Estimating Monetary Value of KHH Ecosystem Services

For the present study, six ecosystem services were evaluated: *provisioning service* of timber; *regulating services* of C stocks and CO_2 sequestration, O_2 production, flood prevention and water supply; and the *supporting service* of biodiversity. The services are those relating to KHH natural forest as well as to its competing use: rubber plantation. Where possible, primary data from field studies was used to estimate the values of the ecosystem services provided by KHH. Secondary data was obtained from a literature review of documents, research reports, journal papers, data collected by governmental offices, statistical data, case studies and so forth. Table 13.2 presents a list of the valuation techniques that were used to estimate monetary values of the six ecosystem services selected for the study of KHH.

Timber Value

This value was estimated using data from another research project conducted by the authors. The gross value of timber was calculated by multiplying the volume of timber by its price. Cost of harvesting was deducted from this value to estimate the net benefit of this service. This timber value was entered in the CBA only in the year during which forest is converted to rubber plantation. NTFPs were not included in the analysis. Total timber value was estimated at 632,974 Bt/ha.

Carbon Dioxide Sequestration

There are several ways to assess the economic value of forest CO_2 sequestration. Guo et al. (2001) suggested three ways, comprising (1) a formula for photosynthesis and respiration, (2) test and survey and (3) a mathematical model. Guitart and Rodriguez (2010) evaluated carbon fixing value using alternative compensation value for silvicultural regimes. Whenever trees are cut, CO_2 kept in the trees or carbon accumulated in the forest in terms of wood, leaves, roots and soil is released back into the atmosphere. This loss of carbon becomes a cost of deforestation. Conversely, a benefit is derived when forest trees are kept intact: they can sequester existing CO_2 from the atmosphere. This is true also for rubber farms which can sequester CO_2 though at a much lower rate.

Anielski and Wilson (2005) suggest methods of valuing carbon store in terms of the damage cost of climate change, carbon fee, replacement cost, carbon credit trading and the cost of timber income foregone in lieu of protecting the carbon stored in the forest ecosystem. In this study, the authors applied a method devised by Yolasiğmaz and Keleş (2009), using the carbon credit trading price to calculate the value of carbon fixing. This approach readily communicates the value to stakeholders. The sequestration rate was obtained by field experiment. The price of carbon was multiplied by the amount of carbon fixed by KHH to obtain an estimate of the value for KHH. The results are shown in Table 13.3.

Oxygen Generation

Similar to the estimation of CO_2 sequestration value, oxygen provision by forest can be determined by the quantity of oxygen provided by the forest multiplied by the price of sold oxygen or the replacement cost of generating oxygen. Studies that have attempted to value oxygen provision services include Xi (2009), Zhang et al. (2009), Jim and Chen (2009), Tong et al. (2007) and Guo et al. (2001).

The annual oxygen generation rate was obtained by taking the CO_2 absorption rate previously calculated and multiplying by 1.2, following the method adopted by Yolasiğmaz and Keleş (2009). The price of oxygen was then multiplied to the amount of oxygen generated by KHH. The outcome of these calculations is presented in Table 13.4.

Water Supply

KHH is a head watershed for the Khlong Rian Canal below the catchment, which then flows to Songkhla Lake Basin. It supplies water to Hat Yai City, servicing in particular the Prince of Songkhla University and Senanarong Military Base. The authors identified 11 reservoirs that receive water from KHH, although only six of them can really be considered to receive their full capacity of water from KHH forest. Without water provided by KHH, the downstream communities would need to find substitute sources for their water supply, such as connecting their inlet pipes to the main pipeline of the municipal waterworks authority. The municipality would itself have to expand the operation of its water supply distribution to cover these communities, involving both capital and operating and maintenance costs.

In order to calculate the amount of water provided by KHH, a Simple Hydrologic System Model was used. A continuity equation was applied to estimate water

Table 13.3 Value of carbon storage service by KHH

Types of ecosystem	Carbon capacity		Carbon price ($/tCO_2$)	Value	
	Carbon storage (ton/ha)	Annual carbon sequestration (ton/ha/year)		Carbon storage (Bt/ha)	Annual carbon sequestration (Bt/ha/year)
Primary forest	218.55	17.01	12.30 Euro or 524.32 Bt	114,595	8,919
Secondary forest	167.97	31.13		88,068	16,317
Average of forest	**193.26**	**24.07**		**101,331**	**12,618**
Rubber plantation	**100.79**	**7.12**		**52,847**	**3,738**

Exchange rate: 1 Euro = 42.6276 Baht

Table 13.4 Oxygen production capacity of KHH from field study

Type of ecosystem	Oxygen production (t/ha/year)	Price of oxygen (Bt/ton O_2)	Value (Bt/ha/year)
Primary forest	39.1145	US$74.31 or 2,328.11 Bt	91,063
Secondary forest	54.1101		125,975
Average of forest	*46.6123*		**108,519**
Rubber plantation	13.3934		**31,181**

flow. The value of this ecosystem service from KHH was assessed by multiplying the modelled volume of water held and released by KHH by the price of water per unit. Unlike CO_2 sequestration and oxygen generation, this ecosystem service was not considered relevant for rubber plantations, because it is known that rubber trees do not hold water well since their roots are shallow. It is often observed that rubber trees fall down when there is runoff water.

Flood Control

Large amount of water is being held by the vegetation cover on KHH. For the past 3 years with KHH being illegally converted to other land uses which have less plant cover area, the flash flood incidence also arises. Losing KHH forest will put pressure on flooding area. Volume, area flooded and flood depth in each scenario are simulated using the hydrological modelling. The water balance study, carried out by our engineering team, indicates how much volume and depth of water will be flooded in which area. The simple model figuring the relationship between forest area, flood depth and flooding area was simulated. Losing more forest means getting more of runoff water which impact the communities with floods. The damage cost corresponding with each change of forest area is then estimated.

The study of flood control benefits of KHH was conducted by the university engineering team. Different scenarios were modelled in terms of flooded areas (which can be shown on a GIS map), flood depth and volume of flood. On GIS map, the flooding area with the same flood depth is categorized into one zone. This enabled flood zones to be identified and mapped. For each flood zone, the number of households, business properties and agricultural land was determined. Average damage cost was then surveyed in these particular areas, according to its flood depth, based on primary data from 100 samples and damage cost estimates for different land-use types. The cost of flood for households includes damaged properties and household items, foregone income resulting from not working due to flooding, cost of illness due to flooding and cost of cleaning after flood. The survey data indicated, as expected, that the average damage cost increases as flood depth increases.

The damage costs for other land-use types are not included in the analysis, such as damage costs relating to infrastructure, governmental offices, educational institution buildings and agricultural production areas. This means the net benefits of flood control provided by KHH or any management regime would be higher than the value incorporated in the CBA.

Biodiversity

Due to limited time, the valuation of benefits from biodiversity could not be performed. In this study, the unit value transfer approach is used. The adjusted benefit estimate B_{p} at the policy site can be calculated as

$$B_{\mathrm{p}} = B_{\mathrm{s}}\left(\frac{Y_{\mathrm{p}}}{Y_{\mathrm{s}}}\right)\beta$$

where B_{s} = the original benefit estimate from the study site
Y_{s} and Y_{p} = the income levels at the study and policy site, respectively
β = the income elasticity of demand for the environmental good in question

Surveys and studies of biodiversity by researchers in the Department of Biology, Faculty of Science, Prince of Songkla University, have been ongoing since 1984. One particular study, carried out during 1984–1986, found 637 species of plants in 130 families, of which 19 species were ferns, 90 species were dicotyledon and 19 species were cotyledon. Due to limited time, the value of biodiversity of KHH could not be undertaken by survey. The benefit transfer method was therefore applied to value this ecosystem service. A literature review revealed that only a few studies have been done on valuing biodiversity in Thailand. More studies have been done in other countries, but according to Navrud and Bergland (2001), the simple unit value transfer approach should not be used for transfers between countries with different income levels and costs of living. This would be particularly true in the case of KHH.

Table 13.5 Value of biodiversity of KHH

Scale of beneficiaries	Average WTP (Bt/person/year)	Total of population	Value (Bt/ha/year)
Hat Yai district only	159	374,891	111,606
Three districts adjacent to KHH		560,336	**163,837**
Songkhla province		1,357,023	326,251

Note: This value is considered underestimate.

Various Thai studies were reviewed as a basis for benefit transfer, including the study of Huai Kha Khaeng (Krasuaythong 2000), Thung Yai Naresuan (Naruchaikusol 2002) and Phu Khieo Wildlife Sanctuaries (Wiwatthanapornchai 2001). They are well known and were all surveyed at the national level. However, the characteristics of those forests and their biodiversity are much different from KHH, so they may give an overestimated benefit for KHH. A contingent valuation study of Pa Krad, a tropical forest in Nathawee District near Hat Yai, was considered more appropriate. The willingness to pay to conserve Pa Krad was estimated at 128 Baht/person/year using 2001 values. Allowing for inflation, the relevant estimate for 2011 was 159 Baht/person/year. This value was transferred to three main groups of beneficiaries: only Hat Yai people, people in three districts around KHH and the population of the whole province. The results are shown in Table 13.5.

Summary of Ecosystem Service Values for KHH

The benefits associated with all six ecosystem services selected for KHH are shown in Table 13.6, and their corresponding estimated values are shown in Table 13.7. These values are used in the CBA of possible management scenarios for KHH, over the next 25 years. Table 13.7 Indicates that the value of water absorption is higher than for the other services, with the exception of timber value. However, the table does not reveal the fact that reducing the damage cost from flooding constitutes the greatest benefit from KHH. If KHH forest is lost, the communities around the area will be affected by worsening floods resulting from excessive runoff. The communities will also incur a reduction in the supply of water for daily consumption, since less water would be made available as the area of forest decreases. Biodiversity value is the next largest value, even though the figure is underestimated, as explained earlier. The oxygen generation value is higher than CO_2 sequestration value due to the higher price of substituted oxygen. All ecosystem service values associated with natural forest are higher than those for rubber plantations.

Table 13.6 Benefits from ecosystem services as basis for CBA

Ecosystem service benefit	Benefit or cost corresponding the type of land use on KHH[1]	
	Forest	Rubber
Timber	–	Benefit from selling forest wood[2]
		Release of CO_2 storage from logged tree and soil[2]
Water absorption	Benefit of water absorbed that will be used throughout the year by the people downstream	Reduced benefit of water absorbed by forest
CO_2 sequestration	Benefit from CO_2 being sequestered at the rate of forest sequestration	Benefit from CO_2 being sequestered by rubber plants
		Release of CO_2 when the rubber plants reach their maximum age and needed to be cleared
Flood prevention	Damage cost of flooding	Damage cost of flooding when forest is converted to rubber
O_2 provider	Benefit from O_2 providing by forest trees	Benefit from O_2 providing by rubber trees
Biodiversity	Benefit of forest biodiversity	–

[1]Benefit and cost is considered per area of land use
[2]Only when the forest is cut down to make rubber farms

Table 13.7 Summary of ecosystem service values of KHH

Ecosystem services	Value at current stage (Bt/ha/)	Explanation
From forest in tact		
Timber, Bt/ha/year	632,974	Will enter into CBA only when the forest is cut
CO_2 sequestration, Bt/ha/year	12,618	Annual benefit
CO_2 storage, Bt/ha	47,983	Will enter into CBA only when the forest is cut
Oxygen generation, Bt/ha/year	108,518	Annual benefit
Water absorption by forest, Bt for total area of forest in that year	118,954,045	This value is at the first year
Flood prevention, Bt for total area of forest in that year	–	Different in each year in each scenario
Biodiversity	163,836	Annual benefit
From rubber plantation		
CO_2 sequestration	3,738	Annual benefit
Oxygen generation	31,181	Annual benefit

Note that rubber farms are reported not to contain significant volume of water or biodiversity, therefore are not accounted hereby

Management Options for KHH

Constructing the Management Scenarios

Information for constructing management options was obtained from face-to-face interviews with stakeholders, focus group discussions and questionnaires. In all, eight focus group discussions were held with representatives of local people living in the forest, local authorities associated with the forest and local academic experts. The number of total representatives in each group was 10, following the number of between 6 and 10 suggested by Bryman (2001). Issues were considered to be important in constructing the management options which included the decline in wildlife species, food sources from the forest, land entitlement, water use and cultural values. From the people's perspectives, KHH is being destroyed by deforestation, illegal encroachment, converting forest to rubber plantations or orchards, buildings and building for tourist attraction.

Three Management Scenarios

The outcome of the consultation process with stakeholders and other interested parties resulted in the following three broad options for managing KHH:

- Base case scenario where the change in land use continues at their usual rate.
- Declaring KHH as a protected area.
- Rehabilitate the forest to its primary stage, with or without payment for ecosystem services to owners who practice agroforestry or replanting forest trees in rubber farms or keeping trees not rubbers for carbon market.

The planning horizon is assumed to be 25 years, equivalent to the expected life of rubber trees from the time of planting.

Scenario I: Base Case Scenario

The first scenario is the base case, where there is no intervention in the existing condition or no new management actions taken. Most stakeholders agreed that if there were no intervention with the current trend in land use, it is assumed that forest will continue to be cut at the current rate of 16.68 ha/year and the rate of forest conversion to rubber plantation is 14.30 ha/year.

Ecosystem services from forest area are therefore expected to be lower the next year than the current year. These include CO_2 sequestration value, O_2 generation value and biodiversity value. However, since rubber farms will be increasing, they also give ecosystem services, although not as much as natural forest. The services from rubber farms considered in the study are in accordance with forest services, i.e. CO_2 sequestration value and O_2 generation value. Biodiversity of rubber farms is far less than forest and thus not usually considered as ecosystem function. The same is true for water retention in rubber plantations.

Assumptions of this scenario include:

- Rate of forest reduction and rubber plantation is linear using current data available.
- Timber benefit is considered only when it is cut.
- Rubber product starts at year 8. Ecosystem services from rubber comprise of CO_2 sequestration and O_2 generation at rubber's rate.
- Current rubber farms are at the average age of 21 years.
- Released CO_2 from harvested forest include CO_2 stored in wood, poles, litter, shrubs and soil, where the CO_2 from wood is assumed to be released only 40 % of carbon storage in wood.

Benefits in the base case include revenue from sales of latex and the value of timber when the forest is cleared to make way for rubber plantations. Costs in the scenario include the benefits foregone from lost services of forest and the cost of rubber farming. Another cost is the value of carbon released from timber when it is harvested. This is assumed to be 40 % of the total carbon storage in wood as most of wood if used for furniture can still store carbon within. The rest of carbon storage (in litter, floor plants, poles and soil) is counted as carbon released back to atmosphere which therefore raises the level of CO_2 in the atmosphere. Table 13.8 shows cost and benefit items appearing in the CBA.

Scenario II: Establishing a Protected Area (PA)

This scenario is possible, as the prospect is currently being explored of including this area as a part of the greater Songkhla Lake Basin Protected Area. There have been some discussions with the Office of Natural Resources and Environmental Policy and Planning (ONEP) and with the movement to bring KHH into the agenda of PA establishment. With this option, all current land-use activities on KHH will have to cease, and no further conversion of forest to other land use will be permitted. Thus, the area of 666.35 ha of forest and 469.78 of rubber farms will remain. The existing condition of the forest will remain, and any further environmental deterioration will be prevented.

Benefits are mainly those yielded by ecosystem services to the community. The cost of this management option includes the cost of establishing the management regime and the budget of local authorities who have the responsibility of protecting and managing this area after being declared protected. Rubber farming costs also have to be included as the current area of rubber farm will still continue production. The relevant information for Scenario II is shown in Table 13.9.

Table 13.8 Scenario I: base case

Items	Year 1	Year 2	Year 3	...	Year 24	Year 25
Area of forest	666.35	649.67	632.99	...	282.74	266.06
Area of rubber plantation	469.78	484.09	498.39	...	798.78	813.08
Cost						
Management cost	0	0	0	...	0	0
Cost of rubber farming	Cost or rubber farming at the age of rubber farms per ha per year					
Ecosystem service losses (benefit foregone from losing forest)						
Loss of annual CO_2 sequestration capacity	The annual CO_2 sequestration of forest per ha × area of forest cut × price of CO_2					
Release of CO_2 from timber harvested into atmosphere	The CO_2 storage in wood and other parts of forest per ha × area of forest cut × price of CO_2					
Loss of O_2 provider	O_2 supplied per ha forest × area of cut forest × price of O_2					
Water supply	Amount of total forecasted lost water in each year × price of water					
Foregone benefit from biodiversity	Foregone benefit of biodiversity per unit area of forest × area of cut forest					
Damage cost of flooding	Total forecasted flood water of this scenario × damage cost					
Benefit						
Benefit from changing forest to rubber						
Harvested timber value	Timber value × area of cut forest					
Rubber latex	For present rubber plantation = net revenue of rubber sold. For newly planted, net revenue will start around year 7 after planting					
Benefit from intact forest						
Value of annual CO_2 sequestration	Amount of CO_2 sequestration by forest per ha × area of forest intact × price of CO_2					
Value of O_2 provider	Amount of O_2 provided per ha of forest × area of forest intact price of O_2					
Value of water supply	Amount of water supplied by forest × price of water					
Value of biodiversity	Value of biodiversity per unit of forest area × area of forest					
Benefit from intact rubber						
CO_2 sequestration	Amount of CO_2 sequestration by rubber farm per ha × area of rubber farm × price of CO_2					
O_2 provider	O_2 supplied by rubber × area of rubber farm × price of O_2					

Assumptions of this scenario include:

- PA saves the land use as it is at present. No further reduction of forest will be possible.
- Costs of establishing PA involves the PA preparation cost (mostly cost of study) and monitoring cost. The preparation cost is derived from other studies of similar nature in the area. The monitoring cost is assumed to be 10 % of study cost each year.

Table 13.9 Scenario II: declaration of KHH as a protected area (PA)

Items	Year 1	Year 2	Year 3	. . .	Year 24	Year 25
Area of forest	666.35	666.35	666.35	. . .	666.35	666.35
Area of rubber plantation	469.78	469.78	469.78	. . .	469.78	469.78
Cost						
Management cost	Establishment cost	Monitoring cost	Monitoring cost	. . .	Monitoring cost	Monitoring cost
Cost of rubber farming	Cost or rubber farming at the age of rubber farms per ha per year					
Damage from flood	Total forecasted flood water of this scenario × damage cost					
Benefit						
Benefit from intact forest						
Value of annual CO_2 sequestration	Amount of CO_2 sequestration by forest per ha × area of forest intact × price of CO_2					
Value of O_2 provider	Amount of O_2 provided per ha of forest × area of forest intact price of O_2					
Value of water supply	Amount of water supplied by forest × price of water					
Value of biodiversity	Value of biodiversity per unit of forest area × area of forest					
Benefit from intact rubber						
Rubber latex	For present rubber plantation = net revenue of rubber sold.					
CO_2 sequestration	Amount of CO_2 sequestration by rubber farm per ha × area of rubber farm × price of CO_2					
O_2 provider	O_2 supplied by rubber × area of rubber farm × price of O_2					

Scenario III: Rehabilitating Forest from Rubber Plantation Area

This option involves introducing a management regime that converts rubber plantations back to native forest and restores the forest to original condition. The Help Conserve KHH Project considers that this may be the most effective option for managing the area. Two cases are considered for converting rubber plantations back to forest: Case A which involves rubber farmers voluntarily handing back areas under rubber production, given the fact that most of the area dedicated to rubber was illegally occupied, and Case B which involves establishing a special fund to buy back the rubber plantation area and rehabilitate it to conservation area. Under this scenario, the area of natural forest will finally reach 100 %.

Assumptions of this scenario include:

- Forest can be fully restored to primary forest after 40 years.
- The growth of carbon sequestration rate from restored rubber farms is 7.3687 tC/ha/year. Oxygen generation is based on this figure multiplied by 1.2 as in other cases.
- The rate of restoration is considered in two cases. The first case assumed the forest is restored at the rate of reduction, namely, 16.68 ha/year. However, if there is some intervention such as purchasing rubber farms to be converted back to forest, it is assumed that 10 % of the rubber production area will be restored per annum.
- Land purchase prices are considered at two levels: full land price (2,500,000 Bt/ha) and half of land price. These represent simply a monetary incentive for rubber growers to relinquish their holdings, rather than real opportunity costs.
- With this management, rubber farms do not lose all benefit from rubber latex, but only part of it. This is because rubber latex can still be collected, but at a lower rate, from rubber trees that are abandoned. However, in the CBA, this benefit is ignored.

Costs in this scenario comprise the restoration expenses for nursery establishment, labour costs and seedlings costs. Since Case A is voluntary, there is no additional cost. In Case B, where the restoration rate is expected to be high, a market mechanism may be needed to accelerate the conversion. Rubber farmers may be paid either full land price or half to relinquish their present use. Details of Scenario III are shown in Table 13.10.

Cost-Benefit Analysis of Management Options

Conducting the Cost-Benefit Analysis (CBA)

All costs and benefits were entered into a spreadsheet model to conduct the BCA and calculate the net present value (NPV) of each management option relative to the base case scenario. The time horizon adopted for all three scenarios is 25 years, matching the cycle of rubber farms. Discount rates ranging from 1 to 7 % are assumed.

Table 13.10 Scenario III: full restoration of area to natural forest

Items	Year 1	Year 2	Year 3	...	Year 24	Year 25
Area of forest	666.35	713.32	760.30	...	1,157.50	1,157.50
Area of rubber plantation	469.78	422.81	375.83	...	0	0
Cost						
Purchasing land	Land price × area of land restored each year (only in Scenario III Case B)					
Cost of restoration	Cost of nursery, seedlings and labour cost for restoration					
Cost of rubber farming	Cost or rubber farming at the age of rubber farms per ha per year					
Ecosystem service losses (benefit foregone from losing forest)						
Damage cost of flooding	Total forecasted flood water of this scenario × damage cost					
Benefit						
Benefit from intact forest						
Value of annual CO_2 sequestration	Amount of CO_2 sequestration by forest per ha × area of forest intact × price of CO_2					
Value of O_2 provider	Amount of O_2 provided per ha of forest × area of forest intact price of O_2					
Value of water supply	Amount of water supplied by forest × Price of water					
Value of biodiversity	Value of biodiversity per unit of forest area × area of forest					
Benefit from intact rubber						
CO_2 sequestration	Amount of CO_2 sequestration by rubber farm per ha × area of rubber farm × price of CO_2					
O_2 provider	O_2 supplied by rubber × area of rubber farm × price of O_2					
Rubber latex	For present rubber plantation = net revenue of rubber sold					

Results for Scenario I

The results for Scenario I are presented in Table 13.11. Although rubber farmers may benefit from large financial benefits, the CBA demonstrates that in the base case, society as a whole itself does not benefit. NPV increases with higher discount rates, indicating that the net benefit is higher in earlier years compared with later years. Indeed, the annual net benefit begins to turn negative from year 15 onwards. This result demonstrates that it would be desirable in terms of the economic welfare of people around KHH to preserve the ecosystem service benefits provided by forest on KHH.

Results for Scenario II

Table 13.12 contains the results for Scenario II. When the current land use is maintained, the NPVs are positive for all discount rates and the BCAs all exceed 2. This means that establishing KHH forest as a protected area and allowing no further land-use change, society would receive significant net benefits for the next 25 years.

Table 13.11 Results of CBA of scenario I

Discount rate (%)	Net present value	Benefit cost ratio	Annual equivalent benefit
1	60,784,334	1.00	607,842
2	130,528,481	1.02	2,610,568
3	183,935,062	1.03	5,518,051
4	224,531,610	1.05	8,981,263
5	255,071,853	1.06	12,753,592
6	277,710,718	1.08	16,662,642
7	294,138,461	1.09	20,589,691

Table 13.12 Results of CBA of scenario II

Discount rate (%)	Net present value	Benefit cost ratio	Annual equivalent benefit
1	4,548,691,971	2.880	45,486,919
2	4,023,692,477	2.878	80,473,849
3	3,581,714,446	2.874	107,451,432
4	3,207,653,492	2.870	128,306,139
5	2,889,404,643	2.867	144,470,231
6	2,617,227,284	2.864	157,033,636
7	2,383,251,622	2.861	166,827,612

Results for Scenario III

The results for Scenario III are shown in Table 13.13. They indicate that positive NPVs are obtained at all discount rates, for both Cases A and B. This shows that the benefit we receive from restoring the forest is exceedingly greater than the cost. The magnitude of NPVs in voluntary case (Case A) is lower than the payment options (both Case Bs). This is because the rate of conservation is lower in Case A than in Case B; thus, the benefits generated from forest are therefore gained more slowly than in Case B. Or the faster we convert area back to forest, the higher benefit we will receive. When compared with the same faster rate of restoration, the cost of payment affects the NPVs. Naturally, the higher we pay for the land price, the lesser benefit we will receive. The details of NPVs are shown in Table 13.13.

Comparison of the Three Scenarios

The results indicate that the NPVs for Scenario III are the highest, in particular Scenario III Case B, where there is a restoration scheme at the rate of 10 % per annum, and payment is made at half the price of land. Scenario II has the next highest NPVs. The lowest NPVs resulted under Scenario I.

Table 13.13 Results of CBA of scenario III

Discount rate (%)	Net present value	Benefit cost ratio	Annual equivalent benefit
Case A			
1	5,666,960,203	3.52	56,669,601
2	4,975,243,925	3.48	99,504,877
3	4,395,466,924	3.45	131,864,007
4	3,907,004,334	3.42	156,280,172
5	3,493,360,286	3.39	174,668,013
6	3,141,288,730	3.36	188,477,323
7	2,840,111,000	3.33	198,807,769
Case B purchasing at full price			
1	5,029,490,834	2.07	50,294,907
2	4,379,783,815	2.05	87,595,675
3	3,835,386,731	2.03	115,061,601
4	3,377,040,814	2.01	135,081,631
5	2,989,311,362	1.98	149,465,567
6	2,659,779,319	1.96	159,586,758
7	2,378,412,682	1.94	166,488,887
Case B purchasing at half price			
1	6,322,756,976	2.86	63,227,569
2	5,526,260,077	2.83	110,525,200
3	4,857,939,295	2.79	145,738,178
4	4,294,416,476	2.76	171,776,658
5	3,816,950,306	2.73	190,847,514
6	3,410,456,557	2.70	204,627,392
7	3,062,746,242	2.67	214,392,236

Conclusions and Recommendations

Research Conclusions

This study has shown that the net present values (NPVs) from the analysis prove that it is highly undesirable on economic grounds to continue losing KHH forest at its current rate. Although there are private benefits for rubber farmers, for society as a whole it is not worth converting natural forest to rubber plantations. Further loss of forest area should therefore be prevented. The analysis shows positive NPVs if the current areas of forest and rubber plantation are frozen, meaning that it is better to stop converting forest to rubber farms now. However, the best option is to convert existing rubber plantation areas back to natural forest. The main benefits arising from this management option take the form of increased value of ecosystem services.

It is worth noting that the benefits associated with ecosystem services evaluated in this study are only a part of the total ecosystem services of KHH. For example, the services such as tourism, preventing soil erosion, soil fertility, pollination and microclimate regulation are not accounted for. These services are more difficult to evaluate because of their complex nature and thus require longer study time, closer observation, more physical data and a fuller understanding of the system.

Policy Recommendations

Table 13.14 compares the advantages and disadvantages of each management scheme. It is clear that to leave things as they are at present without any intervention will be detrimental to society as a whole, especially after year 15. To implement the

Table 13.14 Advantages and disadvantages of the three management schemes

			Scenario III	
Criteria	Scenario I: doing nothing	Scenario II protected area	Restoration Case A	Restoration Case B
Economic feasibility	Positive NPV with annual net benefit turns negative from year 15 onwards	Positive NPV	Positive NPV	Positive NPV
Implementation cost (Present value at 7 % discount rate)	None	3,077,137	6,611,929	1,387,236,025 (Case B1) and 702,902,465 (Case B2)
Impact on communities/ acceptance	Rubber farmers have no impact, but they are illegal farmers	Current rubber farmers have no impact but it may create a problem of fairness with other prospect rubber farmers	Current rubber farmers lose part of their income as the rate of latex collection is not in full as before	Current rubber farmers have no impact as they are compensated
	Other people, e.g. people in Hat Yai, PSU people, media and some businesses enterprise, may not accept this	Other communities can accept this alternative		
Political acceptance	Governmental officers are in the position of taking action to deal with illegal encroachment	Local politicians may lose their political support from rubber farm owners	Local politicians may lose their political support from rubber farm owners	Local politicians may not lose the support as rubber farmers are compensated
Management possibility	Nothing has to be done	Possible and not so complicated once the area has been proclaimed	It is quite uncertain who will voluntary join the restoration	May be complicated as KHH lands are not legally owned and it raises some concerns about justice and fairness
Enforceability	Encroachment is against laws, but lack of enforcement makes it possible	Enforcement should be constant and strong for those who violate the laws	No enforcement needed since it is voluntary	No enforcement needed because the management is based on cooperation

most preferred option – Scenario III – several actions could be taken. The first would be to declare KHH as a protected area. The next phase is to improve the understanding of stakeholders, especially land owners with illegal land holdings, regarding land use on KHH. Raising awareness and implementing programmes of environmental education regarding the importance and value of KHH can assist in encouraging people to appreciate the value of KHH. The Help Conserve Kho Hong Hill Project has been doing that and is gaining more support from the wider community.

Attempting to reduce rubber plantation area and restoring natural forest is more challenging, yet this is the ultimate goal of the Help Conserve Kho Hong Hill Project. It would be desirable to achieve this goal through voluntary action by rubber farmers, but more likely, payments would have to be made to encourage them to relinquish their land holdings. If such a policy were to be announced, would it perhaps result in further encroachment in the short run in order to be paid later? Where might the funding come from to make such payments? Action would also be required to prevent people from illegally harvesting timber from the forest. A careful study of appropriate incentives is clearly warranted to achieve these outcomes, but that is beyond the scope of the present study.

References

Abel N, Cork S, Gorddard R, Langridge J, Langston A, Plant R, Proctor W, Ryan P, Shelton D, Walker B, Yialeloglou Mandy M (2003) Natural values. CSIRO, Canberra, Australia

Anielski M, Wilson S (2005) Counting Canada's natural capital: assessing the real value of Canada's boreal ecosystems. Canadian Boreal Initiative and Pembina Institute, Ottawa

Brown TC, Bergstrom JC, Loomis JB (2007) Defining, valuing and providing ecosystem goods and services. Nat Resour J 47:329–376

Bryman A (2001) Social research methods. Oxford University Press, Oxford

Chee YE (2004) An ecological perspective on the valuation of ecosystem services. Biol Conserv 120:549–565

Curtis IA (2004) Valuing ecosystem goods and services: a new approach using a surrogate market and the combination of a multiple criteria analysis and a Delphi panel to assign weights to the attributes. Ecol Econ 50:163–194

Farber SC, Costanza R, Wilson MA (2002) Economic and ecological concepts for valuing ecosystem services. Ecol Econ 41:375–392

Guitart B, Rodriguez LCE (2010) Private valuation of carbon sequestration in forest plantations. Ecol Econ 69:451–458

Guo Z, Xiao X, Gan Y, Zheng Y (2001) Ecosystem functions, services and their values – a case study in Xingshan County of China. Ecol Econ 38:141–154

Hat Yai Municipality (2010) http://www.hatyaicity.go.th/index.php

Irwin F, Ranganathan J (2007) Restoring nature's capital: an action agenda to sustain ecosystem services. World Resources Institute, Washington, DC

Jim CY, Chen Wendy W (2009) Ecosystem services and valuation of urban forests in China. Cities 26:187–194

Krasuaythong T (2000) A non-use value assessment of natural resources and environment: a case study of Huai Kha Khaeng Wildlife Sanctuary. Master of Science Thesis in Agricultural Economics, Faculty of Economics, Kasetsart University

Kuosmanen T, Kortelainen M (2007) Valuing environmental factors in cost-benefit analysis using data envelopment analysis. Ecol Econ 62(1):56–65

Millennium Ecosystem Assessment (MA) (2005a) Ecosystems and human well-being: biodiversity synthesis. World Resources Institute, Washington, DC

Millennium Ecosystem Assessment (MA) (2005b) Ecosystems and human well-being: a synthesis. Island Press, Washington, DC

Moons E (2002) Cost- benefit analysis of the location of new forest land, Working paper series no. 2002–05. Faculty of Economics and Applied Economic Sciences, Center for Economic Studies, Energy, Transport and Environment, Leuven, Belgium

Murray BC (2000) Carbon values, reforestation, and 'perverse' incentives under the Kyoto Protocol: an empirical analysis. Research Triangle Institute, Center for Economics Research, Research Triangle Park

Naruchaikusol S (2002) Economic valuation of non-use values of natural resource: a case study of Thung Yai Naresuan Wildlife Sanctuary, Kanchanaburi and Tak Provinces. Master of Science Thesis in Resource Management, Faculty of Economics, Kasetsart University

Navrud S, Bergland O (2001) Value transfer and environmental policy. Cambridge Research for the Environment, Cambridge

Olewiler N (2004) The value of natural capital in settled areas of Canada. Ducks Unlimited Canada/Nature Conservancy of Canada, Manitoba, Canada

Ranganathan J, Raudsepp-Hearne C, Lucas N, Irwin F, Zurek M, Bennett K, Ash N, West P (2008) Ecosystem services: a guide for decision makers. World Resources Institute, Washington, DC

Stenger A, Harou P, Navrud S (2009) Valuing environmental goods and services derived from the forests. J For Econ 15:1–14

Stinson G, Freedman B (2000) Potential for carbon sequestration in Canadian forests and agroecosystems. School for Resource and Environmental Studies. Dalhousie University, Halifax, Nova Scotia Canada

Tawna M (2010) Nature's services: ecosystems are more than wildlife habitat. http://www.rand.org/scitech/stpi/ourfuture/NaturesServices/section1.html

Tianhong L, Wenkai L, Zhenghan Q (2008) Variations in ecosystem service value in response to land use changes in Shenzhen. Ecol Econ. doi:10.1016/j.ecolecon.2008.05.018. (in press).

Tong C, Feagin RA, Jianjian L, Zhang X, Zhu X, Wei W, Wenshan H (2007) Ecosystem service values and restoration in the urban Sanyang wetland of Wenzhou, China. Ecol Eng 29:249–258

USDA (2007) Valuing ecosystem services: capturing the true value of nature's capital. The U.S. Department of Agriculture, Washington, DC

USDA (2010) Valuing ecosystem services. http://www.fs.fed.us/ecosystemservices/About_ES/index.shtml

Wiwatthanapornchai S (2001) Non-use value assessment of endangered wildlife species and natural resources of Phu Khieo Wildlife Sanctuary, Chaiyaphum Province. Master of Science Thesis in Agricultural Economics, Faculty of Economics, Kasetsart University

Xi J (2009) Valuation of ecosystem services in Xishuangbanna biodiversity conservation corridors Initiative pilot site, China. Greater Mekong Subregion Core Environment Program, Asian Development Bank, Metro Manila, Philippines

Yolasiğmaz HA, Keleş ASH (2009) Changes in carbon storage and oxygen production in forest timber biomass of Balci Forest Management Unit in Turkey between 1984 and 2006. Afr J Biotechnol 8(19):4872–4883

Zhang M, Wang K, Chen H, Liu H (2009) Impacts of land use and land cover changes upon oxygen regulation values for the Karst ecosystem: a case study of Northwest Guangxi, China, IT in Medicine & Education, 2009, ITIME 09, 14–16 Aug 2009, Jinan China, pp 1276–1280, Jinan China

Chapter 14
Cost-Effectiveness of Policy Options for Sustainable Wetland Conservation: A Case Study of Qixinghe Wetland, China

Jian Wu, Xiaoxia Wang, Kunyu Niu, and Shushan Li

Abstract A primary cause of the loss of wetlands in China is the competition for water between wetlands and their surrounding areas. This study explores cost-effective policy options to reduce off-site water use to support the sustainable conservation of the Qixinghe Wetland in Sanjiang Plain. The cost-effectiveness of four policy options is assessed and compared using multi-criteria analysis. Option I is to reconstruct the irrigation systems in the surrounding areas of the wetland where agriculture competed with the wetland for water. This option is the government's most favored strategy, but only the third most cost-effective. Option II is to construct a dam to store and control floodwaters to relieve seasonal water scarcity. This option is the most reliable in terms of saving water. It was also farmers' most favored strategy, but it imposes a high cost on the local government and therefore did not receive strong support from the authorities. Option III is to promote the adoption of water-saving practices by providing farmers with training courses. This strategy is the most cost-effective, but is less effective in saving water. This option also did not receive strong support from farmers and the government and is therefore not likely to be selected. In Option IV, water saving is achieved by converting some paddy fields to dryland crops. This option turned out to be politically unfeasible because it was the least preferred strategy of the government and farmers. It was also the least effective in saving water. If equal weights are given to all four assessment criteria, Option I would have the best overall performance, while Option IV would be the least preferred strategy. Based on these conclusions, suggestions are offered on how the local government should tackle the wetland's water shortage problem and how the central and provincial governments could tackle the problem at the macro-level.

Keywords Wetland conservation • Multi-criteria analysis • Cost-effectiveness analysis

---

J. Wu (✉) • X. Wang • K. Niu • S. Li
School of Environment & Natural Resources, Renmin University of China, Beijing, People's Republic of China
e-mail: zhxwj@263.net; jianwu@ruc.edu.cn; bwang725@ruc.edu.cn; niukunyu2004@163.com; lishushan.lss@163.com

D. James, H.A. Francisco (eds.), *Cost-Benefit Studies of Natural Resource Management in Southeast Asia*, DOI 10.1007/978-981-287-393-4_14

Introduction

Background

Sanjiang Plain is a vast, low-lying alluvial floodplain in the northeastern segment of Heilongjiang Province. It is situated at the confluence of three rivers: the Heilong (Amur), the Wusuli (Ussuri), and the Songhua.

The plain has a high ecological significance nationally, regionally, and globally. The wetlands perform crucial ecological functions: maintaining the hydrological balance, regulating water flow, mitigating floods, and purifying the water and air. The Sanjiang Wetlands are the most important breeding grounds and migration routes of migratory waterfowls in northeastern Asia and provide habitats for numerous species of wildlife. Thirty-seven wildlife vertebrate species ranked by the World Conservation Union (IUCN) as globally threatened are found in these wetlands, 20 of which are wetland birds. For some of these wetland birds, the plain represents a significant portion of their remaining global habitat area (ADB 2004). However, these wetlands have been dramatically reduced by agricultural development during the last 50 years. In the 1950s, the wetland area was 3.7 million ha, while in 2005 it had declined to 0.92 million ha.

The Sanjiang Plain's high-quality soil and favorable climate for grain production have made it the focus of government agricultural development programs, beginning in the early 1950s. The central government has strongly encouraged settlement in and reclamation of the wetlands and the development of large-scale farming in the plain. This long-term focus on agricultural production in the plain has exacerbated the conflict between ecological conservation and agricultural development.

Water is the key factor affecting the value of wetlands. Recent statistics show that agriculture has consumed more than 75 % of the water resources in the plain, and it is predicted that the irrigated acreage under paddy will double in the next 15 years which will require – under current farming practices – increasing quantities of irrigation water, leading to a sharp drop in the water table in the wetlands (Wang et al. 2005).

The conservation of the Sanjiang wetlands raises typical conservation issues in China. Wetland ecosystems have traditionally encountered threats from the direct consumption of their precious resources by local residents, but the more important and indirect threats come from the water demands driven by the overall development strategy of the country, including agricultural development, urbanization, and industrialization.

Qixinghe Wetland

Qixinghe Wetland is a typical inland freshwater wetland, situated on the right bank of the Qixinghe River in Sanjiang Plain, Heilongjiang Province, northeast China, with a total area of 200 sq. km. It performs crucial ecological functions. For the purpose of wetland ecosystem conservation, the Qixinghe National Nature Reserve (QNNR) was founded in the year 2000. Since then, according to the China Nature Reserve Regulation (1994), activities that directly affect the Qixinghe Wetland have been prohibited, such as the expansion of farming and fishing and the harvesting of reeds and other raw materials, while activities that indirectly affect the wetland still remain, among which competing water use is the most important. The Qixinghe Wetland is already suffering from water shortage. Data collected on surrounding wells show that the water table in the Inner Qixinghe Watershed has dropped by 2.5 m from 1997 to 2005. Some wells even had a decrease of 7–12 m (Xia and Wen 2007). Water conflicts arise especially during the irrigation season.

There have not been any measures taken to deal with the water supply issue for the wetland. Although conservation efforts within the QNNR have been fully implemented, unless there are policies or plans to confront the threats from outside the wetland, the Qixinghe Wetland will not be saved.

Research Objectives

The overall objective of the study has been to explore cost-effective approaches to reducing off-site water use for sustainable wetland conservation within the policy and institutional context in China. The more specific objectives were:

- To establish a profile of water problems encountered in wetland conservation and associated water-use patterns
- To explore and propose alternative policy options to manage off-site water uses for the purpose of wetland conservation
- To analyze and assess the impacts of these options in terms of their ecological effects and economic implications on different stakeholders
- To provide recommendations on how to implement the policy options with acceptable and properly distributed costs of conservation

Methodology

Study Area

The geographic coverage of the study is the Qixinghe Wetland and its surrounding areas where users compete directly with the wetland for water use, which includes the Qixinghe National Nature Reserve, Baoqing County; the Wujiuqi (WJQ) state-owned farm on the right bank of the Qixinghe River; and the Youyi state-owned farm on the left bank.

Data Collection

Information to support the study was obtained from various sources. Data describing land areas and population were provided by the Baoqing Civil Administration and Baoqing Water Resource Authority. Similar data for WJQ farm was collected from interviews, while for the Youyi farm, data was obtained from the Youyi Agriculture Committee and Youyi farm interviews.

Detailed data relating to farmers' awareness, attitudes, and behavior were obtained from a large-scale survey conducted by the research team in July 2008, involving focus group discussions and structured interviews with farmers in Baoqing County and the WJQ state-owned farm, as well as officials of the Heilongjiang Provincial Hydraulic Design Institute, Baoqing Agriculture Technology Promotion Center, Baoqing Water Affairs Bureau, and the Qixinghe National Nature Reserve Management Bureau. Two hundred and one households were interviewed in the survey. The household survey did not include the Youyi state-owned farm due to feasibility constraints. Given that the overall conditions of the two state-owned farms were quite similar, however, the WJQ farm could be taken as representative of both. The sample of 201 households represented about 14 % of the overall number of households (margin of error of 6.5 %) and about 16 % of the total in terms of the paddy acreage.

To take into account the differences in paddy production between the villages and the state farms, a stratified sampling method was applied. The first stratum consisted of irrigation districts in the rural areas in Baoqing County, and the second stratum comprised the WJQ state farm. The samples were allocated according to the area of paddy fields: about 30 % of the samples were from the rural areas and 70 % from the state farm. In each stratum, simple random sampling was applied.

Cost-Effectiveness Analysis

Cost-effectiveness analysis was the technique applied to assess the economic efficiency of options to achieve water savings in competing uses of the resource, designed to improve the prospects for increasing the volume of water allocated to wetland areas. The cost-effectiveness (C/E) of each policy option was measured by the average cost per unit of water saved. To overcome the complication that water savings and their costs would occur at different points in time, discounting was applied to the volumes involved as well as to their costs, following the approach recommended by Warford (2003). The relevant formula is:

$$C/E = \Sigma\big(C_t/(1+r)^t/\Sigma\big(Q_t/(1+r)^t \quad t = 1, 2 \ldots T$$

where the numerator in the above expression measures the present value of the total cost of water saved and the denominator the discounted total quantity of water saved, over the assessment period from year 1 to T. In all the economic assessments, a 30-year time horizon was assumed. Residual values of capital were estimated using the straight-line method of depreciation.

Multi-criteria Analysis

To supplement the cost-effectiveness analysis, a simple multi-criteria analysis (MCA) framework was applied. Based on the consultations conducted with the local officials, four different criteria were chosen through which to compare the trade-offs between the options covered by the study. A stakeholder analysis was incorporated in this exercise.

Water Requirements for Wetland Conservation

Water Availability

For the Inner Qixinghe Watershed (wherein the QNNR is located), the long-term average annual surface water quantity is 351 million m^3 while the groundwater quantity is 246 million m^3. Neglecting the double counting between them, the total long-term average annual water quantity in the watershed is 597 million m^3. The average water resource is about 4,650 m^3 per capita and 5,550 m^3/ha, much higher than the country average (Xia and Wen 2007).

The average surface water resource quantity for the QNNR is 42.9 million m^3 while the average groundwater resources amount to 13.9 million m^3. The long-term average water quantity of the reserve is 56.8 million m^3. The surface water resources of the QNNR come from upstream of the Inner Qixinghe River and from precipitation and its runoff, while the groundwater is recharged mainly by rainfall (Xia and Wen 2007).

The interaction between groundwater and surface water is complicated and difficult to assess due to the lack of essential hydrographic and geological records. In this study, it was presumed that the impact of surface water and groundwater usage was the same. This assumption is reasonable from a long-run hydrological perspective.

Water Utilization

The overall water utilization volume for the whole Qixinghe Watershed is 540 million m^3 in which the surface water utilization volume is about 92 million m^3 with a utilization rate of 18.8 % and the groundwater utilization volume is about 448 million m3 with a utilization rate of 81.4 % (Xia and Wen 2007). Generally, water supply can meet the economic demand, but the ecological water demand has not been taken into account in these calculations.

As for the distribution of water among water-using sectors, taking Baoqing County as an example, in 2007, the total amount of water use in the county was 141 million m^3 in which agriculture used 105 million m^3 of water (74.6 % of total use), industries used 32.7million m^3 (23.1 % of total use), and households used 3.2 million m^3 (2.3 % of total use) (pers. comm. Water Affairs Bureau of Baoqing Country, 2008). Since the WJQ and Youyi state farms are dominated by agriculture, the overall water utilization level for the whole study area is expected to exceed that of Baoqing County itself.

Factors Influencing Wetland Water Shortage

According to the ADB (2007), the factors contributing to a reduced water level in the wetland are as follows:

Increasing Off-Site Water Use Surrounding areas of the QNNR compete with it for water. According to the Baoqing County's planning estimation (Water Affairs Bureau of Baoqing County 2005), from 2006 to 2010 the total water demand is expected to be almost doubled, increasing from 119 million m^3 to 206 million m^3.

Irrigation and Drainage Systems The more important but indirect effect is the development of the drainage and irrigation system driven by agricultural expansion. There is no large hydrological project in the upstream part of the Inner Qixinghe River, but the development of the drainage system along with the expansion of paddy fields has damaged the wetland's capacity to retain water. The wetland is situated in a low-lying area where water resources converge at a slow flow rate. With the expansion of agriculture along the river, however, the water drainage system has been fully developed. The system accelerates the rate at which water is drained from this region, and then irrigation systems divert the water according to various economic uses.

Reduced Water Inflow Water inflows from upstream and precipitation are the only two sources of water supply, but the natural water supply or water inflow from upstream and precipitation has reduced in recent years. As well, the flood control banks or dams that were constructed to protect economic projects prevent small- and medium-scale floodwaters from entering the wetland, and this contributes significantly to the water shortage in the wetland.

Ecological Water Demand for QNNR

In general, the water shortage in the Qixinghe Wetland has been recognized. However, authoritative quantitative estimations of the situation and the wetland water demand are as yet unavailable.

In Sanjiang Plain, some researchers have calculated the ecological water demand based on the geographical and ecological situation of the plain (Guo et al. 2004; Cui et al. 2005; SongLiao Water Resources Commission 2005). Based on their findings, the land types that should be involved in calculating the ecological water demand of the wetland are reservoirs/lakes, water swamps, marshes, and reed wet soil. By using the methodology developed by the study on the Zhalong Wetland (SongLiao Water Resources Commission 2005) and taking into account the rainfall, evaporation rate, soil infiltration capability, and other factors of the Qixinghe River Basin, Xia and Wen (2007) found that the depth of water in the core zone of the wetland should be kept at about 20–25 cm while the depth of water in the buffer zone should be 15–20 cm in order to protect the QNNR and the basin wetland habitat from degradation. According to these requirements and deducting the existing water surface area, the total ecological water demand for the QNNR was estimated to be 38.17 million m^3 (Xia and Wen 2007).

According to satellite images and the QNNR Management Bureau, the amount of water that needs to be recharged to maintain the wetland is 20 % of the total ecological water demand. Accordingly, this study set 8 million m^3 as the target amount of water to be recharged to the wetland through policy intervention.

Water Use by Agricultural Sector

As compared with industry and urban livelihoods, agriculture was the largest water user in the studied area. There were severe conflicts between agricultural water use and wetland ecological water demands. Given the large room left for water saving from the agriculture sector, agriculture has been identified to be the key sector for water saving policy interventions.

Field investigation and a survey that were carried out with farmers revealed their awareness, attitudes, and behavior in relation to water use, as reported below.

Expansion of Paddy Fields

The paddy fields in the surrounding areas of QNNR have developed rapidly in the past 10 years, which accelerated the increase of water consumption by agriculture. Since 1995, paddy field areas in Baoqing County, WJQ state farm, and Youyi state farm have increased significantly by 111 %, 105 %, and 248 %, respectively, mainly due to policies encouraging their development and the high market price of rice (Baoqing Statistics Bureau (1996, 2006); Heilongjiang State Farm Bureau (1996, 2006)). As Baoqing County's planning estimation, agricultural usage will increase from 92 million m^3 to 153 million m^3 during 2006–2010 (Water Affairs Bureau of Baoqing County 2005).

Water Efficiency

Data shows that the water efficiency in agriculture is very low. The water consumption for paddy growing is about 12,000–13,500 m3/ha in the county, whereas the standard for Heilongjiang Province is only 6,750 m3/ha, according to the Water Affairs Bureau of Baoqing County.

Our survey also reveals reasons for this low efficiency dilemma: farmers listed poor hydrological system and land features as the two key factors leading to low efficiency in water use.

Water-Use Behavior

The survey found that 28 % of the people (56/200 households) relied completely on surface water (from rivers and/or the wetland), 31.5 % (63/200 households) used only groundwater, and 40.5 % (81/200 households) used both surface water and groundwater. Generally, the farmers tended to use surface water when it was available because the higher temperature of the surface water was more favorable for rice cultivation.

Under the current irrigation infrastructure, water supply cannot be guaranteed. Farmers cannot get water in a timely manner according to crop growth requirements so they try to get as much water as they can whenever there is water available without considering water saving. Thus, in high water-use seasons, there is usually a struggle to obtain water.

Water-Saving Consciousness

The survey showed that about 54 % (107/198 households) of the people were concerned about the decrease in water resources, 24 % (48/198 households) did not think that there was really a scarcity, and 21.2 % (42/198 households) were not clear about the water situation. This indicated that there was no consensus among the farmers regarding the local water situation. One possible reason was that they live in a low area around the wetland, and it is relatively easy to get water compared with other areas.

The survey also showed that the local farmers had low water-saving skills and many did not think water-saving was necessary. Most of the interviewees (80.6 % or 112/139 households) had never done any trials on water-saving practices, while 89 % (178/200 households) had never had the opportunity to get water-saving training.

Farmers' Awareness

A majority (86.3 %) of the farmers (170/197 households) knew that the QNNR was a wetland. About 46 % (91/198 households) agreed that their livelihoods were related to the wetland, while 41.4 % (82/198 households) denied such a relationship, and 12.7 % (25/198 households) of the people did not know if there was a relationship. About 64 % (63.8 % or 125/196 households) clearly expressed that they were willing to save water for the wetland, while 17.3 % (40/196 households) said that they were not willing to do so. From these information obtained from the survey, it was concluded that some of the local people already had some environmental awareness and recognized that the quality of the environment was linked to their livelihoods. However, their awareness was still too limited to result in water-saving behavior.

Water Resource Management in Heilongjiang Province

Institutional Arrangements for Water Resource Management

The overall institutional arrangement for water resource management in Heilongjiang Province involves a multi-sectoral water administration management system in which the Provincial Department of Water Resources (PDWR) plays the core role. Generally, the PDWR and Heilongjiang State Farm Bureau (HSFB) oversee the management of water resources in Heilongjiang Province. The PDWR manages provincial water issues for the Heilongjiang government and guides the Water Affairs Bureaus in the cities and counties. The Water Affairs Bureaus under the HSFB manage water issues within the state-owned farm areas and implement water administration policies. Water conservation plans in state farms need approval from the PDWR.

At the provincial level, there is also the Provincial Agriculture Development Office (ADO) that provides guidance on water-saving practices in agriculture, but this office has no strong role at the county level. At the county level, it is the Agriculture Development Commission (ADC) of the county that provides education on how to save water.

Under such an institutional framework, water issues are clearly controlled by different agencies and jurisdictions. Agriculture plays a big role in the management of water resources. The Forestry Bureau controls the QNNR and is mandated to address the wetland's water needs; however, it is not closely involved in the institutional arrangement.

Water Policies

Policies and regulations related to water management have been developed at the national and provincial levels to form an overall policy framework which covers water intake permits, water resource plans, water resource fees, water pricing policies, water resource assessments of construction projects, discharge fees, and other issues. In 2006, China issued a regulation on water permits and water resource fees, which required all water users in the various sectors to pay for water resources via the application of a water permit. That delivered a clear signal of the move toward gradually increasing water-use efficiency by strengthening water resource management and charging the use of water resources. In general, however, these policies mainly focused on industrial water users, with an exemption or favorable discount to agricultural water users.

Water Fee Management System

Because land property rights belong to the nation, farmers have to rent land for farming. Water fees are built into the land rental agreements for state farms, and this guarantees a high collection rate. However, the water fee system and policy implementation in the rural areas are weak.

According to the interview conducted with the Baoqing County Water Affairs Bureau, in the rural areas, the present water fee levy system is as follows: the farmers or the whole village pays water fees to the town's fiscal department, which then hands the fees to the county's fiscal department. The county allocates about 60 % of the water fees collected to the city authorities and keeps about 40 %. The water fee for surface water is 300 RMB/ha in Baoqing County, but there is no charge for well water. The county or town provides grants from its budget for the maintenance of the local reservoir. The fee collection rate was very low, below 30 %, showing that the water fee management system was inefficient.

On the WJQ state farm, the Water Affairs Bureau levies water fees for each working district, and the collected fees are handed to the management bureau of the farm. The management bureau from each branch farm buys water from the reservoir (government owned and run by the Reservoir Management Bureau). In practice, the water fees are handed over together with the land taxes, i.e., the water fees are bundled together with land contract fees and collected in advance when farmers rent land. The levy collection rate is as high as 100 %. The water fee standard for surface water is 300 RMB/ha, while well water costs 75 RMB/ha.

Water fees should be linked to water quantity and charged by volume, but the implementation of this policy requires corresponding metering devices which most irrigation districts do not have at present. Charging based on acreage rather than volume cannot establish a nexus between the quantity of water used and the water fee, so the existing water pricing policy will not induce water saving. This is partly due to the poor irrigation infrastructure and lack of water measurement devices.

Prospects for Economic Instruments in the Agricultural Sector

Various studies have discussed mechanisms for water resource allocation including water pricing and water rights markets (Randall 1981; Shen et al. 2001; Jiang 2003; Dietrich and Grossmann 2005). Many researchers have also focused on agricultural water-use efficiency. Amir and Fisher (1999) analyzed the demand for water from agriculture under various price scenarios. Bazzani (2005) applied the DSIRR (decision support system for irrigated agriculture) model to analyze the impact of a water pricing policy on water consumption at the catchment level. Gómez-Limón and Riesgo (2004) developed a methodology to evaluate alternative irrigation water pricing policies, while Varela-Ortega (2003) and Mejias et al. (2004) examined the effect of different policy scenarios for wetland conservation (mainly water pricing policies) on agricultural water use, farmers' incomes, and government revenue. Fu et al. (2002) developed an agricultural production function with water as a factor input for setting up irrigation water quota targets in Sanjiang Plain. Guo and Wang (2004), Mao (2005), Wei et al. (2007), and Su (2003) discussed the effectiveness of a water pricing policy in enhancing agricultural water efficiency in China. Wang and Zhou (1987), Liu et al. (2005), and Zhao (2006) applied an agricultural water-saving decision support model for western China, where water scarcity is a major issue.

In China, the direction of agricultural water price reform has been as follows: first, improving the measurement devices; second, charging on volume instead of acreage; and third, increasing the water price to cover the water supply cost. The goal of the first two is to promote water saving while the last one is to maintain the operation and management of the irrigation services.

But in Sanjiang Plain, generally speaking, administrative mechanisms, such as water resource plans and engineering projects, are still the dominant instruments in water resource allocation (Songliao Water Resources Commission 2005). The use of market mechanisms in water resource allocation is still rare in the region. The existing system, which lacks economic incentives to control water use, is unsustainable.

There are problems that cannot be solved by a water pricing policy alone, for example, the low willingness to pay of farmers; difficulty in collecting water fees, water fee cuts, and embezzlement by the fee collectors; and the poor management and maintenance of the irrigation canals, especially the end parts, the maintenance of which is not funded by the government. These problems must be addressed by a proper reform of the irrigation district management system.

Policy Options

Policy Goal

As noted, for the purpose of this study, to meet the water conservation demands of the Qixinghe Wetland, the policy goal was assumed to be a reduction in off-site water use of about 8 million m^3 (equivalent to about 20 % of the ecological water demand of the wetland), thereby allowing an increased water supply to the wetland without diminishing socioeconomic growth of the surrounding areas. Conservation of the wetland will ultimately improve the water resources and climate in the watershed and support its sustainable development.

Prospective Policy Options

In China, most reforms in water allocation have been initiated by the upper-level government authorities rather than at the local/county levels, so in the policy design in this study, the focus was on what could be implemented at the local level without excessive political or bureaucratic complexity. Through consultation with local officials and experts, policy options to reduce off-site water use were considered that included water-saving measures involving agro-hydraulic engineering, better agricultural skills, and water management reforms.

Before designing policy options for solving the water scarcity problem in the wetland, a preliminary screening was carried out of all potential policy options, based mainly on the general feasibility of the options, including the legal and political feasibility. The results of this assessment are shown in Table 14.1.

Options for Evaluation

Based on the above preliminary screening, the most feasible options selected for detailed evaluation were the following:

Option I: Irrigation System Reconstruction (ISR) Reconstructing the irrigation network for water-saving purposes

Table 14.1 Framework for policy options selection

Problem	Factors contributing to the problem	Strategy	Options
Decreasing water table in wetland	Competitive water use in surrounding area	Reducing off-site agriculture water use	(1) Reconstructing the irrigation system for water conservation
			(2) Water-saving planting practices
			(3) Change from paddy to dry crops
			(4) Management measures like water permits and water fees
	Accelerated water drainage in whole watersheds	Increasing the amount of stored water	(5) Restricting the expansion of paddy fields
			(6) Building new facilities to retain water
	Reduced water inflow	Securing water	(7) Afforestation in upstream areas
		Inflow (by improving the water cycle in the watershed)	

Option II: Ecological Water Control Reservoir (EWC) Constructing a small dam to store and control floodwater to increase the water supply and relieve seasonal water scarcity

Option III: Water-Saving Practices (WSP) Introducing water-saving practices to reduce water usage for paddy planting

Option IV: Switching from Paddy to Dry Crops (PTD) Changing from paddy to dry crops to reduce water use in paddy planting

Each policy option was evaluated in terms of cost (10^4 RMB/year), water saved (m^3/year), farmers' income loss, and government revenue and expenditure.

Implementation Zone and Stakeholders

Based on our local investigation and experts' advice, we chose six zones directly adjacent to and competing in water use with the QNNR in which to implement the four policy options: two small towns in Baoqing County in the rural areas (Qixinghe town (QHT) and Qixingpao town (QPT)) and two branch farms of the WJQ state farm (WJQ-4 and WJQ-5) and two of the Youyi farm (Youyi-6 and Youyi-8.) The main stakeholders potentially affected by the selected policy options at the local level included the government (Water Affairs Bureau, Agriculture Development Commission, Wetland Management Bureau, etc.), enterprises (state-owned farms, water supply enterprises, etc.), and farmers.

The state-owned farms are normally categorized as enterprises, but they function as government units and are operated according to government procedures. Water supply enterprises (such as reservoir and irrigation district enterprises) are not pure enterprises but are part of the Water Affairs Bureau. They have to remain financially viable in terms of their operational costs by providing water supply services and charging water fees, similar to some public service sectors. In the analysis, state-owned farms and water supply enterprises were treated as part of the government with their own specific interests under each policy option.

For each policy option, stakeholders may have different roles and be affected in different ways or to different degrees. They may also attach different degrees of importance to the options or exert varying levels of influence on the decision-making and implementation processes.

Economic Evaluation of Water-Saving Policy Options

Option I: Irrigation System Reconstruction (ISR)

Description

In order to save water to meet the ecological water demand of the wetland, this option involves a reconstruction of the irrigation system in the surrounding areas that compete with the wetland for water by – among other things – sharing the same water sources as the wetland.

These measures would more effectively ensure the supply of water for agriculture, improve the efficiency of the water canal system, and lay the foundation for implementing field water-saving practices and water pricing policies. The option includes designing an appropriate irrigation system to save 8 million m^3 of water to meet the wetland's ecological water needs.

Cost-Effectiveness

The main costs involved in Option I comprise the capital costs of the infrastructure and running costs, with the residual value of capital as a cost offset. The cost-effectiveness results are shown in Table 14.2. They were calculated in terms of the cost of water saving per ha, the water-saving volume per hectare, and cost-effectiveness at discount rates of 2 %, 5 %, and 8 %.

Distributional Effects

In order to understand how the stakeholders would contribute to or gain from Option I, the incidence of costs and effects was assessed, as shown in Table 14.3.

Table 14.2 Cost-effectiveness analysis of Option I (at different discount rates)

r	$E_{ISR}{}^{t}$ (10^6 m^3/year)	TC_{ISR} (10^6 RMB)	TE_{ISR} (10^6 m^3)	*C*/*E* **(RMB/m³)**
2 %	8.00	22.89	179.17	0.128
5 %	8.00	20.14	122.98	0.164
8 %	8.00	18.52	90.06	0.206

Table 14.3 Distribution of the costs and effects of Option I

Stakeholder	Costs	Effects
Farmers	The operation and maintenance of the irrigation system need to be covered by the water fees paid by the farmers, i.e., 552 RMB/ha	–
Government	Irrigation reconstruction investment cost: 19.74 million RMB	Saving 8 million cubic meters of water per year

Table 14.4 Cost-effectiveness analysis of Option II

r	TC_{EWC} (10^6 RMB)	*TE* (10^6 m^3) *EWC*	*C*/*E* (RMB/m³)
2 %	100.97	155.43	0.650
5 %	87.39	106.68	0.819
8 %	79.44	78.13	1.017

Option II: Ecological Water Control Reservoir (EWC)

Description

This option involves building a small dam to store and control floodwater at a suitable location in upstream areas. Such a proposal was raised by the local government years ago. Preliminary investigations were also conducted. The same proposal was used in the present study. It involves building a reservoir midstream along Qixinghe's first branch, the Jinshahe River, located in Qixingpao town in Baoqing County, with the aim of retaining floodwaters and the excess water of the Jinshahe River for irrigation.

Cost-Effectiveness

The costs incurred for Option II are again the capital costs of infrastructure and running costs. The results of the cost-effectiveness analysis are shown in Table 14.4 assuming discount rates ranging from 2 % to 8 %.

Distributional Effects

The investment for this project would be mainly funded by the government, and the government would also have to bear the running costs (see Table 14.5). The water

Table 14.5 Distribution of costs and effects of Option II

Distribution	Costs	Effects
Farmers	–	–
Government	Investment cost, 67.55 million RMB; compensation cost, 19.38 million RMB; running cost, 1.93 million RMB/year	Water saving of 6.94 million m^3/year

saving would mainly serve the purpose of supplying water to the wetland to create ecological benefits. There would be no direct costs and effects for the farmers as there would be no direct impact on yields and water-use methods. However, an indirect benefit could be a yield increase as a consequence of improved local environmental conditions

Option III: Water-Saving Planting Practices (WSP)

Description

In this option, the government provides training in water-saving planting skills for farmers by organizing classes and promoting the implementation of the skills. The Agricultural Techniques Promotion Center (ATPC) of Baoqing County would be the appropriate implementing agency, responsible for engaging agricultural professionals to give water-saving training to farmers at the village level. The training would be held every year during the project period of 30 years. Each class is estimated to have 20 trainees. The ATPC would provide the trainers, training materials, and support services until the farmers could successfully implement the water-saving practices. The ATPC would also monitor and evaluate implementation of the program.

Cost-Effectiveness

For the government, the costs of this option consist of the training and management costs, including monitoring and evaluation. On the farmers' side, the option would involve learning costs, mainly in terms of time, but the benefits would be savings on water fees, water, and energy and increased productivity. The cost-effectiveness analysis results for this option are shown in Table 14.6 at discount rates of 2, 5, and 8 %. Although the cost-effectiveness of this option seems ideal, it could not be achieved without meeting some prerequisites such as irrigation system improvement, installation of water measurement devices, and economic incentives. A high water price would act as an economic incentive for farmers to implement water-saving practices.

Table 14.6 Cost-effectiveness analysis of Option III (at different discount rates)

R (%)	TC_{WSP} (10^6 RMB)	TE_{WSP} (10^6 m^3)	*C*/*E* (RMB/m^3)
2	1.50	179.17	0.008
5	1.03	122.98	0.008
8	0.75	90.06	0.008

Table 14.7 Distribution of the costs and effects of Option III

Stakeholder	Costs (10^3 RMB)	Effects (10^6 m^3)
Farmers	–	–
Government	Training cost: 51,940 RMB/year	Saving 8 million m^3 of water per year
	Management cost: 14,840 RMB/year	

Distributional Effects

This proposal assumes that the government would provide the water-saving training at no charge, which means that the government would bear the main cost (Table 14.7).

Option IV: Switching from Paddy to Dry Crops (PTD)

Description

Switching from paddy to dry (or dryland) crops is another potential choice to achieve water saving. In the targeted policy implementation areas, dry crops are completely reliant on rainfall and do not require irrigation. Thus, switching from paddy to dry crops is expected to save a large amount of water. Changing from rice to dry crops is not unusual in local areas. In the survey, it was found that farmers chose to change crops for many reasons such as crop price fluctuation and conditions of land and water resources.

The roles of the government in this option include being the provider of information, helping farmers identify appropriate farmland, and promoting the change from paddy to dry crops. Farmers are the most important actors in this option. Changing from paddy to dry crops will impact farmers' net incomes. Significant price gaps between paddy and dryland crops would decrease farmers' willingness to switch, so the net income change caused by this option should be carefully considered.

Table 14.8 Cost-effectiveness analysis of the switch from paddy to soybean (at different discount rates)

Paddy to soybean (%)	TC_{PTD} (10^6 RMB)	TE_{PTD} (10^6 m^3)	C/E (RMB/m^3)
2	15.53	179.17	0.087
5	10.90	122.98	0.089
8	8.18	90.06	0.091

Table 14.9 Cost-effectiveness analysis of the switch from paddy to corn (at different discount rates)

Paddy to corn (%)	TC_{PTD} (10^6 RMB)	TE_{PTD} (10^6 m^3)	C/E (RMB/m^3)
2	17.34	179.17	0.097
5	12.13	122.98	0.097
8	9.07	90.06	0.101

Cost-Effectiveness

The main costs of this option involve short-term costs associated with reduced productivity and the cost of leveling land. Long-term costs would comprise losses of farmers' income resulting from the switch from rice to dryland crops, for which only a lower price can be obtained. The results of the cost-effectiveness analysis are presented in Table 14.8 for a switch from paddy to soybeans and in Table 14.9 for a switch to corn.

The cost of changing from paddy to corn is a little higher than that of changing from paddy to soybean, and the higher one was chosen to compare with the cost-effectiveness of the other three options.

Distributional Effects

Table 14.10 shows how farmers absorb both the short-term and long-term costs of Option IV. However, as the representative of public welfare, the government can be considered the main beneficiary from meeting the water-saving target.

Economic Comparison of Options

The overall results of the option analysis are shown in Table 14.11. Option III is the most cost-effective. The next most cost-effective is Option IV and then Option I. The least cost-effective is Option II. Regarding the distribution of costs, Option I distributes the costs between farmers and the government. Option II imposes a high cost on the government. Option III involves a low cost to the government. Option IV has a medium-level cost, borne by farmers.

Table 14.10 Distribution of the costs and effects of Option IV

Stakeholder	Costs		Effects
	Paddy to soybean	Paddy to corn	
Farmers	Short-term cost:	Short-term cost:	
	1,518,055.8 RMB in the first year	1,559,975.8 RMB in the first year	
	Long-term cost:	Long-term cost:	
	655,569.75 RMB/year	738,035.6 RMB/year	
Government			Saving 8 million m^3 of water per year

Table 14.11 Comparison of the cost-effectiveness and distributional cost effects of options (at a discount rate of 5 %)

Options	Costs (10^6 RMB)	Effects (10^6 m^3)	C/E (RMB/m^3)	Distributional Cost Effects: Government revenue and expenditure	Distributional Cost Effects: Farmers' income
I (ISR)	20.14	122.98	0.164	19.7 million RMB for irrigation reconstruction – a one-off investment cost	552 RMB/ha/year for the running cost of the irrigation system
II (EWC)	87.39	106.68	0.819	67.7 million RMB one-off investment cost	–
				19.4 million RMB compensation cost	
				1.9 million RMB running cost per year	
III (WSP)	1.03	122.98	0.008	51,940 RMB/year training cost	–
				14,840 RMB/year management cost	
IV (PTD)	10.90	122.98	0.089	–	Paddy to soybean as example: short-term cost = 1.52 million RMB in the first year; long-term cost = 0.66 million RMB/year

Regarding the sensitivity of the cost-effectiveness (C/E) ratio to different discount rates, it was determined that Option II is the most sensitive, followed by Option I and then Option IV. Option III is insensitive to the discount rate. The comparative relationship between the four options, however, does not change under discount rates ranging from 2 to 8 %. The only exception is the comparative relationship between Option I and Option IV, which changes under very low discount rates.

Trade-Off Between Policy Options

Multiple Criteria for Comparing Options

Based on consultations with the local officials, four different criteria were chosen to compare the trade-offs between the four options: cost-effectiveness, reliability in achieving the water-saving effect, government's attitude, and acceptance by farmers. Cost-effectiveness was quantified in terms of RMB per m^3 of water saved, while the other three were scored on a scale of 1–4. Reliability and government's attitude were investigated by interviewing technical experts and relevant government agencies, respectively. Farmers' attitudes were surveyed, during which farmers were required to choose only two preferred options from the four, and a higher score means the options were chosen as the best two more often. Trade-offs between the four options are summarized in Table 14.12.

Ranking of Options

The final ranking of options appears in Table 14.13. If an option ranked first, it was allocated a score of 10. If it ranked second, third, and fourth, it received a score of 8, 5, and 0, respectively.

It can be seen that each of the options has pros and cons. Option III is the most cost-effective, while Option II has the highest reliability in terms of water saving

Table 14.12 Trade-offs between the four policy options

Criteria	Option I: ISR	Option II: EWC	Option III: WSP	Option IV: PTD
Water saved	8 million m^3	6.94 million m^3	8 million m^3	8 million m^3
Cost-effectiveness	0.164 RMB/m^3	0.819 RMB/m^3	0.008 RMB/m^3	0.089 RMB/m^3
Reliability in achieving the effect	High	High	Medium	Low
	Needs good management	Needs mechanism to guarantee supply to wetland	Needs incentive	Needs incentive
	Needs mechanism to guarantee supply to wetland			Unsustainable in the long run
Government's attitude	High	Medium	High	Low
	Matches policy priority and funding mechanism	Has technical support but no financial capacity	Not independent	
Acceptance by farmers	High	High	Medium	Low
	Not sure about cost burden	No cost		

Note: For cost-effectiveness, the lower the value, the more cost-effective the option

Table 14.13 Ranking of the four policy options

Criteria	Option I: ISR	Option II: EWC	Option III: WSP	Option IV: PTD
Cost-effectiveness	5 (3)	0 (4)	10 (1)	8 (2)
Reliability in achieving the effect	8 (2)	10 (1)	5 (3)	0 (4)
Government's attitude	10 (1)	5 (3)	8 (2)	0 (4)
Acceptance by farmers	8 (2)	10 (1)	5 (3)	0 (4)
Total score (ranking)	31 (1)	25 (3)	28 (2)	8 (4)

Note: The rank of the options is given in parenthesis (1 = best option; 4 = worst option)

and is the most supported by the farmers, but the cost is extremely high for the local government. Option I, on the other hand, has the government's greatest support. Given the same weighting for all four criteria, Option I is the best overall policy, while Option III ranks second, Option II ranks third, and Option IV is the least feasible.

Conclusions and Policy Recommendations

General Conclusions

Among the four policy options designed at the local level to manage the off-site agricultural water use, taking consideration the reliability of supplying water to the wetland, cost-effectiveness, and the attitudes of the government and farmers, we concluded that Option I (irrigation system reconstruction or ISR) was the optimal option. Option III (the promotion of water-saving planting practices or WPP) was the second best and could in fact be complementary to Option I. As for the implementation, both the ISR and WPP options require reform of the water pricing policy and improvement of the irrigation district management system.

The current water pricing system (charging according to acreage and having a low charge rate) has brought about low collection of water fees, to the extent that the water fees collected are insufficient to cover the running costs of the irrigation system. Even if the irrigation project were to be put in place, without strengthening the implementation of the water pricing policy, it would still result in an ineffective irrigation system.

The same is the case for water-saving planting practices. Under the low water pricing policy, there is no incentive for farmers to take water-saving action, so even if the irrigation project were put in place, incentives would still be needed to encourage farmers to adopt these practices to create real water savings.

In order to ensure the effectiveness of the proposed measures, an effective pricing policy should be established that provides economic incentives to ensure implementation. Increasing the water price is a logical *must* step in this direction.

The irrigation district management system should also be reformed in order to create an institutional environment to support a new water pricing policy.

Other mechanisms are required to ensure that the water-saving effects are converted to ecological benefits in terms of wetland conservation. The options above presume that the water saved will not be diverted to other water users but will be allocated to the wetland. However, the analysis revealed that whether the water saved will be allocated to the wetland depends mainly on the water resource management authorities.

Policy Recommendations

Given the findings of the study, several policy recommendations can be made. In order to tackle the wetland water shortage problem, the local government could reconstruct the irrigation system in the surrounding area of the Qixinghe Wetland for water-saving purposes as soon as possible, while restricting further expansion of paddy fields in the Qixinghe River basin.

Water-saving practices should be continually promoted by providing regular training to farmers. The option of water-saving planting practices was found to be the most cost-effective, so it should be applied as fully as possible. The training is also necessary because the study found that local farmers were not very open to water saving.

Speeding up water pricing reforms and irrigation district management system reforms is important. The implementation of hard measures needs to be strengthened by using economic incentives.

Reconstructing the irrigation system and promoting water-saving planting practices should be conducted at the same time as irrigation management and water pricing reforms to accelerate the whole process. The direction of agricultural water pricing reform should be to make economic benefits play a role in water resource allocation.

The key constraint for wetlands in competing with other water users is the lack of funding. Making the wetland a competitive water user, by arranging for funding for reliable water supply under the National Wetland Protection Program, is necessary. In view of the wetland's spill-over benefits, such funds should be guaranteed by the government. The current funding mechanism for wetland conservation does not take full account of the wetland's water resource demands. It is suggested that the central government set up a special budget for dealing with the problem of water shortage in wetland conservation.

For the provincial government, it is recommended that a water resource plan for the whole Qixinghe River basin be made as soon as possible, taking account of the real ecological water demands of the wetland.

Appropriate institutional arrangements should also be made at the provincial level by involving representative wetland departments in the decision-making process for water allocation and agricultural development.

References

ADB (2004) Initial environmental examination report: Sanjiang Wetland Protection Project (TA 3998-PRC), Asian Development Bank, Manila

ADB (2007) Inception report: Sanjiang Wetland Protection Project (ADB Loan 2157 PRC and GEF Grant 4517-PRC), ADB internal report, Asian Development Bank, Harbin

Amir I, Fisher FM (1999) Analyzing agricultural demand for water with an optimizing model. Agric Syst 61:45–56

Baoqing Statistics Bureau (1996, 2006) Baoqing Statistics Yearbook. Baoqing County. Shuangyashan City. Heilongjiang Province, China

Bazzani GM (2005) A decision support for an integrated multi-scale analysis of irrigation: DSIRR. J Environ Manag 77(4):301–314

Cui LJ, Bao DM, Xiao H, Lei K (2005) Calculation of wetland ecological water and case study. J Soil Water Conserv 19(2):147–151

Dietrich O, Grossmann M (2005) Global change impacts on wetlands water balance and economic value – water management strategies for the protection of the Spreewald Wetland.In: ICID 21st european regional conference, 15–19 May 2005, Frankfurt (Oder), Germany; Slubice, Poland, International Commission on Irrigation and Drainage (ICID), New Delhi, pp 1–5

Fu Q, Wang LK, Men BH, Liang C (2002) Study on water production function model and variable law of sensitivity index of rice in Sanjiang Plain, Water Saving Irrigation, No. 4, pp 81–83

Gómez-Limón JA, Riesgo L (2004) Irrigation water pricing: differential impacts on irrigated farms. Agric Econ 31(1):47–66

Guo S, Wang R (2004) Analysis of agriculture water price policy effects. The agriculture economic issue, no. 7, pp 41–44

Guo YD, He Y, Deng W, Zhang MX, Zhang GX (2004) Study on ecological water demands of Zhalong Wetland. J Soil Water Conserv 18(6):393–398

Heilongjiang State Farm Bureau (1996, 2006) Statistics Yearbook of Heilongjiang Reclamation Area, Heilongjiang Province, Harbin

Jiang WL (2003) Studies of bearing capacity of agricultural water pricing. China Water Res 11:41–43

Liu XL, Huang GB, Zhang R (2005) Demand and function analysis of decision support system on water-saving agriculture. J Gansu Agric Univ 40(6):817–821

Mao C (2005) Analysis of the relationship between agricultural water price reform and water-saving effect. China rural water and hydropower, no. 4, pp 2–4

Mejias P, Varela-Ortega C, Flichman G (2004) Integrating agricultural policies and water policies under water supply and climate uncertainty. In: Proceedings of the 25th international conference of agricultural economists (IAAE), 16–22 August 2003, Event Dynamics, Durban, South Africa, pp 1039–1047

Randall A (1981) Property entitlements and pricing policies for a maturing water economy. Aust J Agric Econ 25:195–220

Shen DJ, Liang RJ, Wang H, Yang XL (2001) Water pricing theory and practice. Science Press, Beijing

SongLiao Water Resources Commission (2005) Report of water resources plan for Zhalong Wetland, Changchun, Jilin Province

Su YX (2003) Analysis of agriculture water price and water use amount in central Gansu Province. Gansu Water Hydropower Technol 39(3):244–245

Varela-Ortega C (2003) Assessment of agricultural policy options for sustainable ground-water management: a case study of wetland conservation in Spain. CIHEAM –Option

Wang YL, Zhou XH (1987) The multi-objective planning model of water resources optimal allocation in the HeiheMain River System. J Desert Res 7(4):23–33

Wang B, Zhang Z, Yu H, Wen J, Chen J (2005) Subject report on water resources exploitation and utilization of Sanjiang Plain, Technical report, Heilongjiang Provincial Institute of Hydraulic and Hydropower Reconnaissance and Design, Heilongjiang Province, pp 28–36

Warford J (2003) Marginal opportunity cost pricing for municipal water supply, EEPSEA special paper, EEPSEA, Singapore

Water Affairs Bureau of Baoqing County (2005) The 11th five-year plan report of Baoqing County on water conservation development, Baoqing County, Shuangyashan City, Heilongjiang Province

Wei YP, Chen DL, Li BG (2007) Effectiveness of agricultural water pricing on solving the water shortage of North China Plain: a case study in Fengqiu County of Henan Province. Res Sci 29 (2):40–45

Xia GL, Wen JJ (2007) Water resources assessment report on Qixinghe National Nature Reserve and Qixinghe Watershed, Sanjiang Wetland Protection Project, Baseline Survey Report, (ADB Loan 2157-PRC and GEF Grant 4517-PRC), Harbin, Heilongjiang Province, China

Zhao Z (2006) Study on the mode and application of agricultural water saving at the middle reaches of Hei River. China rural water and hydropower, No. 8, pp 62–64

Part V
Adaptation to Climate Change

Chapter 15
Economic Vulnerability and Possible Adaptation to Coastal Erosion in San Fernando City, Philippines

Jaimie Kim B. Arias, Moises A. Dorado, and Rowena Dorado

Abstract Coastal erosion or shoreline retreat, resulting from natural and human activities, is currently affecting many Philippine coastal areas. This research evaluates the adaptation strategies of planned protection (i.e., construction of bulkheads and revetments complemented by bioengineering measures) and relocation relative to the business-as-usual strategy (the construction of bulkheads) in San Fernando Bay in La Union, Philippines. Study results show that about 300 structures, 283,085 m^2 of land, and 123,033 m^2 of beach along San Fernando Bay will be lost to coastal erosion/shoreline retreat by the year 2100. Planned protection, which is a combination of hard and soft measures, yields the highest net present value (NPV) of about Php 148.63 million under the assumption that beaches are not resilient (Scenario A) and about Php 126.78 million under the assumption that beaches are resilient (Scenario B), at a discount rate of 6 %. This strategy is supported by both the government and the locality. The key issue is where to get the funding to support the project.

Keywords Philippines • Adaptation • Coastal erosion • Cost-benefit analysis • Sensitivity analysis

---

J.K.B. Arias (✉) • R. Dorado
Department of Economics, College of Economics and Management, UPLB, Laguna, Philippines
e-mail: jkbayani@gmail.com; jbarias@up.edu.ph; wengdorado@yahoo.com; radorado@up.edu

M.A. Dorado
Agrometeorology and Farm Structures Division, Institute of Agricultural Engineering, College of Engineering and Agro-industrial Technology, University of the Philippines Los Baños (UPLB), Laguna, Philippines
e-mail: doradoma@yahoo.com

D. James, H.A. Francisco (eds.), *Cost-Benefit Studies of Natural Resource Management in Southeast Asia*, DOI 10.1007/978-981-287-393-4_15

Introduction

Background

Coastal erosion or shoreline retreat is currently affecting many coastal areas in the Philippines. It is also prevalent in other countries in Southeast Asia including Malaysia (Zamali and Lee 1991), Vietnam (Mazda et al. 1997; Ngo et al. 2006), Indonesia (Prasetya and Black 2003), and Thailand (Prinya 1996). Natural factors such as wind and waves, long-shore currents, and tectonic activities, as well as anthropogenic factors such as dam construction, sand mining, coral reef destruction, groundwater extraction, wetlands conversion, dredging of inlets for navigation, and boat traffic, have been identified as the factors contributing to the hazard. The impact of this coastal hazard is expected to become more widespread due to climate change and sea-level rise as well as with the continuing urbanization and development of coastal communities in the country. The hazard can inflict serious adverse impacts on society. Land, properties, infrastructure, and natural resources, such as sandy beaches, can be destroyed.

It is not surprising that huge efforts are being exerted to mitigate the impacts of the hazard. Typical responses usually involve protection activities or retreat. These adaptation options, however, entail large investments and sometimes even cause undesirable impacts. It is important, therefore, to carefully evaluate and assess the feasibility of these options before an action is taken. Studies have pinpointed that sea-level rise can exacerbate the extent of coastal erosion/shoreline retreat as low-lying areas become inundated (Mimura and Harasawa 2000; Hareau et al. 1999). With the archipelagic nature of the country, many areas in the Philippines are prone to coastal erosion/shoreline retreat. In fact, it has been documented in several areas of the country, including La Union (Salvador et al. 1997; Siringan et al. 2004), Bataan (Perez et al. 1999), and Leyte (Balce et al. 1999).

When coastal erosion/shoreline retreat occurs, adverse impacts on society, the economy, and the environment are to be expected. These include the loss of beaches, loss of land, loss of livelihood, displacement of people, and destruction of property and infrastructure. As a response, adaptation strategies are usually undertaken to address the hazard. These adaptations take the form of either protection or retreat/relocation which entail huge investments and sometimes even have undesirable impacts and consequences. It is necessary, therefore, that an evaluation of adaptation strategies be undertaken to ensure efficient coastal erosion/shoreline retreat management.

Recognizing this need, this study evaluated three adaptation strategies to coastal erosion/shoreline retreat in one of the coastal areas in the country identified to be experiencing the hazard, i.e., San Fernando Bay in San Fernando City, La Union Province. In support of this goal, the areas at risk to coastal erosion/shoreline retreat until 2100 were first delineated, and the economic vulnerability of these areas was then quantified.

Research Objectives

The general objective of this study was to estimate the economic vulnerability of San Fernando Bay in San Fernando City, La Union, to coastal erosion/shoreline retreat and identify and evaluate various adaptation options to address the hazard. The specific objectives were as follows:

- To identify and delineate critical areas at risk to coastal erosion/shoreline retreat until 2100
- To prepare an inventory and estimate of the value of resources, properties, structures, and economic activities at risk to coastal erosion/shoreline retreat
- To identify possible adaptation strategies to address coastal erosion/shoreline retreat
- To conduct a cost-benefit analysis on the identified adaptation strategies
- To evaluate the different adaptation strategies based on social, administrative, and legal/political feasibility

Methodology

Study Site

The selected study site, San Fernando Bay, is located in the northwestern part of San Fernando City, La Union. San Fernando City is a densely populated coastal city and the provincial capital of La Union Province. It is an important area not only because of its function as the administrative seat of the provincial government but also because of its economic and strategic significance. Lying about 270 km northwest of Manila, cradled by the South China Sea in the west, and the mountainous boundaries of the Cordillera Mountain Range in the east, the city serves as the gateway to the northern Philippines, the Ilocos Region. San Fernando Bay was chosen because it was identified as an area where coastal erosion was prevalent (Siringan et al. 2004) and which was densely populated. The area was already built-up; hence, there was no need to project the trends in the future development in the bay.

The study covers approximately 7 km of the 9-km coastline. The total population of the coastal barangays in San Fernando Bay is about 25,235 with households numbering 5,520. The average household size is five, and the average household income is Php 132,460 per annum. All the eight barangays are built-up and considered as 100 % urban by the City Planning and Development Office (CPDO). With an estimated poverty threshold of about Php 77,350 per annum, an estimated 44 % of the total population can be considered as poor. The unemployment rate is also very high, averaging about 57 %. With only half of the population having completed secondary education, most of the residents are engaged in service-related occupations. The primary occupations are fishing, fish vending, driving, construction labor, teaching, stevedoring, and overseas employment (San Fernando CPDO 2006a).

Factors Affecting Coastal Erosion

Siringan et al. (2004) identified five factors that contributed to coastal erosion along San Fernando Bay: (a) sea-level rise – global tectonics which cause inundation of low-lying areas, (b) climate change which causes changes in precipitation and storminess, (c) land cover changes which affected sediment yields of the river, (d) shifting river mouth positions which lead to sediment budget deficits along the coast, and (e) human activities which include mining, construction of seawalls and ripraps, and destruction of coral reefs, mangroves, and sand dunes.

Identification of Areas at Risk

The analytical methods applied in the prediction of the possible change in San Fernando Bay coastline included Markov chain analysis (MCA) (Anderson and Goodman 1957) and the cellular automata (CA). The MCA is a convenient tool for modeling change particularly when changes and processes are difficult to describe. The Markovian process enables modeling of a future condition purely from the immediately preceding condition, which means that the effects of the different factors are taken collectively. The MCA, however, lacks the ability to show where possible changes may occur. It quantifies the probability, but it does not show the spatial distribution of the predicted changes. The CA was therefore used in tandem with the MCA to introduce the spatial aspect to the modeling of change.

The 1966 San Fernando Harbor map was used as the initial state for the MCA. The present condition of the shore was developed from 2005 aerial photos and the GPS survey. This present condition was used as the second state for the MCA. The data from the 1977 topographic maps and SRTM were used as transition states/conditions.

Economic Vulnerability and Risk Assessment

Economic vulnerability, as used in this study, refers to the potential damage from coastal retreat under the "no action" assumption. The "no action" assumption is different from the "business as usual" in that the latter takes into consideration current adaptation measures already being undertaken.

From the “areas at risk” projections, the threatened lands, threatened buildings, and threatened beaches were first identified and delineated. An inventory of the buildings and infrastructures was then undertaken using aerial photographs from Google (2005) and the spot maps collected from the barangay offices. On-site validation was also conducted. Then economic vulnerability (EV) was computed based on three sources: (a) the value of threatened buildings and infrastructures, (b) the value of threatened lands, and (c) the value of environmental services from threatened beaches.

In the estimation of economic vulnerability, the following assumptions were made: (a) the current land use will prevail in the future (up to 2100); (b) the potential increase in land prices, due to population pressure and increase in demand, will be offset by the potential decrease in prices due to coastal erosion; (c) no new construction and real estate development will occur within the study period since open lands (undeveloped lands) comprise only a small portion of the total land area, while most of the vulnerable areas are already built-up; and (d) the depreciation of structures will be offset by repairs and renovation so that the houses will retain their current market values (up to 2100).

Value of Threatened Buildings and Structures

Data on property values was obtained by undertaking the following steps: (a) examination of tax maps to get the property identification number of the buildings within the areas at risk, (b) opening of books at the Assessor’s office to get the tax declaration numbers based on the property identification numbers, and (c) checking the tax declaration forms to get information on building characteristics, land use category, and market values. Properties within 100 m from the coastline were isolated, which served as the sample for computing the per square meter market value of the buildings. The 100-m delineation was used since the projected coastline retreat was estimated to average 100 m inward. The market values of buildings quoted in the books were then adjusted to 2006 price levels, and a statistical analysis was conducted to estimate the per unit values. To estimate the market value of buildings, the estimated per unit value was multiplied by the total building floor area. The cost of construction, on the other hand, was used in estimating the value of threatened public infrastructure, the data for which was sourced from the City Planning Office.

Value of Threatened Lands

For the value of the threatened lands, current market prices were used based on the average prices quoted by real estate brokers. The market value was estimated at Php 3,200 per square meter.

Value of Environmental Services from Threatened Beaches

Beaches provide important regulatory, ecological, and economic functions. One regulatory function of beaches is that they act as a natural protection and armor for coastal properties against storm surges and waves (US Army Corps of Engineers 1981). They also serve as habitats for diverse biological species and provide recreational services. To estimate the recreational value of the beaches in San Fernando Bay, the study applied the benefit-transfer method. In this study, the recreational values of beaches estimated by Colgan and Lake (1992) applied in a US Environmental Protection Agency research project on sea-level rise in Maine, USA, were used. The values were translated into local currency using the shadow exchange rate, and further adjustments were made based on the prevailing recreational use of the beaches in San Fernando City.

To approximate the economic value of docking services, the producer surplus of fishermen in San Fernando Bay was estimated. The producer surplus was calculated by deducting the costs incurred from the fishing activity from the gross income received from fishing. Secondary data from the San Fernando LGU Planning Office was used.

Benefits and Costs of Adaptation

In the CBA, a total of six scenarios (S1–S6) were considered (Table 15.1). The scenarios were based on two assumptions about the resilience of the beaches in San Fernando Bay (Scenario A, the beaches are not resilient; Scenario B, the beaches are resilient) combined with three assumptions about the level of erosion impacts – low, average, and high.

The estimation of the costs and benefits of the three adaptation strategies ("business as usual," planned adaptation, and planned retreat) was undertaken from a local perspective. The stakeholders included in the study were fishermen, households, and businesses situated in the areas at risk to coastal erosion/shoreline retreat in the study site as well as the local government of San Fernando City.

The benefit of an adaptation option is essentially the potential damages avoided as a result of undertaking the adaptation activity. This was computed as a fraction of economic vulnerability. Suppose, for example, a certain adaptation option is able to save x% of the total buildings and x% of the total land area at risk, then the benefit is computed as x% of the total value of threatened buildings and x% of the total value of land. The cost estimates, on the other hand, include construction and maintenance costs as well as the value of the potential losses resulting from undertaking the adaptation option. In the calculation of the stream of benefits and costs, it was assumed that the rate of coastal erosion/shoreline retreat was constant (i.e., the loss of land was spread evenly over time).

Table 15.1 Scenarios used in the study

Scenario	*Scenario A:* the beaches are not resilient and will be eroded away with shoreline retreat	*Scenario B:* the beaches are resilient and will remain in spite of shoreline retreat (even without applying beach erosion mitigation)
Low impact	S1	S2
Average impact	S3	S4
High impact	S5	S6

Note: S1–S6 in the table refer to different scenarios based on coastal erosion impact and beach resilience

To determine the economic acceptability of the adaptation options, the net present value (NPV) of each option was computed using the usual formula. The rule applied was to adopt the project with the largest NPV.

A sensitivity analysis was carried out with varied discount rates to determine the sensitivity of NPV calculations to the rates. The appropriate discount rate used to evaluate the options was assumed to range from 3 to 6 %, but a 15 % discount rate was also included in the analysis since this was the rate that the National Economic Development Authority (NEDA) was using in evaluating the acceptability of investment projects in the Philippines.

Results and Discussion

Areas at Risk

The current and projected shorelines by 2100 were superimposed so as to delineate the areas at risk. The projections are based on the cumulative effects of all factors including sea-level rise. There are no areas where accretion is projected. The possible reason for this is because of the assumption that all of the shoreline will be inundated by one meter because of sea-level rise by 2100, and this has created a future condition of guaranteed erosion along the shore. Coastal retreat was projected to range between 30 and 140 m inland. In all, the total land area at risk was estimated to be 283,085 m^2, while the total beach area at risk was 123,033 m^2 (Table 15.2).

Table 15.2 Summary of projected land and beach loss by 2100

Barangay	Land loss (sq. m.)	Beach loss (sq. m.)
Dalumpinas Oeste	59,107	38,435
Lingsat	89,693	17,375
Carlatan	30,117	8,482
Pagdaraoan	30,495	4,534
Ilocanos Norte	15,297	–
Ilocanos Sur	22,021	4,939
Catbangen/Poro	36,355	49,268
Total	283,085	123,033

Economic Values at Risk

Value of Threatened Lands

The impacts of coastal erosion/shoreline retreat are primarily felt through the loss of land resources. Based on the projections of shoreline retreat, the total land area at risk was segregated according to the actual use. Figure 15.1 shows that the largest proportion is currently allocated for residential use, followed by institutional, commercial, and open lands.

Institutional lands are those that are occupied by school and government buildings, while commercial lands are for restaurants, hotels, and other commercial establishments. It was estimated that the current total value of threatened lands was about Php 932.5 million or USD 21 million.

Value of Threatened Buildings and Structures

The number of residential structures at risk adds up to almost 300. These houses were categorized according to size: small for houses with a floor area of less than or equal to 50 m^2, medium for houses with 51–100 m^2 of floor area, and large houses with floor areas greater than 100 m^2 (Table 15.3).

Apart from the residential structures, there were institutional and commercial buildings situated within the areas at risk to coastal erosion. These included four educational establishments, three churches, a plant, two restaurants, a hotel, and four government-owned buildings. The total current value of the structures or capital threatened by coastal erosion was estimated to be Php 112.1 million or USD 2.5 million.

Value of Environmental Services from Threatened Beaches

The recreational value of beaches adopted by the US Environmental Protection Agency (EPA) in their 1995 research project on sea-level rise planning in Maine

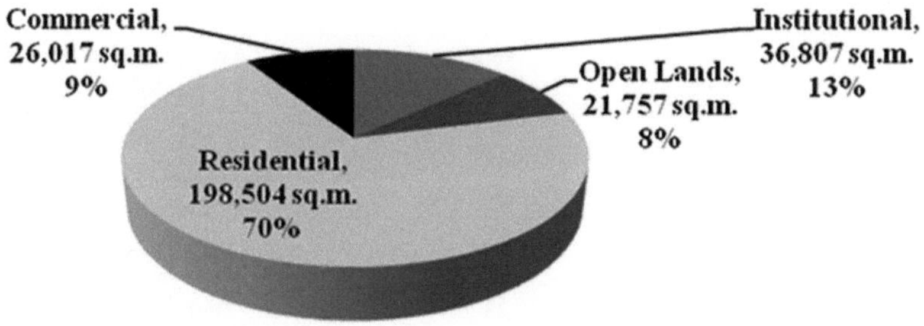

Fig. 15.1 Land areas at risk

Table 15.3 Number of residential structures at risk in San Fernando Bay

Barangay	Less than 50 m^2	51–100 m^2	Above 100 m^2	Total
Dalumpinas Oeste	4	10	12	26
Lingsat	6	51	18	75
Carlatan	0	0	1	1
Pagdaraoan	33	29	2	64
Ilocanos Norte	2	40	0	42
Ilocanos Sur	0	31	1	32
Catbangen	15	18	0	33
Total	60	179	34	273

(based on the 1992 Colgan and Lake valuation study in Casco Bay) ranged from USD 6 (low) to USD 54 (high) per person per day. In this study, the low estimate of USD 6/person/day was used. The simple benefit-transfer method was applied wherein the estimate was converted into local currency using the shadow exchange rate.

The results of the household survey showed that 79 % of the respondents utilized the beaches for recreational purposes. The respondents visited the beaches an average of five days per week for about two hours per day. As such, corresponding adjustments were made resulting in an estimated annual recreational value of Php 4.54 million, using the equation below:

$$\text{Annual Recreational Value of Beaches} = \text{USD6} \times \text{Shadow Exchange Rate (SER)} \times 0.79 \times \text{ San Fernando Bay coastal population of } 273 \times 260 \text{ days} \times (2\ \text{h}/24\ \text{h}).$$

Apart from recreational services, the beaches along San Fernando Bay also provide other use benefits. As of 2006, there were 130 registered fishermen and about 300 registered boats that docked along San Fernando Bay. From the secondary data collected from the CPDO, the average net income of these fishermen was about Php 566 per week (Table 15.4).

Table 15.4 San Fernando Bay fishing sector statistics

Barangay	No. of registered fishermen	Number of registered boats	Average net income per week (Php)
Dalumpinas Oeste	0	0	–
Lingsat	17	28	350.00
Carlatan	3	22	650.00
Pagdaraoan	9	19	434.00
Ilocanos Norte	13	23	350.00
Ilocanos Sur	49	82	477.00
Catbangen	21	63	1,470.00
Poro	25	72	1,539.00
Average	137	312	566.00

San Fernando City Planning and Development Office (2006b)

It was assumed that when the beaches along San Fernando Bay are completely eroded, access to the beach/sea will be lost, and fishermen living in the area will have no place to dock their boats. From an interview conducted with 10 fishermen, the respondents claimed that they had no other docking alternatives if the beaches along San Fernando Bay were to become completely eroded. Moreover, due to high unemployment in the city, the probability of changing jobs was very low. As such, it is expected that the loss of beaches will result in a net welfare loss to society. This welfare loss was approximated from the current net income that fishermen earned from fishing. Using the formula below, the docking value of the beaches was estimated at Php 7.99 million per year:

$$\text{Annual Docking Value of Beaches} = \sum \left(\text{No. of Boats}_j \times \text{Average Net Income}_j \times 52 \text{ weeks}\right)$$

where j = barangay.

Costs of Adaptation Strategies

For the "business-as-usual" strategy, under Scenario A, wherein the beaches are assumed to be lost as a result of coastal erosion, the costs cover only the construction and maintenance costs of building the bulkheads. While for Scenario B, wherein it is assumed that the beaches are resilient, the costs involve the construction and maintenance costs as well as the value of lost social services (recreation and docking) from the beach. Under this scenario, it is asserted that the main factor causing the loss of the beach's social services is the construction of bulkhead structures.

The construction and maintenance costs used in the calculations were based on the appraisal made by an engineering expert, which amounted to about Php 5,000

Table 15.5 Breakdown of the construction costs of the adaptation options (2006 prices)

Costs		Bulkheads	Revetments	Vegetation	Revetments + vegetation
Cost per meter (Php/m)	Labor	1,750	260	12	300
	Materials	3,250	1,040	88	1,200
	Total	5,000	1,300	100	1,500
Coastal length protected (meters)		800	3,500	1,000	1,500
Maintenance cost		40 % every 5 years after the 25th year			
Average lifespan (years)	50	50	50		50

per linear meter. The structures are expected to last about 50 years, after which the bulkheads will be replaced completely. Regular maintenance after the 25th year and every 5 years thereafter was deemed to be undertaken, the cost of which was estimated to be equivalent to 40 % of the construction cost (Table 15.5). In the estimation, the schedule of investments was projected based on the assumption that the households will start to erect structures only when their properties start to be encroached by the sea. The costs associated with lost recreational and docking services, on the other hand, were assumed to be incurred only after the beaches had been completely eroded.

For the planned retreat strategy, both scenarios entail costs associated with the loss of land, loss of buildings, and retreat costs. Retreat costs were assumed to be shouldered by the government and were estimated to be equal to the cost of procuring the properties. To provide an incentive for property owners to sell, a premium was assumed to be given by the government on top of the current market value. Thus, the acquisition cost was equal to the property value plus a 10 % premium. It was also assumed that relocation would be gradual. Only when the property was already within the salvage zone area would the government purchase the property. For this strategy, the major cost component is the purchasing or acquisition cost.

Comparing the three strategies, the highest cost is from planned retreat, followed by the "business-as-usual" strategy, while the least cost strategy is planned protection (Table 15.6).

Benefits of Adaptation Strategies

For the "business-as-usual" strategy, the benefits are attributed to the value of land and buildings saved in both Scenarios A and B. A large component of the benefit

Table 15.6 Present values of the costs of the "business-as-usual," planned protection and planned retreat strategies (in Php millions)

		Scenario A			Scenario B		
Business as usual	Discount Rate	Planned retreat	Business as usual	Planned protection	Planned retreat		
Low	1 %	82.40	37.92	672.31	284.54	59.88	672.31
	3 %	38.38	21.32	311.73	86.10	26.50	311.73
	4 %	29.06	17.60	233.09	52.80	20.18	233.09
	5 %	23.34	15.25	183.26	35.33	16.56	183.26
	6 %	19.70	13.72	150.02	25.84	14.39	150.02
	10 %	13.63	11.14	86.99	14.11	11.19	86.99
	15 %	11.67	10.40	58.12	11.70	10.40	58.12
Average	1 %	82.40	37.92	2,016.93	439.60	76.72	2,016.93
	3 %	38.38	21.32	935.20	146.78	33.10	935.20
	4 %	29.06	17.60	699.28	91.20	24.35	699.28
	5 %	23.34	15.25	549.77	59.81	19.22	549.77
	6 %	19.70	13.72	450.06	41.55	16.09	450.06
	10 %	13.63	11.14	260.98	16.97	11.50	260.98
	15 %	11.67	10.40	174.36	12.09	10.44	174.36
High	1 %	82.40	37.92	1,344.62	628.82	97.28	1,344.62
	3 %	38.38	21.32	623.47	256.39	45.00	623.47
	4 %	29.06	17.60	466.19	175.34	33.50	466.19
	5 %	23.34	15.25	366.52	124.75	26.27	366.52
	6 %	19.70	13.72	300.04	91.95	21.57	300.04
	10 %	13.63	11.14	173.98	36.17	13.59	173.98
	15 %	11.67	10.40	116.24	18.43	11.13	116.24

estimation comes from the value of saved lands, contributing about 65 % (at the 15 % discount rate) to 87 % (at the 1 % discount rate) of the total cost.

With the planned protection strategy, the benefits under Scenario A cover the value of land and buildings as well as the value docking and recreation from the saved beaches. The stream of benefits derived from the preservation of beach services was projected to begin only in the year when the beaches were expected to be complete (the 40th year). Also, since the protection strategy alters the natural landscape of the beach, it was assumed that only 70 % of the current recreational benefits will be saved by adopting the strategy, but 100 % of the docking benefits will be retained. For Scenario B, on the other hand, since beaches are assumed to be resilient and would not erode even without mitigation, the benefit of the strategy includes the value only of the saved land and buildings. The component that has the greatest weight in the benefit estimation is the value of saved lands which covers about 58–64 % of the total benefits under Scenario A and about 64–87 % under Scenario B for all discount rates.

For the planned retreat strategy, no benefits are expected to be derived under Scenario A since without protection, beaches are assumed to erode along with the

Table 15.7 Present values of the benefits of the "business-as-usual," planned protection and planned retreat strategies (in Php millions)

		Scenario A			Scenario B		
Business as usual	Discount Rate	Planned retreat	Business as usual	Planned protection	Planned retreat		
Low	1 %	320.15	500.32	–	320.15	320.15	202.14
	3 %	148.44	190.98	–	148.44	148.44	47.72
	4 %	111.00	132.16	–	111.00	111.00	23.74
	5 %	87.27	97.95	–	87.27	87.27	11.99
	6 %	71.44	76.91	–	71.44	71.44	6.14
	10 %	41.42	41.85	–	41.42	41.42	0.48
	15 %	27.68	27.70	–	27.68	27.68	0.03
Average	1 %	640.30	958.69	–	640.30	640.30	357.20
	3 %	296.89	393.52	–	296.89	296.89	108.41
	4 %	221.99	277.38	–	221.99	221.99	62.14
	5 %	174.53	207.03	–	174.53	174.53	36.46
	6 %	142.88	162.35	–	142.88	142.88	21.85
	10 %	82.85	85.82	–	82.85	82.85	3.33
	15 %	55.35	55.72	–	55.35	55.35	0.41
High	1 %	960.44	1,447.49	–	960.44	960.44	546.41
	3 %	445.33	639.66	–	445.33	445.33	218.01
	4 %	332.99	463.38	–	332.99	332.99	146.28
	5 %	261.80	352.19	–	261.80	261.80	101.41
	6 %	214.32	278.72	–	214.32	214.32	72.25
	10 %	124.27	144.36	–	124.27	124.27	22.54
	15 %	83.03	89.05	–	83.03	83.03	6.76

retreat of the coastline. For Scenario B, the benefit is the value of the social services derived from the saved beaches wherein the bulk of the benefit estimate is associated with docking services.

Table 15.7 shows the present values of the benefits that will be generated from the three adaptation strategies. Under Scenario A, the planned protection strategy generates the highest present value of benefits followed by the "business-as-usual" option and planned retreat. Under Scenario B, planned protection and "business as usual" generate equal benefits which are higher than the present value of the benefits from the planned retreat option.

Cost-Benefit Analysis Results and Sensitivity Analysis

The costs and benefits considered in the CBA calculation are summarized in Table 15.8, while the results of the net present value estimation are shown in Table 15.9.

Table 15.8 Cost-benefit framework

Adaptation option		Scenario A (beaches are not resilient)	Scenario B (beaches are resilient)
Business as usual (hold the line)	Benefit	Value of saved lands + value of saved buildings	Value of saved lands + value of saved buildings
	Cost	Construction cost of bulkheads	Construction cost of bulkheads + value of lost economic and recreational benefits from beaches
Planned protection	Benefit	Value of saved lands + value of saved buildings + value of economic services from saved beaches + 70 % (value of recreational benefits from saved beaches)	Value of saved lands + value of saved buildings
	Cost	Construction cost of bulkheads, revetments and planting of vegetation	Construction cost + 30 % (value of lost recreational benefits from beaches)
Planned retreat	Benefit	None	Value of economic and recreational benefits from saved beaches
	Cost	Value of lost lands + value of lost buildings + acquisition cost of properties	Value of lost lands + value of lost buildings + acquisition cost of properties

Regardless of the scenario, consistently positive net benefits for the planned protection and the "business-as-usual" strategies were derived, while the planned retreat option yielded consistently negative net benefits (Table 15.9). It is seen that at the 6 % discount rate under the "average" scenario, the "business-as-usual" option has an NPV of Php 123.18 million under Scenario A and Php 101.33 million under Scenario B. The planned protection strategy, on the other hand, has an NPV of Php 148.63 million under Scenario A and Php 126.78 million under Scenario B. Lastly, the planned retreat option garnered a negative NPV of Php 300.04 million under Scenario A and negative Php 278.19 under Scenario B.

Also, for all levels of discount rates and for the low and average scenarios, the NPV estimates from the planned protection strategy were consistently higher than the "business-as-usual" option. Furthermore, the ratio of the NPV of planned protection vis-à-vis "business as usual" is higher under Scenario B (where the beaches are assumed to be resilient) than under Scenario A. Moreover, the difference between the NPVs is more pronounced at lower discount rates than at higher ones. Under Scenario A, the gap between the NPVs of the two options can be largely attributed to the higher benefits derived from planned protection while there are only minimal differences in the costs. Under Scenario B, on the other hand, the difference comes primarily from the higher costs associated with "business as usual."

The planned retreat strategy produced negative NPV estimates because the value of the saved beaches could not offset the cost of relocation (purchasing properties at

Table 15.9 Net present values of adaptation strategies (in Php millions)

		Scenario A			Scenario B		
Business as usual	Discount Rate	Planned retreat	Business as usual	Planned protection	Planned retreat		
Low	1 %	237.74	462.41	(672.31)	35.61	260.27	(470.17)
	3 %	110.07	169.67	(311.73)	62.34	121.94	(264.01)
	4 %	81.94	114.56	(233.09)	58.20	90.81	(209.35)
	5 %	63.92	82.70	(183.26)	51.93	70.71	(171.27)
	6 %	51.74	63.19	(150.02)	45.60	57.05	(143.88)
	7 %	27.79	30.71	(86.99)	27.31	30.23	(86.51)
	10 %	16.00	17.30	(58.12)	15.98	17.28	(58.10)
	15 %	237.74	462.41	(672.31)	35.61	260.27	(470.17)
Average	1 %	557.89	920.77	(1,344.62)	200.69	563.57	(987.42)
	3 %	258.51	372.20	(623.47)	150.11	263.79	(515.06)
	4 %	192.94	259.78	(466.19)	130.80	197.64	(404.05)
	5 %	151.19	191.78	(366.52)	114.72	155.32	(330.05)
	6 %	123.18	148.63	(300.04)	101.33	126.78	(278.19)
	10 %	69.22	74.68	(173.98)	65.88	71.35	(170.65)
	15 %	43.68	45.32	(116.24)	43.27	44.91	(115.83)
High	1 %	878.04	1,409.57	(2,016.93)	331.63	863.16	(1,470.52)
	3 %	406.96	618.34	(935.20)	188.95	400.33	(717.19)
	4 %	303.93	445.77	(699.28)	157.65	299.50	(553.00)
	5 %	238.45	336.93	(549.77)	137.04	235.53	(448.36)
	6 %	194.62	265.00	(450.06)	122.37	192.75	(377.81)
	10 %	110.64	133.22	(260.98)	88.11	110.68	(238.44)
	15 %	71.36	78.65	(174.36)	64.60	71.90	(167.60)

10 % premium) and the value of lost properties. Even if it is assumed that the coastal erosion impact on properties is low, the NPV estimates for planned retreat are still negative.

Sensitivity analysis can determine whether the same conclusion will hold even if the impacts are significantly lower or higher than the projections. Under the low impact scenario wherein it is assumed that the rate of erosion of beaches is slower by 50 % compared to the average, the NPV of the planned protection strategy is still higher than the "business-as-usual" option by a ratio of 1.22 for Scenario A and 1.25 for Scenario B, using a discount rate of 6 %. This is equivalent to a difference of about Php 11.45 million in absolute terms (for both Scenarios A and B). Under the high impact scenario, the ratio increases to 1.36 for Scenario A and 1.58 under Scenario B, equivalent to a difference of about Php 70.4 million in absolute terms (Table 15.10).

Table 15.10 Ratios of and difference between the benefits and costs of the planned protection strategy (PP) vs. the "business-as-usual" strategy (BU)

		Net benefits				Benefits				Costs			
		Ratio: PP/BU		Difference: PP/BU		Ratio: PP/BU		Difference: PP/BU		Ratio: BU/PP		Difference: BU/PP	
		A	B	A	B	A	B	A	B	A	B	A	B
Low	1 %	1.94	7.31	224.66	224.66	1.56	1.00	180.17	–	2.17	4.75	44.49	224.66
	3 %	1.54	1.96	59.60	59.60	1.29	1.00	42.54	–	1.80	3.25	17.06	59.60
	4 %	1.40	1.56	32.62	32.62	1.19	1.00	21.16	–	1.65	2.62	11.46	32.62
	5 %	1.29	1.36	18.77	18.77	1.12	1.00	10.68	–	1.53	2.13	8.09	18.77
	6 %	1.22	1.25	11.45	11.45	1.08	1.00	5.47	–	1.44	1.80	5.98	11.45
	10 %	1.10	1.11	2.92	2.92	1.01	1.00	0.43	–	1.22	1.26	2.49	2.92
	15 %	1.08	1.08	1.30	1.30	1.00	1.00	0.02	–	1.12	1.12	1.28	1.30
Ave	1 %	1.65	2.81	362.88	362.88	1.50	1.00	318.39	–	2.17	5.73	44.49	362.88
	3 %	1.44	1.76	113.69	113.69	1.33	1.00	96.63	–	1.80	4.44	17.06	113.69
	4 %	1.35	1.51	66.85	66.85	1.25	1.00	55.39	–	1.65	3.74	11.46	66.85
	5 %	1.27	1.35	40.59	40.59	1.19	1.00	32.50	–	1.53	3.11	8.09	40.59
	6 %	1.21	1.25	25.46	25.46	1.14	1.00	19.48	–	1.44	2.58	5.98	25.46
	10 %	1.08	1.08	5.46	5.46	1.04	1.00	2.97	–	1.22	1.47	2.49	5.46
	15 %	1.04	1.04	1.64	1.64	1.01	1.00	0.37	–	1.12	1.16	1.28	1.64
High	1 %	1.61	2.60	531.53	531.53	1.51	1.00	487.05	–	2.17	6.46	44.49	531.53
	3 %	1.52	2.12	211.38	211.38	1.44	1.00	194.32	–	1.80	5.70	17.06	211.38
	4 %	1.47	1.90	141.84	141.84	1.39	1.00	130.39	–	1.65	5.23	11.46	141.84
	5 %	1.41	1.72	98.48	98.48	1.35	1.00	90.39	–	1.53	4.75	8.09	98.48
	6 %	1.36	1.58	70.38	70.38	1.30	1.00	64.40	–	1.44	4.26	5.98	70.38
	10 %	1.20	1.26	22.58	22.58	1.16	1.00	20.09	–	1.22	2.66	2.49	22.58
	15 %	1.10	1.11	7.30	7.30	1.07	1.00	6.02	–	1.12	1.66	1.28	7.30

Conclusions and Policy Implications

Scenario Outcomes

San Fernando Bay is vulnerable to the impacts of coastal erosion/shoreline retreat. By 2100, it is projected that about 300 structures, 283,085 m^2 of land, and 123,033 m^2 of sandy beaches will be lost due to the hazard. The current value of these capital and land resources is estimated at Php 112.1 million and Php 932.5 million, respectively, while the annual benefits from the threatened sandy beaches are approximated at Php 4.5 million for recreation and Php 8.0 for docking services.

The study identified three adaptation strategies/options to address the problem: "business as usual," planned protection, and planned retreat or relocation. A cost-benefit analysis was conducted to determine the economic feasibility of these strategies, while social, legal/political, and administrative feasibility assessments were also undertaken. Because the resilience of the beaches in San Fernando Bay is yet to be known, two scenarios were analyzed. In the first scenario, it was assumed that the beaches (in the study site) were not resilient and would be lost as a consequence of coastal erosion/shoreline retreat (Scenario A). The second scenario assumed that the beaches were resilient and would retreat inland (Scenario B). A sensitivity analysis was carried out to account for uncertainties by varying the discount rates (from 1 to 15 %) and the coastal erosion impact projections (low, average, and high).

In the "business-as-usual" framework, the study used the results of the survey to predict the future responses of property owners. From the survey and site visits, it was found that most of the household properties had started to be encroached by the sea and they were adopting the hold-the-line strategy, employing bulkheads/seawalls. It was predicted that this would be the prevailing strategy under this option. This strategy, however, has an undesirable consequence in that it will lead to the complete loss of the beach and restrict public access to the sea.

For the planned adaptation option, it was assumed that the government would intervene and implement hard (bulkheads and revetments) and soft (vegetation) protection along the coast. The goal of these interventions is to maintain public access and preserve the beaches while at the same time protecting properties and infrastructure along the coast. The last option, planned retreat, assumed that the government would prohibit any protection activities by property owners so that the shoreline was allowed to retreat. In the meantime, the government would gradually purchase properties situated in "risky" areas to ensure that these areas were vacated over time.

Under the "business-as-usual" strategy, there would be potential losses in terms of public access to the beaches and other benefits provided by the beaches. These externalities would arise from the autonomous protection activities of property owners, whose decisions are based upon expected private benefits and costs. When a shoreline property owner decides to construct a bulkhead to protect his property, the external costs (potential impacts on neighboring properties) are not considered in his decisions. A bulkhead can effectively protect the property immediately behind it, but it can also accentuate the erosion of the beach in front of it. If all shoreline property owners built bulkheads to protect their land, this would mean the subsequent loss of the beach resource and its services. In San Fernando Bay, the main use of the beach is for recreation by nearby residents and for docking by local fishermen. They are, therefore, the sectors that would be most affected if the "business-as-usual" scenario were to prevail.

From the analysis, government-financed planned retreat is not a viable option to pursue, basically because it would be very costly. However, it should be emphasized that the analysis considers solely the impacts of coastal erosion/shoreline retreat. If the damage costs of other coastal hazards such as tsunamis and typhoons, which can potentially cause harm to human health and life, are considered, the benefits of the retreat option may be significantly higher.

It can be concluded that among the three strategies evaluated, the planned protection strategy is the most rational option to adopt along the coast of San Fernando Bay. Such a strategy protects the welfare of property owners as well as satisfying the goal of preserving the beaches and the environmental services derived from using them. This can be seen from the results of the CBA, wherein the planned adaptation strategy consistently yielded the highest NPV estimates under all the scenarios assumed in this study and at varying discount rates. This adaptation strategy, which combines hard and soft protection, is also socially feasible with 65 % of the property owners interviewed expressing agreement with the implementation of the options. It is also politically feasible with 82 % of the city government officials interviewed expressing a willingness to support the strategy. Administratively, it is also relatively easier to implement than planned retreat/relocation and has a lower investment requirement of about Php 57 million compared with Php 1.15 billion for the latter. However, the "business-as-usual" option is more desirable in terms of administrative feasibility as it will entail no investment from the government.

Limitations of Study

It is important to note that the conclusions and recommendations derived from this study are based on projections covering a relatively longtime period of about 100 years. The implication of the long-term frame of analysis is that it became necessary to adopt certain assumptions particularly with regard to the future rate of erosion, future use of threatened lands and beaches, and the future response of the next generation in dealing with coastal erosion/shoreline retreat. The results of the analysis and conclusions are, therefore, contingent upon these assumptions.

There is also a need to emphasize that the conclusions are restricted to only the three adaptation options which were identified, based on a stakeholder's workshop and appraisals made by our technical expert. Moreover, since no specific engineering plan was drafted, the calculated construction and maintenance costs should be considered as rough estimates. Lastly, the study also did not include the potential impacts of pursuing protection activities in San Fernando Bay on the coastline of adjacent municipalities, particularly San Juan, La Union.

References

Anderson TW, Goodman LA (1957) Statistical inference about Markov chains. Ann Math Stat 28:89–110

Balce C, Leones J, De Leon P, Geosciences Bureau Tacloban City, Philippines (1999) Coastal erosion in Eastern Leyte Island, Philippines. In: Proceedings of the Committee for Co-ordination of Joint Prospecting for Mineral Resources in Asian Offshore Areas (CCOP) 35th session, Part II, Technical Report, Regional Offshore Prospecting in East Asia, UNDP, Bangkok

Colgan C, Lake F (1992) The economic value of Casco Bay, final report, Maine State Planning Office, Maine

Google (2005) Google Earth aerial photos. http:www.googleearth.com. Downloaded on 1 Aug 2007

Hareau A, Hofstadter R, Saizar A (1999) Vulnerability to climate change in Uruguay: potential impacts on the agricultural and coastal resource sectors and response capabilities. Clim Res 12:185–193. http://www.int-res.com. Downloaded on 11 Jan 2007

Mazda Y, Magi M, Kogo M, Hong NP (1997) Mangroves as a coastal protection from waves in the Tong King Delta. Viet Nam, Mangroves Salt Marshes 1:127–135

Mimura N, Harasawa H (2000) Data book of sea level rise 2000, Center for Global Environmental Research, National Institute for Environmental Studies, Environment Agency of Japan, Tsukuba, Ibaraki

Ngo N, Tien C, Pham HT, Do DS, Nguyen NB (2006) Status of coastal erosion of Vietnam and proposed measures for protection. http://www.fao.org/forestry/media/11286/1/0/. Downloaded on 11 Jan 2007

Perez RT, Amadore LA, Feir RB (1999) Climate change impacts and responses in the Philippines coastal sector. Clim Res 12:97–107. http://www.int-res.com. Downloaded on 11 Jan 2007

Prasetya GS, Black KP (2003) Sanur and Kuta Beaches in Bali: case studies for replacing traditional coastal protection with offshore reef. In: Proceedings of the third international surfing reef symposium, 23–25 June 2003

Prinya N (1996) Coastal erosion in the Gulf of Thailand. GeoJournal 38(3):283–300

Salvador JH, Kanbara GH, Sato T, Machida S, Almeda R, Garelan S, Laserna S (1997) Satellite monitoring of coastal topography in La Union, Northwestern Luzon, Philippines. Poster presented in GIS development conference 1997. GIS Development Network. http://www.gisdevelopment.net/aars/acrs/1997/ps2/ps4024.asp. Downloaded 11 Jan 2007

San Fernando City Planning and Development Office (2006a) Population survey of San Fernando City, La Union, Philippines

San Fernando City Planning and Development Office (2006b) Survey of the fisheries sector in San Fernando City, La Union, Philippines

Siringan F, Berdin R, Maria YS (2004) Coastal erosion in San Fernando, La Union: trends, causes and possible mitigation measures. Terminal report, Marine Geology Laboratory, National Institute of Geological Sciences, University of the Philippines, Diliman

US Army Corps of Engineers (1981) *Low Cost Shore Protection: A Guide for Local Government Officials*, US Army Corps of Engineers.

Zamali BM, Lee SC (1991) Analysis of coastal protection work along the Southwestern Coast of Johore, Malaysia. In: Chou LM, Chua T-E, Khoo HW, Lim PE, Paw JN, Silvestre GT, Valencia MJ, White AT, Wong PK (eds) Towards an integrated management of tropical coastal resources, International Center for Living Aquatic Resources Management (ICLARM) Conference Proceedings 22, October 28–31, 1988, National University of Singapore, National Science and Technology Board, Singapore, and International Center for Living Aquatic Resources Management, Philippines. http://test.esmology.com/water/images/pdf/did3.pdf. Downloaded on 17 Nov 2007

Chapter 16
Adaptation to Sea Level Rise in the Vietnamese Mekong River Delta: Should a Sea Dike Be Built?

Vo Thanh Danh

Abstract Vietnam is one of the five countries in the world most likely to be seriously affected by climate change, most particularly from sea level rise (SLR). This study was conducted in Tra Vinh Province, which borders the Southeast Asia Sea and therefore represents an area of the Vietnam Mekong Delta (VMD) that would typically be affected by SLR. A risk cost-benefit analysis of the proposed construction of a concrete sea dike system to protect from SLR is conducted. The study develops five dike options associated with three hypotheses on the scale of different sea dike systems. Option 1 represents a dike that could withstand the severity of a storm that occurs once every 20 years. Options 2 and 3 represent a dike that could withstand the severity of a storm that occurs once every 50 years. Lastly, options 4 and 5 represent a dike that could withstand the severity of a storm that occurs once every 100 years. Based on the NPV decision rule, the results indicate that dike options should be recommended as an appropriate adaptation measure for the VMD, and the larger in scale the dike system options are, the higher are the ENPVs. The study recommends constructing a small dike system first, with the future option of increasing its height. The findings support the necessity of constructing a concrete sea dike system in the VMD given the context of global climate change.

Keywords Vietnam • Adaptation • Sea level rise • Cost-benefit analysis • Sensitivity analysis

V.T. Danh (✉)
School of Economic and Business Administration, Can Tho University, Can Tho City, Vietnam
e-mail: vtdanhctu@yahoo.com.vn

D. James, H.A. Francisco (eds.), *Cost-Benefit Studies of Natural Resource Management in Southeast Asia*, DOI 10.1007/978-981-287-393-4_16

Introduction

Background

Vietnam is one of the five countries that may be the most seriously affected by global climate change and a consequent rise in sea level (SLR). If the sea level rises by between 0.2 and 0.6 m, 100–200 thousand ha of Vietnamese plains will be submerged. A 1-m rise would result in 0.3 to 0.5 million ha of the Red River Delta being submerged, and 90 % of the Vietnam Mekong Delta (VMD) would be flooded. The SLR scenarios released by the Ministry of Natural Resources and Environment (MONRE) in 2009 were constructed with two levels of environmental emergency in mind – a high level and a medium level. The results showed that compared to 1980–1999, on average, the SLR would measure between 30 and 33 cm and between 74 and 100 cm by the middle and end of this century, respectively. Table 16.1 presents the SLR scenarios for Vietnam.

According to the MONRE's forecast, due to the impact of a global rise in sea levels, 15,000–20,000 km^2 of the VMD's coastal areas would be inundated. Nine of its thirteen provinces would be completely below water. The current sea dike system in coastal areas cannot effectively protect people and the land when storms and high tides occur at the same time. The construction of a sea dike has to be considered as a potential solution to a rise in sea level.

Sea Dikes in Vietnam

The dike system in the VMD is 1,400 km long, with nearly 620 km of sea dikes. The current sea dike system in the VMD is made out of earth, set into earth. These dikes are not able to cope with the level eight storms that come from the Southeast Sea during the flood season. The sea dike system in the VMD has a unique characteristic – the sea dikes were created in conjunction with planting mangrove forests, to further protect the dike.

Some VMD provinces have no sea dikes or not enough sea dikes to prevent seawater intrusion. For example, in Ca Mau Province, along a coastline of 254 km, there is only 93 km of earth-built sea dikes in the west, and there are no sea dikes in the east. Every year sea dikes are destroyed by tides, floods, and storms. The repair and maintenance of sea dikes requires a great deal of money, which is supplied by the national dike management budget.

In May 2009, the government issued Decree No. 667/QĐ-TTg regarding sea dike maintenance and upgrading. The implementation programme is divided into three periods:

2009–2012, 2013–2016, and 2017–2020. From 2009 to 2012, mangrove forests will be planted parallel to the sea dike system. From 2013 to 2016, the sea dike system will be upgraded and developed alongside the road network. From 2017 to

Table 16.1 Forecast sea level rise in Vietnam, 2020–2100, compared to 1980–1999 (MONRE 2009)

	Time								
Scenario	2020	2030	2040	2050	2060	2070	2080	2090	2100
High	12	17	24	33 cm	44	57	71	86	100 cm
Medium	12	17	23	30 cm	37	46	54	64	74 cm

2020, a sluice system will be constructed so that the sea dike system can be operated for the purposes of both adapting to an SLR and for transportation. However, up to the year 2020, the sea dike system from the centre of Vietnam to the south will still be an earth-built one. The main objective of the programme is to establish the sea dike system in preparation for the impacts of a rise in sea level. The total budget for the programme is 19.5 thousand billion VND (more than 1 billion USD).

The necessity of investing in a concrete sea dike system in the VMD is the subject of an ongoing policy debate in Vietnam. Some think that the government should not build a concrete sea dike system for the VMD. The necessity of investing in a concrete sea dike system in the VMD is the subject of an ongoing policy debate in Vietnam.

Some think that the government should not build a cement sea dike system for the VMD. The reason given for such a view is that a concrete sea dike system will need billions of USD of investment and will not be effective. An alternative solution is proposed, which combines policies of moving people in the affected areas during a natural disaster and adapting life in coastal areas (by increasing collective action and public awareness of the measures necessary for living with SLR). On the other hand, proponents of the sea dike system think that the VMD needs a large sea dike system, like the Netherlands, because the VMD is surrounded by sea and it faces a high risk of SLR due to global climate change. The national budget would not be sufficient for such a big investment.

In summary, there are two different points of view: one is an adaptation policy and the other is a coping policy. The question of whether a sea dike system needs to exist or not needs to be answered. The VMD's agriculture-based economy would certainly be affected by a rise in sea level, and the region has to prepare for future changes. This study proposes an economic valuation of a concrete sea dike system as an adaptation to the impacts of a rise in sea level. The study uses a risk cost-benefit analysis (CBA) framework.

Research Objectives

The overall objective of this study was to assess the viability of sea dikes as a structural response to a rise in sea level caused by climate change and also to investigate the acceptability of such a project in the Vietnamese Mekong Delta. The specific objectives were:

- To develop a risk assessment and cost-benefit analysis framework specific to SLR in the VMD
- To assess the viability of building sea dikes as an infrastructure response to the impacts of climate change
- To identify the levels of risk, cost, and benefit associated with the sea dike option

Methodology

Study Location

The selected study area was Tra Vinh Province. Tra Vinh Province is located at the southeast end of the VMD, between the Tien River (Co Chien River) and Hau River. The climate of the province is tropical monsoon. The eastern border of the province sits on the South China Sea. The province's natural area measures 223,000 ha, and the seashore has a length of 65 km. The entire coastal area of Tra Vinh is affected by high tides and seawater intrusion. More than 90 % of the total agricultural land area of 90,000 ha suffers from seawater intrusion. Salinity usually begins in December at Hung My, at the Co Chien River, and Tra Kha, at the Hau River.

Data Collection

Data collection was conducted in Cau Ngang, Duyen Hai, Cang Long, and Cau Ke districts of Tra Vinh Province. Cau Ngang and Duyen Hai districts are in coastal areas and Cang Long and Cau Ke are not. These districts were chosen to assess the impact of salinity and seawater intrusion. With the assumption that rice production in Tra Vinh has homogeneous characteristics, a production function with the salinity impact dummy variable allows measurement of marginal productivity loss due to salinity. A sample of 115 rice farmers from Cau Ngang and Duyen Hai districts was taken. These areas are affected by salinity. A sample of 118 rice farmers was taken from parts of Cang Long and Cau Ke districts. These areas are not affected by salinity.

To prepare for the calculations of the CBA model, socio-economic data was collected. Table 16.2 shows forecast indices for 2010–2020. Baseline calculations were implemented based on these values.

Table 16.2 Forecast indices for Tra Vinh Province, 2010–2020 (Vietnam Master Plan 2011)

Item	Unit	2010	2015	2020
1. Natural area	km2	2,292.8	2,292.8	2,292.8
2. Population, in which:	1,000 persons	1,005	1,026	1,046
Urban population	1,000 persons	150.8	266.6	314
3. GDP (1994), in which:	Bil. VND	8,211	14,470	30,430
Industry/construction	%	1,710.9	4,236.4	11,308.2
Agriculture/forestry/aquaculture	%	3,629.7	4,467.0	5,617.8
Services	%	2,870.4	6,426.6	13,504.0
4. Percentage of urbanization	%	15	26	30
5. GDP growth rate, in which:	%	11.64	13	15
Industry/construction	%	18.50	19.88	21.70
Agriculture/forestry/aquaculture	%	4.66	4.24	4.69
Services	%	20.87	17.49	16.01
6. GDP (2005)	Bil. VND	11,681	23,451	48,688
	Bil. USD	0.65	1.31	2.71
7. GDP percentage (2005)	%	100.0	100.0	100.0
Industry/construction	%	23.59	28	36
Agriculture/forestry/aquaculture	%	43.85	40	30
Services	%	32.56	32	34
8. GDP per capita (2005)	Mil. VND	11.6	22.8	46.5

Note: exchange rate – USD 1 = VND 18,000

Theoretical Framework

The study used a risk CBA framework that considers the likelihood of an extreme storm event and SLR. In a traditional CBA, all the variables in the model are non-random and have single values. However, in the risk CBA framework, critical variables relating to the probability of an extreme event (storm) are random. This allows consideration of both the range of values of the variables and the method of measuring the values of variables in the context of the likelihood of an extreme storm event. To do the risk analysis, one needs an assessment of probability with which changes in critical variables may occur. By assigning appropriate probability distributions to the critical variables, probability distributions for the economic indicators can be estimated. For the critical variables relating to the extreme event, a binomial distribution function is built. A simulation model is then used to obtain the expected/forecast values for the risk CBA calculations.

Following Boardman et al. (2001) and introducing risk analysis into the CBA study, a risk assessment and cost-benefit analysis framework specific to the SLR sea dike options in the VMD can be described as in Fig. 16.1.

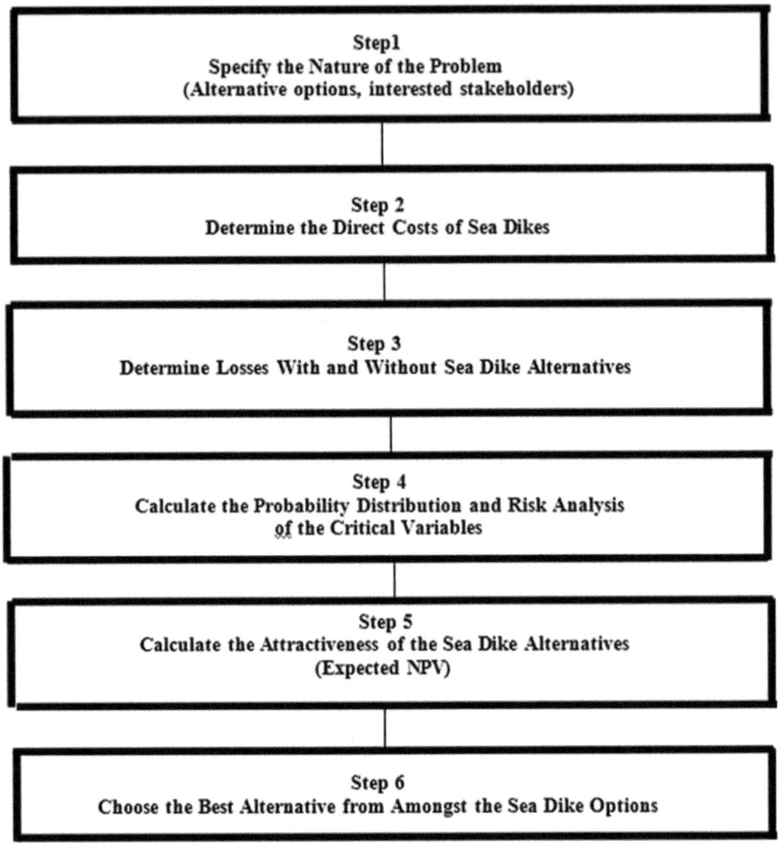

Fig. 16.1 The risk CBA framework

Problem Description

As previously mentioned, the sea dike system runs along the coastal areas of the VMD. Most of the existing sea dikes in Tra Vinh Province are made of earth, except for one 615-m concrete sea dike at Bao Village, Hiep Thanh Commune, Duyen Hai District. In this CBA study, the base scenario is the status quo, in which there are no concrete sea dikes. A permanent concrete sea dike system is a principal potential adaptation strategy for mitigating the impacts of SLR. The alternative options to concrete sea dikes also need to be specified, and the interested parties need to be identified.

Identifying alternatives to sea dikes depends on timing, size, and construction materials. Firstly, the time factor chosen in this study is long term – a concrete sea dike system must survive for a very long time, for instance, 100 years. The proposed lifespan of a permanent concrete sea dike network for the VMD is 100 years. Secondly, the scale of the sea dike system depends on the SLR scenario that is selected, the probability of the occurrence of an extreme storm event, and which safety standards are specified. Thirdly, the construction costs vary depending on the types of materials chosen for the dike.

Three scales of sea dike system can be considered: the first is small in scale (a dike that can withstand a storm that occurs once every 20 years), the second is medium scale (one that can withstand a storm that occurs once every 50 years), and the third is large scale (one that can withstand a storm that occurs once every 100 years). The time frame given to the construction of the sea dike is important. Technically, the dike may be constructed in one operation or in sequential periods. This study examines five different dike options (see below) associated with different scales, construction phases, and lifespans:

Option 1: A small-scale dike 2 m high, lasting 50 years.

Option 2: A medium-scale dike 3 m high that is constructed all-in-one operation with no plans for future upgrading. The lifespan of this dike is 100 years.

Option 3: A medium-scale dike with a height of 3 m that is constructed over two time periods, with the initial investment in a dike on as small a scale as option 1 (at a height of 2 m) but with the body of the dike constructed on the medium scale; at the second phase of construction, the dike would be upgraded to a height of 3 m. The lifespan of this dike is 100 years.

Option 4: A large-scale dike constructed all at once, with a height of 4 m and no plans for upgrading in the future. The lifespan of this dike is 100 years.

Option 5: A large-scale dike constructed in two phases, with the initial investment of an option 1 dike (height of 2 m) but with the body of the dike constructed on a large scale; at the second phase, additional investment would raise the dike to 4 m. The lifespan of the dike is 100 years.

Using a base scenario of "no concrete sea dike system" allows one to compute the differentials between with and without alternative option values. These values are the costs and benefits used in the CBA calculations.

Quantifying the Costs

The cost of sea dikes depends on the safety standards that are adhered to and their scale. In Vietnam, dike costs vary because of the differing prices of materials, land use, and revetments. The cost of labour is highly variable but constitutes a relatively small percentage of the total cost (Hillen et al. 2010; Mai et al. 2008). Because information about the cost of dikes was not available from the local dike management authority, the cost category in this study uses the dike cost calculations given by Mai et al. and by Hillen et al. for a typical sea dike in rural Vietnam. Mai et al. (2008) determined the cost of dike heightening with a comparable probabilistic approach to ascertain the safety standards of the sea dike system. Mai et al. used both outer- and inner-slope protections and included the costs of maintenance in the dike costs category. Because dike costs data are estimated at different levels, in the risk CBA framework, the probability distributions of the construction costs, maintenance costs, and dike heightening variables were assigned to have a uniform distribution with the minimum values of the Hillen et al. estimations and the maximum values of the Mai et al. estimations.

Quantifying the Benefits

Using sea dikes as a coastal defence avoids damage to the VMD. In this study, two types of damage were avoided: (1) loss of life, homes, infrastructure (roads, electricity network, water connections, etc.) due to storms and flooding and (2) loss in yields of rice and aquaculture due to salinity. The measurement of each type of benefit was calculated by the methods described in the following sections.

Avoidance of Storm Damage

Storm damage can incur loss of life, homes, and infrastructure (roads, electricity network, public facilities, etc.) and the destruction of rice and aquaculture production. According to the National Centre for Hydrometeorological Forecasting (NCHMF), MONRE, from 1961 to 2010, 258 storms hit Vietnam, 17 of which were in the south of the country. While many strong storms (level 11 and above, >103 km/h) have visited other parts of Vietnam, the VMD has rarely been a victim of this type of natural disaster. During this period, 43 level 11 (and above) storms hit Vietnam (or 16.7 % of the total), and 9 storms (or 3.5 % of the total) reached level 13 and above (>133 km/h). From 1961 to 2010, only 17 storms, or 6.6 % of the total number of storms across Vietnam, hit the MRD, and only one of these storms reached level 11, with an additional one attaining level 13.

The frequency of storms in the VMD follows a pattern: once every 4 years, there is a level 6 storm (39–49 km/h); once every 10, years there is a level 8–10 storm (62–102 km/h); once every 20 years, there is a level 11 storm (103–117 km/h); and once every 50 years, there is a level 13 storm (>133 km/h). However, for a project as huge as the sea dike system, the probability distribution of storms (and floods) needs to simulate beyond the 1961–2010 time frame.

An alternative is the World Bank's (2010) simulation of the economic losses caused by storm and flood events with different return period assumptions. The World Bank assessed economic losses caused by storms that take place once every 10 years (0.013 % of national GDP), once every 50 years (0.023 % of national GDP), and once every 100 years (0/03 % of national GDP).

The estimated economic losses caused by storms in the VMD, based on the Vietnam Central Committee for Flood and Storm Control (CCFSC) storm cost report, were consistent with the World Bank's estimates. For example, with the scenario of a "once-every-40-years storm", the percentage of economic loss of the

VMD's GDP was 0.016 %, compared to 0.023 % for the "once-every-50-years storm" scenario of the World Bank's projection. The benefit due to the avoidance of losses due to storms was estimated as follows:

$$\text{Strom loss avoided} = \sum_{i=1}^{100} \sum_{k=1}^{5} \%\ \text{strom loss in GDP RPk} \times \text{GDP}_\text{i} \quad (16.1)$$

where RPk: return period k ($k = 1$–5)
GDP$_i$: GDP at time i ($i = 1$–100, i.e. 2010–2110)

Avoidance of Flood Damage

The VMD is an area familiar with flooding. Flooding occurs frequently and brings much damage to the region. According to a Central Committee for Flood and Storm Control (CCFSC) report (2011), from 1991 to 2005, the VMD suffered eight floods, and each one brought significant economic losses. Similar to the storm loss estimations, the flood scenarios used in this study were based on a combination of World Bank simulations (2010) and calculations by the CCFSC. There are four flood scenarios for the VMD: flooding once every 2 years, flooding once every 10 years, flooding once every 50 years, and flooding once every 100 years. The benefit due to the avoidance of losses due to flooding was estimated as follows:

$$\text{Flood loss avoided} \sum_{i=1}^{100} \sum_{k=1}^{4} \%\ \text{flood loss in GDP RPk} \times \text{GDP}_\text{i} \quad (16.2)$$

where RPk: return period k ($k = 1$–4)
GDP$_\text{i}$: GDP at time i ($i = 1$–100, i.e. 2010–2110)

Reduction of Damage from Seawater Intrusion

Sea and river dike systems help to protect the land from seawater intrusion and salinity, which reduce yields of the rice and cash crops that are the main agricultural products of coastal areas. The benefit gained from avoiding salinity is at least the cost of building the sea dike system. In order to measure the value of losses in agriculture and aquaculture production due to salinity, a damage function was designed. Damage was defined as a loss of productivity due to salinity. Hypothetically, as the degree of salinity increases, the productivity of rice farming and fishing decreases.

The total losses are the product of the marginal loss of productivity (per ha) and the area affected. To estimate the impacts of salinity on farming yields in the affected areas, a simple production function was specified:

$$Q_i = a_0 + a_1 L_i + a_2 K_i + a_3 S_i + ui \tag{16.3}$$

where

Q_i: yield of product *i* (rice (tonne/ha)
L: labour (kg/ha)
K: capital (fertilizer) (kg/ha)
S: dummy variable (1 for salinity; otherwise 0)
a*i*: regression's coefficients

Equation 16.3 includes two types of explanatory variables: yield-increased variables (labour and capital) and yield-decreased variables (salinity and distance).

A decomposition analysis was used to measure the impact of salinity on productivity. Production function decomposition analysis allows decomposition of the difference in the change in farming productivity between land affected by salinity and land unaffected by salinity. That is, the changes are decomposed into two components: changes due to the effects of salinity and input reallocation. A production function in a log-linear form is shown below:

No salinity:

$$LnQn = LnA_n + a_1 LnL_n + a_2 LnK_n \tag{16.4}$$

With salinity:

$$LnQs = LnA_S + b_1 LnL_S + b_2 LnK_S + b_3 LnS \tag{16.5}$$

Taking the difference between Eq. 16.5 and Eq. 16.4 and rearranging terms result in the following:

$$\begin{aligned} Ln(Qs/Qn) = {} & Ln(As/An) + (b_1 - a_1)Ln(L_S/L_n)) + (b_2 - a_2)Ln(K_S/K_n) \\ & + b_3 LnS. \end{aligned} \tag{16.6}$$

The coefficient b_3 in Eq. 16.6 implies the marginal loss of productivity due to salinity impact separately, while other coefficients, (b_1–a_1) and (b_2–a_2), show the impact of differences in labour and capital, respectively. It is expected that the sign of coefficient b_3 will be negative in the estimation.

The benefit of avoiding the negative impact of salinity, due to a dike system, is measured as follows:

$$\text{Salinity loss avoided} = \text{marginal productivity loss} \times \text{total areas affected} \tag{16.7}$$

Probability Distributions and Risk Analysis of Critical Variables

The risk analysis in Step 4 is central to the risk CBA framework. Methodologically, a risk CBA not only considers the range of values of the variables but also attaches to these values a measure of the likelihood of their occurrence. Two uncertainty variables need to be taken into account in the sea dike projection: storms and flooding. Estimation of these critical values must be implemented via the risk analysis framework. In this study, a simulation analysis is applied to obtain the expected values of these uncertainty variables.

In the VMD, storms are not an annual weather phenomenon – they occur rarely in the region. However, when storms do occur, losses are usually large. As storms are a discrete variable, a certain probability distribution function is specified in order that, based on the type of probability distribution function, a simulation model can be run to estimate the expected value of the critical variable. Storm records for 1961–2010 were used to predict the form of probability distribution. Five return periods were specified with a binomial probability distribution function for the storm variable: once every 4 years, once every 10 years, once every 20 years, once every 50 years, and once every 100 years. The value of economic damage associated with each storm frequency was estimated by simulation analysis using Crystal Ball® software.

In contrast, flooding in the VMD is usually riverine in nature, rather than flash flooding, as in other parts of Vietnam. Flooding in the VMD causes significant economic damage. Four flood scenarios were selected: once every 2 years, once every 10 years, once every 50 years, and once every 100 years. In order to estimate the expected values of this critical variable, a binomial probability distribution function was used. Similarly, the values of economic damage associated with each flood frequency were estimated by simulation analysis using Crystal Ball® software.

The exact costs of dike construction and dike heightening are unknown. Some studies (Hillen et al. 2010; Mai et al. 2008) have estimated the typical costs for a typical sea dike in Vietnam but differences in technical specifications, location, region, etc., make accurate costings of dike construction and heightening problematic. In order to overcome these estimation difficulties, expected values have been calculated using Crystal Ball® software based on the assumption of a uniform probability distribution function, with maximum and minimum values given.

Finally, the area of agricultural land affected by salinity cannot be accurately measured. At Tra Vinh, more than 90 % of the total agricultural land area is affected by salinity[1], but the salinity status of other salinity-affected regions differs. Therefore, estimating the area affected by salinity needs to be done with the uncertainty condition.

[1] 2020 Socio-economic Master Plan, Tra Vinh province

In this study, a uniform probability distribution form was assigned for this variable using Crystal Ball® software to predict the percentage of agricultural land area affected by salinity.

Baseline Alternatives

In order to construct the baseline for the risk CBA assessment, the study used secondary data from Tra Vinh's Master Plan for 2010–2020 and also the 2025 Vision. Based on the projections in the Master Plan, further calculations for 2030 were done in order to construct a baseline for 2010–2030. A simple regression model was used to forecast values of indicators (e.g. GDP values of the areas bounded by the dike system). It is plausible that the whole of Tra Vinh Province could be protected by the river dike and sea dike systems. The value of flood and storm losses avoided is based on the proportion of losses per GDP value (see Eqs. 16.1 and 16.2). Based on the development indices shown in the Tra Vinh's Master Plan, the following situation is forecast for 2010–2110 (Table 16.3).

Calculating Desirability of Sea Dike Alternatives

In this step, all positive and negative impacts were calculated in monetary terms. A social discount rate of 3 % was assigned to calculate the ENPV for all benefits and costs at different times.

Choice and Sensitivity Analysis of Best Alternative

Based on the project evaluation criteria, the best sea dike alternative from the options is proposed. Before the final recommendations are made to policy makers, a further sensitivity analysis will be conducted.

Findings and Discussion

Cost Measurement

Dike Construction Costs

Five dike options were considered in this risk CBA study. Construction costs and maintenance costs were projected across all the options, but the costs of dike

Table 16.3 Forecasts of GDP values and rice and aquaculture areas in period of 2010–2110 (Vietnam Master Plan 2011)

Forecasted value	2010	2020	2030	2040	2050	2060	2070	2080	2090	2100	2110
GDP (billion US$)	0.65	2.02	5.24	13.58	26.72	52.56	103.39	203.38	301.05	445.6	659.6
Annual increase in GDP (%)	12	12	10	10	7	7	7	7	4	4	4
Rice area (ha)	90,000	89,000	89,000	89,000	89,000	87,000	87,000	87,000	87,000	85,000	85,000
Aquaculture area (ha)	55,000	55,000	55,000	55,000	55,000	55,000	55,000	55,000	55,000	55,000	55,000
In which:											
Fresh-water areas	10,000	15,000	15,000	15,000	15,000	15,000	15,000	15,000	15,000	15,000	15,000

heightening were only applied to options 3 and 5. The estimations of dyke costs were based on the dike cost projections in Hillen et al. (2010) and Mai et al.'s (2008). According to Hillen and Mai, construction costs include the cost of creating the body of the dike, land use, berm, outer protection, and inter protection or revetments. These construction costs vary because of the differing costs of materials and land use and the application of inner and outer protection or revetments. Although labour costs are important, they were relatively small in the overall scheme of dike construction costs. In these two studies, dike construction costs were projected for the dike system in rural areas.

Dike Maintenance Costs

The annual dike maintenance cost comprises a small amount of total dike capital budgeting. Based on the dike department and ministry budgets, Hillen et al. (2010) estimated the yearly dike maintenance cost for 1 km of dike as USD 27,000. Table 16.4 shows the construction costs and maintenance costs of dikes at the heights estimated by Hillen et al. (2010) and Mai et al. (2008).

Dike Heightening Costs

In order to calculate the dike heightening costs at the second phase for option 3 and option 5, the unit cost price standards given by IPCC CZMS (1990), cited by Hillen et al. (2010), are applied in this study. Table 16.5 shows the unit cost prices of coastal defences with the assumption of "all-in" costs for dike construction. Since most of the sea dikes in the VMD, and in Tra Vinh Province in particular, are in rural areas, the dike heightening costs chosen for calculations in this study were in the range of USD 0.702–1.404 million per km.

Cost Simulations

Cost estimations by Hillen et al. (2010) and Mai et al. (2008) were dependent on various assumptions regarding safety standards and the frequency of natural disasters such as storms and floods. These assumptions vary under different uncertainty conditions. The cost estimations by Hillen et al. (2010) were lower than those of Mai et al. (2008). Based on this, the probability distributions of construction costs, maintenance costs, and the costs of dike heightening were assigned uniform distributions with minimum values and maximum values. Table 16.6 shows the simulation values of the cost variables used in the CBA calculations.

Some assumptions regarding the longevity of the dikes were made for each option: option 1 assumed a life of 50 years, and all the other dikes were assumed to have a life of 100 years. In the case of dike heightening, after a period of 20 years,

Table 16.4 Decomposition of dike construction and maintenance costs (per km) in rural Vietnam (Hillen et al. 2010)

	Height at 2 m			Height at 3 m			Height at 4 m		
	Hillen		Mai	Hillen		Mai	Hillen		Mai
Cost category	(M€)	(M$)	(M$)	(M€)	(M$)	(M$)	(M€)	(M$)	(M$)
Dike body	0.286	0.386	0.92	0.471	0.636	1.38	0.729	0.984	2.00
Land use	0.071	0.096	0.31	0.229	0.309	0.72	0.400	0.540	1.00
Berm	0.014	0.019	–	0.014	0.019	–	0.014	0.019	–
Revetment	0.486	0.656	–	0.714	0.964	–	0.929	1.254	–
Outer/inter protection	–	–	0.57	–	–	0.58	–	–	0.71
Maintenance	–	–	0.02	–	–	0.02	–	–	0.04
Total cost	0.857	1.157	1.82	1.429	1.929	2.70	2.071	2.796	3.75

Table 16.5 Unit cost prices of dike heightening (USD/km) (Hillen et al. 2010)

Type of coastal defence measure	Unit cost IPCC CZMS (1990) (2009 price level, USD)
New 1-m-high sea dike	0.55
New 1-m-high sea dike with regular maintenance	0.84
Raising low sea dike by 1 m in rural areas	0.70
Raising high sea dike by 1 m in rural areas	1.40
Raising sea dike by 1 m in urban areas	14.03

Table 16.6 Simulation results, dike costs, CBA analysis

Cost component	Distribution	Minimum value	Maximum value	Simulated value	Skewness
Construction cost					
2-m-high dike	Uniform	1.16	1.82	1.49	0
3-m-high dike	Uniform	1.93	2.7	2.31	0
4-m-high dike	Uniform	2.8	3.75	3.27	0
Maintenance cost	Uniform	0.027	0.04	0.3	0
Heightening cost					
By 1 m	Uniform	0.702	1.404	1.05	0
By 2 m	Uniform	1.41	2.81	2.11	0

the dikes in options 3 and 5 were heightened to the level of the dikes in options 2 and 4, respectively. The total length of the dikes in Tra Vinh Province is 147 km. Table 16.7 shows the total costs of the proposed dike options.

The dike cost values were used to calculate the present values of dike options. Table 16.8 shows a summary of results of present values of dike costs with a discount rate at 3 % and a timeline of 100 years. Results showed that option 4 had

Table 16.7 Costs of dike options (million USD)

Cost category by option	2010	2020	2030	2040	2050	2060	2110
Option 1	221.97	2.94	2.94	2.94	2.94	221.97	2.94
Construction cost	219.03	0	0	0	0	219.03	0
Maintenance cost	2.94	2.94	2.94	2.94	2.94	2.94	2.94
Heightening cost	0	0	0	0	0	0	0
Option 2	5.25	2.94	2.94	2.94	2.94	2.94	2.94
Construction cost	2.31	0	0	0	0	0	0
Maintenance cost	2.94	2.94	2.94	2.94	2.94	2.94	2.94
Heightening cost	0	0	0	0	0	0	0
Option 3	221.97	2.94	157.29	2.94	2.94	2.94	2.94
Construction cost	219.03	0	0	0	0	0	0
Maintenance cost	2.94	2.94	2.94	2.94	2.94	2.94	2.94
Heightening cost	0	0	154.35	0	0	0	0
Option 4	486.57	5.88	5.88	5.88	5.88	5.88	5.88
Construction cost	480.69	0	0	0	0	0	0
Maintenance cost	5.88	5.88	5.88	5.88	5.88	5.88	5.88
Heightening cost	0	0	0	0	0	0	0
Option 5	221.97	2.94	313.11	5.88	5.88	5.88	5.88
Construction cost	219.03	0	0	0	0	0	0
Maintenance cost	2.94	2.94	2.94	5.88	5.88	5.88	5.88
Heightening cost	0	0	310.17	0	0	0	0

Note: Maintenance cost is yearly annuity cash flow

Table 16.8 Costs of dike options (discount rate = 3 %, unit: million USD)

Dike option	Construction cost	Maintenance cost	Heightening cost	Total cost
Option 1	268.99	92.901	–	361.893
Option 2	339.57	92.901	–	432.471
Option 3	219.03	92.901	85.460	397.391
Option 4	480.69	185.802	–	666.492
Option 5	219.03	142.062	171.734	532.825

the highest cost, at USD 666.492 million, and option 1 had the lowest cost, at USD 361.893 million. Generally, the dike options that included the flexibility to heighten the dikes at a later date incurred higher costs.

Benefit Measurement

To estimate the benefit of flood and storm losses avoided for the whole of Vietnam, return periods of 10, 50, and 100 years were used (Table 16.9), while for the Vietnamese Mekong Delta, return periods of 2, 4, 10, 20, and 50 years were assumed (Table 16.10).

Table 16.9 Indicative annual aggregated probable maximum flood and storm losses for Vietnam (% of 2008 GDP): (a) World Bank 2010); (b) author's calculation)

	Flood		Typhoon	
Return period	Value[a)] (USD million)	Percentage per GDPb) (%)	Value[a)](USD million)	Percentage per GDPb) (%)
10 years	1,093	0.013	1,095	0.013
50 years	2,225	0.026	1,913	0.023
100 years	2,781	0.033	2,290	0.027

Table 16.10 Probable maximum flood and storm losses for the Mekong Delta (CCFSC 2011; author's calculation)

Storm/flood	Storm (mil. VND)	Flood (mil. VND)	Frequency
Linda storm 1997	7,179,615	–	1/50
Storm 1998	317,055	–	1/20
Tropical depression 1999	300	–	1/4
Flood 1996	–	2,571,223	1/10
Flood 1994	–	2,283,858	1/10
Flood 2001	–	1,535,910	1/10
Flood 1991	–	590,000	1/2
Flood 2002	–	456,831	1/2
Flood 1995	–	383,752	1/2
Flood 2000	–	302,069	1/2
Flood 1997	–	67,496	1/2
Average loss per year:	2,498,990	1,023,892	
in million USD	138.83	56.88	
% loss in regional GDP[a]	0.037	0.018	
% loss per event in regional GDP[a]			
Storm once every 4 years	0.000004	–	
Storm once every 20 years	0.005	–	
Storm once every 40 years	0.106	–	
Flood once every 2 years	–	0.006	
Flood once every 10 years	–	0.037	

Note: aMedian 1998 GDP and 1997 GDP for storms and floods, respectively

To estimate productivity losses avoided for rice and aquaculture, a uniform distribution form was assigned for the salinity-affected area variable. Using the probabilities of disaster events and their corresponding damage, the values of the economic loss per event were measured.

Results showed that the benefits of avoiding storm losses for corresponding return periods measured in terms of percentage per GDP were 0.000004 %, 0.013 %, 0.005 %, 0.023 %, and 0.027 %, respectively. By the same method, the benefits of floods avoided for corresponding return periods were 0.006 %, 0.037 %, 0.026 %, and 0.033 %, respectively.

Table 16.11 shows the monetary benefits of the different dike options over different periods of time, classified by storms, floods, and salinity. To calculate

Table 16.11 Benefits of dike options (unit: million USD)

Benefit category by option	2010	2020	2030	2040	2050	2060	2070	2080	2090	2100	2110
Option 1	36.7	129.7	842.9	688.4	4,151.0	2,558.8	15,958.3	9,799.3	46,398.7	21,427.0	101,620.6
Storm	0.0	10.1	581.2	68.0	2,965.7	263.0	11,476.2	1,017.7	33,417.2	2,229.9	73,221.3
Flood	3.9	86.8	225.2	584.0	1,148.8	2,259.9	4,445.6	8,745.1	12,944.9	19,161.7	28,364.0
Salinity	32.8	32.8	36.5	36.5	36.5	35.9	36.5	36.5	36.5	35.4	35.4
Option 2	36.7	129.7	842.9	688.4	4,151.0	5,712.2	15,958.3	9,799.3	46,398.7	21,427.0	141,198.2
Storm	0.0	10.1	581.2	68.0	2,965.7	1,682.0	11,476.2	1,017.7	33,417.2	2,229.9	91,031.2
Flood	3.9	86.8	225.2	584.0	1,148.8	3,994.3	4,445.6	8,745.1	12,944.9	19,161.7	50,131.7
Salinity	32.8	32.8	36.5	36.5	36.5	35.9	36.5	36.5	36.5	35.4	35.4
Option 3	36.7	129.7	842.9	688.4	4,151.0	5,712.2	15,958.3	9,799.3	46,398.7	21,427.0	141,198.2
Storm	0.0	10.1	581.2	68.0	2,965.7	1,682.0	11,476.2	1,017.7	33,417.2	2,229.9	91,031.2
Flood	3.9	86.8	225.2	584.0	1,148.8	3,994.3	4,445.6	8,745.1	12,944.9	19,161.7	50,131.7
Salinity	32.8	32.8	36.5	36.5	36.5	35.9	36.5	36.5	36.5	35.4	35.4
Option 4	36.7	129.7	842.9	688.4	4,151.0	5,712.2	15,958.3	9,799.3	46,398.7	21,427.0	185,393.3
Storm	0.0	10.1	581.2	68.0	2,965.7	1,682.0	11,476.2	1,017.7	33,417.2	2,229.9	110,820.0
Flood	3.9	86.8	225.2	584.0	1,148.8	3,994.3	4,445.6	8,745.1	12,944.9	19,161.7	74,537.9
Salinity	32.8	32.8	36.5	36.5	36.5	35.9	36.5	36.5	36.5	35.4	35.4
Option 5	36.7	129.7	842.9	688.4	4,151.0	5,712.2	15,958.3	9,799.3	46,398.7	21,427.0	185,393.3
Storm	0.0	10.1	581.2	68.0	2,965.7	1,682.0	11,476.2	1,017.7	33,417.2	2,229.9	110,820.0
Flood	3.9	86.8	225.2	584.0	1,148.8	3,994.3	4,445.6	8,745.1	12,944.9	19,161.7	74,537.9
Salinity	32.8	32.8	36.5	36.5	36.5	35.9	36.5	36.5	36.5	35.4	35.4

Note: Losses due to salinity avoided is yearly annuity cash flow

Table 16.12 Present values of the benefits of the different dike options (discount rate = 3 %, units: million USD)

Dike options	Storm losses avoided	Flood losses avoided	Salinity losses avoided	Total benefit
Option 1	10,509.2	7,192.8	1,094.6	18,796.7
Option 2	11,759.6	8,721.1	1,094.6	21,575.3
Option 3	11,759.6	8,721.1	1,094.6	21,575.3
Option 4	12,789.3	9,991.0	1,094.6	23,874.9
Option 5	12,789.3	9,991.0	1,094.6	23,874.9

Table 16.13 Cost-benefit analysis of sea dike options with uncertainty conditions (discount rate = 3 %)

Category	Option 1	Option 2	Option 3	Option 4	Option 5
Benefit	18,796.7	21,575.3	21,575.3	23,874.9	23,874.9
Cost	361.9	432.5	397.4	666.5	532.8
ENPV	18,434.8	21,142.8	21,177.9	23,208.4	23,342.1

the monetary values of storm losses avoided and flood losses avoided, the disaster losses in terms of percentage of GDP were multiplied by the corresponding GDP values projected in the baseline scenario. To measure the monetary values of salinity damages avoided, the values of marginal productivity losses for rice and for aquaculture were multiplied by the salinized areas projected in the baseline scenario and the simulation value of 83 % of areas invaded by salinity. While the cash flow for storms and floods was assumed at the end of the period of the events, the cash low for salinity takes the form of annuities.

The values of losses for each event (storms, floods, and salinity) were used to calculate the present values of benefits in the CBA calculation. Table 16.12 shows a summary of the results of the present values of the benefits of dikes, with a discount rate of 3 % and a timeline of 100 years. The results showed that option 4 and option 5 had the highest benefits, of USD 23,875 million, and option 1 had the lowest benefit, of USD 18,797 million.

Cost-Benefit Analysis

Cost estimations (Table 16.7) and benefit estimations (Table 16.12) under uncertainty conditions are jointly presented in Table 16.13. Based on the NPV decision rule, the results indicated that all the dike options could be recommended as appropriate dike adaptation measures. The larger in scale the dike systems were, the higher the ENPVs were. Among the dike alternatives applicable to the VMD, the initially small-scale dike options (options 3 and 5) that have subsequent heightening as a built-in feature are the most appropriate choices if the impacts of rises in sea level are mainly storms, floods, and increased salinity.

Table 16.14 Present values of costs of dike options (discount rate = 6 %)

Dike options	Construction costs	Maintenance costs	Heightening costs	Total cost
Option 1	230.9	48.9	–	279.8
Option 2	339.6	48.9	–	388.4
Option 3	219.0	48.9	48.1	316.0
Option 4	480.7	97.7	–	578.4
Option 5	219.0	64.0	96.7	379.7

Table 16.15 Present values of benefits of dike options (discount rate = 6 %)

Dike option	Storm losses avoided	Flood losses avoided	Salinity losses avoided	Total benefit
Option 1	1,409.9	1,048.5	593.9	3,052.3
Option 2	1,539.4	1,206.8	593.9	3,340.1
Option 3	1,539.4	1,206.8	593.9	3,340.1
Option 4	1,597.7	1,278.8	593.9	3,470.4
Option 5	1,597.7	1,278.8	593.9	3,470.4

Table 16.16 Cost-benefit analysis of sea dike options with uncertainty (discount rate = 6 %)

Category	Option 1	Option 2	Option 3	Option 4	Option 5
Benefit	3,052.3	3,340.1	3,340.1	3,470.4	3,470.4
Cost	279.8	388.4	316.0	578.4	379.7
ENPV	2,772.5	2,951.7	3,024.1	2,892.0	3,090.7

Sensitivity Analysis

In order to ensure that the selected sea dike options were assessed at the appropriate levels, sensitivity analyses of negative effects of changes in the discount rate and salinity were prepared in this section.

Change in Discount Rate

As the selected discount rate for a CBA calculation increases, the present values decrease. This causes changes in ENPV that provide benchmarks for selecting the best dike options. Tables 16.14, 16.15, and 16.16 show the CBA results in terms of present values of costs, present values of benefits, and ENPVs, respectively, if the discount rate is 6 %. The results showed that the ENPVs of dike options were very sensitive to an increase in discount rate. The uncertainty of the socio-economic environment is a potential factor leading to changes in discount rate.

Table 16.17 Sensitivity analysis of present values of benefits of different dike options (discount rate = 3 %, salinized areas = 50 %)

Dike options	Storm losses avoided	Flood losses avoided	Salinity losses avoided	Total benefit
Option 1	10,509.2	7,192.8	566.6	18,268.6
Option 2	11,759.6	8,721.1	566.6	21,047.2
Option 3	11,759.6	8,721.1	566.6	21,047.2
Option 4	12,789.3	9,991.0	566.6	23,346.8
Option 5	12,789.3	9,991.0	566.6	23,346.8

Table 16.18 Cost-benefit analysis of different dike options with sensitivity analysis(discount rate = 3 %, salinized areas = 50 %)

Category	Option 1	Option 2	Option 3	Option 4	Option 5
Benefit	18,268.6	21,047.2	21,047.2	23,346.8	23,346.8
Cost	361.9	432.5	397.4	666.5	532.8
ENPV	17,906.7	20,614.8	20,649.9	22,680.3	22,814.0

Change in Salinized Areas

The CBA calculations showed that productivity losses avoided provided the highest proportion of total benefits. Values from salinity-affected areas were used to project values for all rice and aquaculture land. According to the Socio-economic Master Plan of Tra Vinh Province, more than 90 % of the total natural area of the province could become salinized. The natural area currently affected in six permanently salinized regions is about 75 %. It is reasonable to assume that these salinity-affected areas depend on the effectiveness of the river dike and sea dike systems. In order to assess the effect of this important variable on the ENPVs, a further analysis was conducted. Tables 16.17 and 16.18 show the present values of the benefits of ENPVs, assuming that 50 % of the rice and aquaculture land is salinized. The sensitivity analysis showed that compared to the initial CBA assessment, the ENPVs were still robust.

Conclusions and Recommendations

The impact on the Vietnamese Mekong Delta of a rise in sea level has been discussed in this report. Various sea dike options were proposed as appropriate adaptation measures. The study developed five dike options associated with three hypotheses regarding different scales of sea dike systems: option 1 would be suitable for a storm that occurs once every 20 years, option 2 and option 3 were suitable for a storm that occurs once every 50 years, and option 4 and option 5 were appropriate for a storm that occurs once every 100 years. It was assumed that the lifespan of the dike in option 1 was 50 years. The lifespan of dike options 2, 3, 4, and 5 was set at 100 years.

The study used the risk CBA framework, involving simulation analysis using Crystal Ball®, to assess the dike options proposed for the VMD. The baseline was derived from Tra Vinh Province's Socio-economic Master Plan. There were three cost components in the cost category: construction costs, maintenance costs, and dike heightening costs. To calculate the uniform probability distribution of dike costs, minimum values from Hillen et al. (2010) and maximum values from Mai et al. (2008) were used to estimate the simulated values of dike costs. The benefit category was defined as economic damage avoided in the VMD because of the protection offered by the dike system. These include (1) losses from storms and floods sustained by houses, infrastructure such as roads, electricity supplies, water connections, crops destroyed, etc., and (2) the avoidance of productivity losses due to salinity.

To estimate the economic losses caused by storms, the study proposed five scales of storms corresponding to different return periods: once every 4 years, once every 10 years, once every 20 years, once every 50 years, and once every 100 years. There were also four flood scenarios: once every 2 years, once every 10 years, once every 50 years, and once every 100 years. The economic damage due to each event was projected from the World Bank's projections (2010) and CCFSC's data. In order to estimate the rice yield loss due to salinity or seawater intrusion, a salinity dummy variable was introduced into the production function. The values of productivity losses were also calculated using the simulation procedure.

The risk CBA results showed that applying option 4 to the entire length of the dike system (147 km) incurred the highest present value (PV) of USD 666.5 million. Option 1 had the lowest PV costs, at USD 361.9 million. In general, dike options with built-in subsequent heightening incurred lower PV costs than alternative options that built to full height from the start. The results also showed that the benefits of losses avoided due to storms and floods were important. In the case of salinity, annual rice and aquaculture productivity losses avoided were USD 331.25 per hectare and USD 915 per hectare, respectively. Based on the NPV decision rule, results indicated that all the dike options should be taken into account if dike adaptation measures were to be considered for the VMD. The larger in scale the dike systems were, the higher were the ENPVs. Of the dike alternatives applicable to the VMD, the small-scale dike options – option 1, option 3, and option 5 – should be chosen, as the impacts of sea level rise are associated with storms, floods, and salinity.

The sensitivity analyses showed that the ENPVs of dike options were very sensitive to changes in discount rate. Also, if the salinity-protected area is 50 % of the total land area, the CBA results were not significantly altered. Although arguments regarding the feasibility of a concrete sea dike system for coping with climate change impacts are still the subject of policy debates, the CBA results in this study have found initial evidence to support the construction of a concrete sea dike system for the VMD.

References

Boardman AE, Greenberg DH, Vining AR, Weimer DL (2001) Cost-benefit analysis: concept and practice. Prentice House, Upper Saddle River

Central Committee for Flood and Storm Control (2011) http://www.ccfsc.gov.vn/KW367A21/Trang-chu.aspx/. Downloaded on 11 Jan 2011

Hillen MM, Jonkman SN, Kanning W, Kok M, Geldenhuys MA, Stive MJF (2010) Coastal defence cost estimates case study of the Netherlands, New Orleans and Vietnam, http://www.google.com.vn/search?hl=vi&source=hp&q=Hillen+M.M.%2C+S.N.+Jonkman%2C+W.+Kanning%2C+M.+Kok%2C+M.A.+Geldenhuys%2C+and+M.J.F.+Stive.+2010.+Coastal+defence+cost+estimates+case+study+of+the+Netherlands%2C+New+Orleans+and+Vietnam&btnG=T%C3%ACm+v%E1%BB%9Bi+Google&aq=f&aqi=&aql=&oq=. Downloaded on 18 Jan 2011

Mai VC, Tri MC, Quy NB, Vrijling JK (2008) Risk-based approach for safety standard of coastal flood defences in Vietnam. J Water Resour Environ Eng 23

The Ministry of Natural Resources and Environment (MONRE) (2009) Climate change and sea level rise scenarios. MONRE, Hanoi

Vietnam Master Plan (2011) Tra Vinh province 2020 Socio-economic Master Plan, Vietnam

World Bank (2010) Weathering the storm: options for disaster risk financing in Vietnam. http://www.reliefweb.int/rw/RWFiles2010.nsf/FilesByRWDocUnidFilename/MYAI-84W88V-full_report.pdf/$File/full_report.pdf\. Downloaded on 11 Jan 2011

Chapter 17
Exploring the Potentials of a Technology-Based Flood Early Warning System in Laguna, Philippines: A Cost-Benefit Analysis

Jaimie Kim B. Arias, Rowena A. Dorado, Maria Emilinda T. Mendoza, and Vicente G. Ballaran Jr

Abstract This study covers the second phase of a 3-year research project on flood risk management in Laguna, Philippines. The first part of the study focused on vulnerability assessment. The second phase of the study entails an economic analysis of flood adaptation options. Adaptation options were formulated using a participatory framework and watershed approach. The study site covers three lakeshore municipalities in the Sta. Cruz River Watershed: Pila, Victoria, and Sta. Cruz. Three sub-studies are undertaken: (1) direct damage cost estimation for floods, (2) cost-benefit analysis of a technology-based early warning system, and (3) cost-effectiveness analysis of relocation versus evacuation and building modification. Typical damage costs per hectare for two types of land used are estimated for different flood depths. For built-up land, the typical direct damage cost per hectare is about Php 410,000, Php 215,000, and Php 85,000 for neck-level, waist-level, and knee-level floods, respectively. Farming land (rice), on the other hand, incurs a damage cost of Php 29,600/ha for a flood level of 2.5 ft and above. The authors found that the respondents in the study site have a positive willingness to pay for a technology-based early warning system. The estimated mean willingness to pay is between Php 127 and Php 152 per household per month. Gender seems to play a significant role in predicting the mean willingness to pay. The analysis also

J.K.B. Arias (✉) • R.A. Dorado
Department of Economics, College of Economics and Management, UPLB, Laguna, Philippines
e-mail: jkbayani@gmail.com; jbarias@up.edu.ph; wengdorado@yahoo.com; radorado@up.edu

M.E.T. Mendoza
Department of Social Development Services, UPLB, Laguna, Philippines
e-mail: makilinggazer@gmail.com; makilinggazer@yahoo.com

V.G. Ballaran Jr
Agrometeorology and Farm Structures Division, Institute of Agricultural Engineering, College of Engineering and Agro-industrial Technology, UPLB, Laguna, Philippines
e-mail: vgballaran@yahoo.com; vdballaran@up.edu.ph

D. James, H.A. Francisco (eds.), *Cost-Benefit Studies of Natural Resource Management in Southeast Asia*, DOI 10.1007/978-981-287-393-4_17

shows that an early warning system would result in large social net benefits, with a benefit-cost ratio reaching as much as 33. One of the sites, the municipality of Sta. Cruz, is exposed to long-term floods which can last as long as 6 months. The cost-effectiveness of three options for responding to such types of flooding is accordingly assessed: relocation, the use of evacuation centers, and building modification. Risk analysis is undertaken to test the robustness of the results using Monte Carlo simulation. The analysis indicates that building modification is the most cost-effective option among the three, followed by evacuation centers.

Keywords Philippines • Flood • Vulnerability assessment • Adaptation options • Risk analysis

Introduction

The Philippine archipelago is blessed with natural beauty – a tropical climate, sprawling beaches, and rolling terrains. Unfortunately, these same geographical features contribute to the country's vulnerability to climate change-related hazards. The most devastating of which are typhoons and floods. Sitting along the Pacific typhoon belt, an average of 20 typhoons hit the country each year. A number of these typhoons have been so destructive causing huge losses of properties, disruption of economic activities, injuries, and even deaths. Table 17.1 shows the deadliest typhoons which hit the country over the last two decades.

The Sta. Cruz River Watershed in Laguna, which covers the lakeshore municipalities of Sta. Cruz, Pila, and Victoria, is one of the areas frequented by heavy rains and typhoons. Flash floods as well as slow-onset, long-term floods occur in the area. Knowing the potential economic and social consequences of the hazard, the local government is keen on mainstreaming adaptation in its policies and programs especially on those that focus on risk reduction, instead of disaster response which just relies on post-flood recovery and rehabilitation. One of the options that they intend to pursue is the establishment of a technology-based early warning system. However, considering that financial resources are constantly a limiting factor, the implementation of an early warning system should be backed by sound decision-making. It is in this regard that a research which focuses on the economic analysis of the adaptation option was found necessary.

Why is economic analysis important? Economics provides a framework by which one can assess the feasibility of a project, with the whole society's welfare in mind. It is different from financial analysis, wherein the only concern is whether

Table 17.1 Deadliest typhoons that hit the Philippines (DOST-PAGASA)

Name	Dates of impact	Deaths
Thelma/Uring	November 1991	5,000–8,000
Freshen/Frank	June 2008	1,410
Durian/Reming	November 2006	1,399

the project is profitable from the perspective of an investor or from fiscal analysis where the only consideration is the impact of the project on the budget of the government (Boardman et al. 2006). Economics lends a broader picture, for which one can have a transparent decision-making. It is deemed that a project is worthwhile doing, only if it promotes the welfare of the society as a whole. If the project is found to be welfare improving, the implementation of the project promotes the efficient allocation of resources.

Research Objectives

The overall objective of the research was to determine the economic feasibility of establishing a technology-based flood early warning system along the Sta. Cruz River Watershed. In line with this objective, it tried to answer the following research questions:

1. Using the contingent valuation method (CVM), what is the estimated maximum willingness to pay for households along the lakeshore municipalities of the Sta. Cruz River Watershed for a technology-based early warning system?
2. What are the costs associated with the implementation of a technology-based early warning system?
3. Is the technology-based early warning system a good project on the basis of economic efficiency?

Description of Technology-Based Early Warning System

The term early warning system (EWS) is defined by the United Nations International Strategy for Disaster Reduction (UNISDR) as "the set of capacities needed to generate and disseminate timely and meaningful warning information to enable individuals, communities and organizations threatened by hazards to take necessary preparedness measures and act appropriately in sufficient time to reduce the possibility of harms or losses." All over the world, there has been evidence that EWS is an effective strategy to minimize loss of lives as well as damage to properties (Basha and Russ 2007; Mercy Corps Practical Action 2010). In the Philippines, it was possible to prove its effectiveness in the case of Bulacan, wherein the establishment of a community-based early warning system saved the aquaculture industry from experiencing losses during the typhoon Marce in 2006, while in Dumangas, Iloilo, the presence of EWS reduced casualties of typhoon Frank in 2008 to zero (Pagulayan 2012).

The institution at the forefront of promoting and establishing the use of early warning systems in the country is the Department of Science and Technology (DOST). One of its flagship programs is the Nationwide Operational Assessment of Hazards (NOAH). It also administers the Advanced Science and Technology Institute (DOST-ASTI) and the Philippine Atmospheric, Geophysical, and Astronomical Services Administration (PAGASA), the two agencies involved in the establishment of EWS in the country.

For the Sta. Cruz River Watershed, the EWS being proposed is technology-based, which means that it will utilize meteorological monitoring equipment like that developed by DOST-ASTI. The technology includes an automated weather system (AWS) and a water level monitoring system (WLMS). Both sets of equipment collect real-time data and through the cellular network send these data to a central server maintained by the DOST. The AWS collects information on wind speed and direction, air temperature, air humidity, air pressure, and rain amount, duration, and intensity, while the WLMS collects information on water levels. This information can be accessed over the Internet and can be used for predicting flood occurrences along the Sta. Cruz River Watershed.

The installation of the system will involve several steps which include (1) the acquisition and setting up of the AWS and WLMS stations, (2) training and capacity building of personnel, (3) research and field validation to establish thresholds for meteorological parameters, (4) establishment of a database containing contact information of households, (5) setting up of the monitoring and communications office, and (6) actual EWS operations.

One of the salient features of the proposed system is the use of *text* messaging or short messaging service (SMS) for issuing warnings. Given that there are 101 mobile cellular subscriptions for every 100 people in the country (Information and Communications for Development: Maximizing Mobile 2011, cited by Roa 2012), *texting* is the easiest way and probably the most efficient way of issuing warnings to households. For the proposed project, computer software that maintains contact numbers and locations of households, as well as one that manages bulk text messaging, will be acquired. To establish the thresholds as a basis for the warning, experts will be commissioned to do research and field validation.

Study Site

The locale of the study covers the municipalities of Sta. Cruz, Pila, and Victoria. Sta. Cruz is the capital of the province of Laguna and has a land area of 3,860 ha comprising 29 barangays. The topography is level or nearly level with slope ranging from 0 to 2.5 %. In 2010, the estimated population was about 127,320 or about 26,150 households. Although, Sta. Cruz serves as the commercial and service hub of eastern Laguna, the municipality is still dependent on agriculture, with rice, coconut, and vegetables as the main crops. Pila is similar in size and topography to Sta. Cruz. The total land area covers 3,120 ha with 17 barangays. Majority (67 %) of the population are considered rural. The estimated population in 2010 was 51,300 with about 10,260 households. The municipality is considered agriculture based. Major crops include rice, coconut, and fruit trees. The municipality of

Victoria has a total land area of 2,235 ha covering 9 barangays. Population as of 2010 was estimated to be 34,600 or about 6,900 households. Out of the 9 barangays, only 2 are considered urban. The major contributor to the economy is the agricultural sector. The municipality is well known for duck raising.

Methodology

Data Collection

A household survey was conducted to collect data on household profile and characteristics, impacts and damages from flooding, and willingness to pay for early warning system. Respondents were chosen using a two-stage stratified random sampling procedure. First, the population of households was stratified by municipality (i.e., Sta. Cruz, Pila, and Victoria) and second, by the household's location, whether in a rural-high vulnerability barangay index, rural-low vulnerability index, urban-high vulnerability index, or urban-low vulnerability index. A stratified random sample of 500 households was then obtained using proportionate sampling, i.e., the size of each stratum in the sample is proportionate to the size of the stratum in the population. The households making up the sample were selected from a list of households provided by the LGUs and the Barangay Integrated Development Approach for Nutrition Improvement of the University of the Philippines (BIDANI-UPLB). Whenever a household included in the sample could not be located or refused to participate, a replacement household was selected randomly from the same stratum.

Cost-Benefit Analysis

The general steps for conducting a CBA recommended by Boardman et al. (2006) were undertaken for the economic assessment. The discount rate used in the study is 15 % comprising the discount rate used by the National Economic and Development Authority (NEDA) in evaluating local projects. Only one option was subjected to CBA, and the benefit-cost ratio (BCR) was calculated. The internal rate of return (IRR) was also estimated. As recommended by Margulis et al. (2008), sensitivity analysis was performed.

Estimating the Benefits of EWS

To capture the benefit of the technology-based early warning system project, the contingent valuation method (CVM) was used. CVM is a survey-based valuation technique which directly asks individuals or households about their willingness to pay (WTP) for a good or service. The procedure is considered a direct method as opposed to methods which infers values from observed behavior (Carson 2000). In a CV study, a hypothetical market is presented to the respondents, after which the value of their willingness to pay is elicited. Leary (1999) defined WTP as the amount paid by a person which would have an equivalent effect on the person's welfare as would the policy. It is referred to as "contingent" since valuation is dependent on the information provided to the respondent during the survey.

CV Theoretical Framework

The theoretical framework of the contingent valuation method described below was adopted from Bateman et al. (2002).

Basic utility theory suggests that individuals (or households) maximize their utility subject to their budget constraint. Following the standard economic theory, an indirect utility function, V, can be defined which shows the maximum amount of utility that a household can derive from their income Y, given prices of goods P, and the level of provision of a nonmarketed good Z (which in this case is the early warning system). It is also assumed that the household's utility depends upon their sociodemographic characteristics S.

It is expected that the higher the income or the lower the prices, the higher the utility level. It is also presumed that an increase in provision of the nonmarketed good, say from Q^0 to Q^1, will increase the household's utility, such that

$$V(Y,P,S,Q^0) < V(Y,P,S,Q^1).$$

When households are answering a CV question, they are essentially comparing their utility at Q^0 vis-à-vis Q^1. At Q^1, they are experiencing higher levels of utility; hence, it is reasonable to assume that they would be willing to pay an amount to obtain Q^1. However, as they increase their payment to obtain higher provision of Q, their utility diminishes. Hence, the maximum amount that they would be willing to pay is defined where

$$V(Y,P,S,Q^0) = V(Y-C,P,S,Q^1)$$

– that is, when the household is indifferent between its original position, when the provision of the nonmarketed good is Q^0, and the position wherein Q^1 is provided but it has to pay for the provision of the nonmarket good. If payment exceeds C, then the household is better off at Q^0. In the equation, C is the compensating surplus measure of a change in welfare or the household's maximum WTP to achieve the increase in provision of the nonmarket good.

From the previous equation, a bid function $C()$ can be defined as a function of the other parameters in the model. It is also assumed that maximum WTP is equal to C and should be less than income Y.

$$C = C(Q^0, Q^1, Y, P, S) = \text{WTP} \leq Y.$$

Parametric Estimation of WTP

In the binary choice format, households simply state whether they are willing to pay a certain amount (B) or not. In response to the CV question, they can either answer

$$\text{No if } C_i < B \text{ or}$$
$$\text{Yes if } C_i \geq B.$$

The probability that the household is willing to pay can be estimated using Hanemann's binary logit model wherein the dichotomous-choice WTP responses are regressed on bid values and a vector of socioeconomic and awareness/attitudinal variables (X_t) (Hanemann 1994):

$$\Pr(\text{Yes}) = E\left(Y = \frac{1}{X_i}\right) = 1/(1 + e - [\beta_0 + \beta_1 X_i]).$$

The mean willingness to pay is then estimated using the formula:

$$\text{WTP} = -\alpha/\beta_1$$

where α is the sum of the constant term and the coefficients of all explanatory variables multiplied with the mean values of explanatory variables (except the variable bid price).

β_1 is the coefficient of the bid price.

Empirical Model

The dependent variable takes on the value of 1 if the household is willing to pay the bid amount and takes on the value of 0 if not. The explanatory variables used in the analysis are summarized in Table 17.2.

Table 17.2 Explanatory variables used in logit regression

Explanatory variable	Notation	Description	Expected sign	Rationale
Bid	BID	Takes on the values of Php25, Php50, Php100, Php200, and Php300	–	As the price of the good increases, the less will be the likelihood of paying
Household income per capita	INC	Monthly household income divided by the household size	+	As the income increases, the greater the likelihood of paying
Exposure	EXP	Number of floods inundating house in the last 10 years	+	As exposure to flooding becomes greater, the greater the likelihood of paying
Knowledge about early warning system	KNOW	Dummy, if self-rating on knowledge is above 2, the variable takes on the value of 1; 0 if below 3	+	The more aware and knowledgeable a household is about the early warning system, the greater the likelihood of paying
Autonomous adaptation	RISK	Dummy, if household has undertaken any form of autonomous adaptation, the variable takes on the value of 1; 0 if otherwise	+	Autonomous adaptation may be an indicative of risk aversion. Risk-averse households will be more likely to be willing to pay
Control variables reflecting respondent characteristics				
Respondent's sex	RSEX	Dummy, if male, the variable takes on the value of 1; 0 if female		
Respondent's education	REDUC	Number of years of schooling		
Respondent's age	RAGE	Age as of last birthday		
Control variables reflecting geographical location				
Pila	PILA	Dummy, if household is located in Pila, the variable takes on the value of 1; 0 if otherwise		
Sta. Cruz	SCRUZ	Dummy, if household is located in Pila, the variable takes on the value of 1; 0 if otherwise		

Nonparametric Estimation

To obtain a nonparametric estimate, the lower-bound Turnbull mean willingness-to-pay equation was used (Haab and McConnell 2002). The Turnbull estimator is a distribution-free estimator relying on asymptotic properties. The proportion of "no" responses is higher as the bid price increases, while the survival function is decreasing in bid. The mean is given by

$$E_{\mathrm{LB}}(B) = \sum_{j=0}^{M} B_j \left(F_{j+1} - F_j\right)$$

with variance

$$V(E_{\mathrm{LB}}(B)) = \sum_{j=1}^{M} F_j(1 - F_j)\ \left(B_j - B_{j-1}\right)^2 / T_j$$

where

M is the number of bids.
B is the bid level.
T_j is the number of respondents offered the bid price β_j.
F_j is the proportion of no responses to the bid price β_j.

CV Survey Design

To obtain reliable estimates of maximum willingness to pay, the CVM survey was carefully designed, from the drafting of questionnaire to the sampling process and survey implementation. The following are the key features of the CV survey and questionnaire as applied in the study:

- The mode of the survey was a face-to-face interview as recommended by Arrow et al. (1993).
- A probabilistic sampling technique was applied. The population from which the sample was drawn includes all households within Pila, Victoria, and Sta. Cruz, which are the lakeshore municipalities belonging to the Sta. Cruz River Watershed. Specifically, the researchers conducted multistage stratified random sampling.
- A total of 497 completed household questionnaires were used in the analysis, 147 of which were replacement households. Of the total nonresponse, 67 households were not located, 37 already moved, 32 refused/could not be interviewed, while 7 did not have electricity connection. The total population of households in the study site was 41,869.
- Most of the contents of the questionnaire used in the survey were written in Tagalog, the local dialect, except for some terminologies that do not have a direct translation in the vernacular. The questionnaire was 22-page long, covering an Introduction/Screener Page and 6 Modules:
  - Module 1: Experience and Perceptions about Flooding Risk
  - Module 2: Knowledge and Perceptions about Early Warning Systems

  - Module 3: CV Scenario and Willingness to Pay
  - Module 4: Household Profile and Characteristics
  - Module 5: Livelihood Characteristics
  - Module 6: Flood Damage Experience and Adaptation Activities

- As an input to the drafting of the CV scenario, focus group discussions were undertaken to obtain information on the possible range of WTP bids, payment vehicle, and local experiences on floods and early warning systems. Also, a technical expert was consulted to provide the description of the early warning system (EWS) technology.
- Before finalizing the questionnaire, a pretest was conducted in Los Baños, a lakeshore municipality in Laguna near the study site. The pretest site was selected because of its similar socioeconomic characteristics and also because of its similar exposure to floods.
- Training was given to the enumerators prior their field work.
- The tailored design method (Dillman 2000) was adopted in the survey. The aim was to encourage respondents to participate in the survey by establishing trust and lowering perceived costs while increasing perceived rewards of joining the survey. First, the purpose of the study as well as the institutions involved was relayed to the respondent. Strict confidentiality of the responses was assured. Respondents were informed that they will be given a short report if they wish to receive an update of the study. Also, a token (a fabric reusable shopping bag) was given to all the respondents. The token was carefully thought out, so as not to encourage biased responses. The token was simple and inexpensive, yet useful. It also bears the name of the project and the name of the institutions undertaking the project so as to reinforce the legitimacy of the survey. Visual aids were also used to help respondents in the cognitive task of understanding the good that they are being asked to value.
- A good CV questionnaire has a clear and easily understood scenario. The scenario used in the study provided a full description of the proposed project – a technology-based flood early warning system (EWS) which will be installed along the Sta. Cruz River Watershed. To aid in explaining the project, a pretested brochure was used. It explained that the proposed EWS technology utilizes an automated weather system (AWS) and a water level monitoring system (WLMS) for collecting data. These data will be used as basis for issuing warnings. The warning will be sent through a text message to households subscribed in the service. The lead time for issuing the warning is between 2 and 4 h. The institution responsible for the project is the local government of the three municipalities. The duration of the project is 10 years which covers the life span of the equipment (2014–2023).
- The payment vehicle used was mandatory, to be collected through an additional charge in the electricity bill. This is collected on a monthly basis for a period of 10 years (the duration of the project). During the FGD, it was found that people were less inclined to pay additional taxes. Also, a water bill surcharge was not

suitable because a significant proportion of households were not connected to the water district. Instead, they sourced their water from private wells. An additional surcharge to the electric bill was accordingly deemed acceptable.

- The elicitation technique was single-bounded dichotomous choice. Although there are other elicitation techniques such as open-ended, bidding game, payment card, and double-bounded dichotomous choice, the single-bounded dichotomous choice was chosen for its simplicity and incentive compatibility. The bid levels used were 25, 50, 100, 200, and 300 pesos per month, paid for 10 years.
- To minimize strategic bias, a provision point was stated, wherein it was explained that the project will push through only if more than 50 % of the respondents vote "yes" in a barangay referendum. To discourage "yea saying" and minimize warm-glow effect, cheap talk script was included. Specifically, the respondents were reminded of their income constraint as well as the existence of alternative projects.
- Debriefing questions were also included to identify invalid and protest votes. Table 17.3 categorizes responses according to whether they are valid or invalid answers. If "yes" answers are invalid, they are converted to "no." If "no," votes are classified as protest votes; these could be dropped from the observations or retained. In this study, the protest "no" votes were retained to maintain conservative estimates.

Table 17.3 Classification of answers to the WTP debriefing question

	Valid answers	Invalid (bias or protest)
Reasons why respondents are willing to pay	My household will benefit from the project	We will not really be made to pay
	I have confidence in the effectiveness of the technology that will be used	We are happy that we are able to help
	I have confidence in the government's ability to execute this project	
Reasons why respondents are not willing to pay	We cannot afford to pay any amount	The government should be made to pay
	Our electricity bill is already too high	Only the rich should be made to pay
	There are other more important problems that should be given priority	I don't have confidence on the ability of the government to implement this project
	Early warning systems are not beneficial	I don't have confidence on the technology
	The lead time is not sufficient	
	The local government gives ample warning	
	No one in my household owns a cell phone	

- A certainty question was included to further verify validity of yea answers. In the survey, the respondents were asked about how certain they are that they will have the same vote if they were asked to vote in a real barangay referendum. With 1 being the lowest score (not sure) and 5 the highest (very sure), those that answered 3 or better were considered as legitimate "yes" votes. Those that answered 2 or below were further classified based on their reasons for being uncertain. If the reason for their high uncertainty is because they want to know first the responses of the majority, then these were still considered as valid "yes" votes. However, if their reason is because they want to consult first with their spouses or because they are unsure about their income in the future, these were converted into "no" votes.

Results and Discussion

Description and Characteristics of Survey Respondents

About 64 % of the survey respondents were female and the majority (76 %) were married. In terms of occupation, about 28 % were housewives, 21 % were small business owners, and 17 % were laborers or unskilled workers. A majority (60 %) of the households sourced their livelihood mainly from employment in the services sector. Twenty-seven percent were engaged in small nonagriculture-related business, while 12 % were agriculture based (Table 17.4).

The mean age of the respondents was 48, while the mean years of schooling was 9. The average individual income was Php 5,125 per month, while the mean monthly income of the whole household was Php 17,000. The average household size was 5. Female respondents had a significantly lower mean individual income compared to male respondents.

Flood Exposure Indicators

In terms of flood exposure, the mean number of flood events that respondents have experienced over the last 10 years is 5. An average of three flood events inundated their homes over the same period. The mean highest flood height is 1 foot, while the mean longest duration of the flood is 16 days (Table 17.5). Fifteen percent

Table 17.4 Composition of respondents based on the main source of livelihood

Household's main source of livelihood	Frequency	%
Farming	36	7
Livestock/poultry	26	5
Small business	135	27
Services sector	301	60

Table 17.5 Flood exposure indicators

Flood exposure	Mean	S.D.
Frequency of floods in the last 10 years	5	13
Frequency of floods inside house in the last 10 years	3	9
Highest flood height in ft	1	2
Longest duration of flood in days	16	34

Table 17.6 Composition of households based on reported maximum flood height inside their house

Flood height	Frequency	%
0	250	50
1 ft	42	8
2 ft	34	7
3 ft	43	9
4 ft	27	5
5 ft	15	3
6 ft	8	2
>6 ft	6	1
Total	250	50

Table 17.7 Composition of households based on reported longest duration of flood inside their house

Duration	Frequency	%
<1 day	246	49
1 day	108	22
1 week	39	8
1 month	24	5
2 months	29	6
3 months	41	8
4 months	7	1
5 months	3	1

experienced knee-deep flood (about 1–2 feet), 9 % waist-deep (3 feet), 5 % about neck-deep (4 feet), and 6 % above the head (5 feet and above) (Table 17.6). In terms of duration, 21 % reported experiencing long-term flooding which corresponds to at least a month of inundation (Table 17.7).

Willingness to Pay for a Technology-Based EWS

A substantial proportion of the respondents revealed that they had little knowledge about early warning systems prior to the survey. When asked to rate their knowledge, with 1 being the lowest and 5 the highest, 45 % rated their knowledge as below 3. This proportion is about the same regardless of whether the respondent is male or female (Table 17.8). The brochure explaining the details of the early warning system proved to be useful in the conduct of the willingness-to-pay survey.

Table 17.8 Knowledge about early warning systems

Rate	All	%	Male	%	Female	%
Rate 1 (lowest)	152	30	53	29	99	31
2	77	15	28	16	49	15
3	132	26	47	26	85	27
4	64	13	20	11	44	14
5 (highest)	75	15	32	18	43	13

Table 17.9 Distribution of responses to willingness-to-pay question by bid level

	Frequency					%				
Bid level	25	50	100	200	300	25	50	100	200	300
Yes	64	67	54	37	31	65	66	53	36	31
No	34	34	46	64	69	34	33	46	63	68
Total	99	102	101	102	101	100	100	100	100	100

Table 17.10 Distribution of reasons for "yes" response

Reasons for yes response	Frequency	%
To ensure the readiness of my community during floods	173	73
I have faith in the technology that will be used	9	4
I have faith in the ability of the local government to implement this project	8	3
I feel happy knowing that I could help others	23	10
The fee is affordable and our area is always at risk	23	10

The responses to the "willingness-to-pay" question which have been corrected for uncertainty are summarized in Table 17.9. It can be seen that the proportion of "yes" votes decline as the bid level increases. For all bid levels, the total "yes" votes were 252 out of the total 500 responses.

A majority (73 %) of those who answered "yes" stated that their primary reason is to ensure the readiness of their community during floods. Ten percent stated that the fee was affordable; 4 % claimed that they have confidence in the technology that will be used, while 3 % stated that they are confident about the capacity of the local government to carry out the project. Ten percent claimed that they feel happy knowing that they could help others through the project (Table 17.10). This latter reason, however, signals the presence of a "warm-glow" effect. Hence, if this was the sole reason for voting "yes," then the votes were converted into "no."

The reasons for "no" responses were varied. A significant percentage (39 %) considered that the fee is unaffordable and that their electricity bill is already too high (22 %). Eight percent thought that there are more important problems that need to be prioritized, while 6 % did not think that the early warning system is useful. Three percent claimed that they have no confidence in the local government and (2 %) have no confidence in the technology that will be used in the project (Table 17.11). Nineteen percent stated that they are not willing to pay because it

Table 17.11 Distribution of reasons for "no" response

Reasons for no response	Frequency	%
The fee is unaffordable	105	39
Our electricity bill is already too high	58	22
There are other more important problems that must be prioritized	22	8
The government should pay for the project	48	18
Only the rich should pay for the project	4	1
I do not have confidence on the capability of the local government	7	3
The early warning system is not useful	17	6
I do not have confidence on the technology that will be used	5	2
We are already being warned by our local officials	3	1

Table 17.12 Results of the logit regression analysis

	Coef.	Std. err.	*z*	$P > \|z\|$	[95 % Conf. interval]	
BID***	−0.0057	0.00	−5.85	0.00	−0.01	0.00
INC*	0.0001	0.00	1.91	0.06	0.00	0.00
RSEX**	0.4865	0.21	2.36	0.02	0.08	0.89
REDUC	−0.0124	0.04	−0.35	0.72	−0.08	0.06
RAGE**	−0.0163	0.01	−2.16	0.03	−0.03	0.00
EXP**	0.0973	0.03	2.81	0.01	0.03	0.17
KNOW	0.3752	0.22	1.67	0.10	−0.07	0.82
RISK***	0.6333	0.21	3.09	0.00	0.23	1.04
PILA	0.1756	0.29	0.60	0.55	−0.40	0.75
SCRUZ	−0.2039	0.27	−0.77	0.44	−0.72	0.32
CONSTANT	0.8698	0.60	1.46	0.15	−0.30	2.04

* Significant at 1 %
** Significant at 5 %
*** Significant at 10 %
No. of observations 497
Likelihood ratio 75.09
Prob > chi2 0.00
Pseudo R2 0.11

should be the government or the rich households who should shoulder the cost of the project. This reason signals the tendency for free riding. However, it was decided to retain these observations in order to arrive at a conservative estimate of mean willingness to pay.

Table 17.12 shows the result of the logit regression analysis. Of the 10 explanatory variables, 6 were found to be statistically significant, which include the bid level, income, respondent's sex and age, flood exposure, and attitude toward risk. The results conform to economic theory in that the bid level has a negative coefficient while income has a positive coefficient.

Table 17.13 Parametric and nonparametric mean willingness to pay

	Parametric	Nonparametric
Mean WTP per HH per month	Php 140	Php 128
(in US Dollars)	US$ 3.5	US$ 3.2
Confidence interval for mean WTP (at 99 % confidence level)	[Php 127, Php 152]	
(in US Dollars)	[US$ 3.175, US$ 3.8]	

Exchange rate: US$ 1 = Php 40

It was also found that the male respondents are more likely to be willing to pay while age reduces the likelihood of paying. The higher propensity of male respondents to pay may be explained by the patriarchal nature of households in the Philippines wherein the males usually take on the lead role of ensuring the safety and well-being of their families. The attitude toward risk wherein the indicator used is the conduct of autonomous adaptation showed positive impact on willingness to pay. That is, those who have undertaken some forms of autonomous adaptation were found to be more likely to be willing to pay.

Based on the logistic regression, the mean willingness to pay was estimated to be Php 140 per household per month. The confidence interval is between Php 127 and Php 152 at a 99 % confidence level. The nonparametric lower-bound mean willingness-to-pay estimate falls within these limits, averaging Php 128 per household per month. The mean willingness to pay is about 0.8 % of the average household income of the respondents (Table 17.13).

Cost of Establishing a Technology-Based EWS

The estimated initial investment needed to establish the EWS is about Php 2 million or US$ 50,000. This covers the acquisition and installation of the equipment, capacity building and training, and setting up of the monitoring and warning dissemination office. The annual operations cost, on the other hand, is about Php 1.7 million per year or US$ 42,500 (Table 17.14).

Using the lower-bound estimate of mean willingness to pay equal to Php 128 per household per month, the estimated benefit for the first year is about Php 65.6 million or Php 1.64 million. Projections were carried out to estimate the benefit for the succeeding years, considering the growth of the population. It was assumed that the population growth rate is 1 % per year, and the number of households was calculated by dividing the total population by the average household size of 5. Table 17.15 shows the baseline number of households corresponding to the year 2012.

Table 17.14 Technology-based early warning system cost items

Cost item	Amount in Php
Initial investment	
a. Technology installation cost	
Machine and equipment purchase	
Automated weather station	120,000
Automated water level monitoring system	260,000
Computers	360,000
Installation (20,000 @ 3 units)	60,000
Shipping (2000 @ 3 units)	6,000
Fencing, polemask, and other civil works (30,000 per monitoring site)	90,000
b. Research to establish thresholds	200,000
c. Capacity building/training	150,000
d. Software acquisition	150,000
e. One-time registration	42,288
f. Communication expense	72,000
g. Internet	72,000
h. Overhead cost	360,000
Operation and maintenance	
a. Personnel	1,152,000
b. One-time registration	423
c. Communication expense	72,000
d. Internet	72,000
e. Overhead cost	360,000

Table 17.15 Baseline household population (2012)

Municipality	Number of households
Pila	9,331
Sta. Cruz	23,550
Victoria	8,988

Using a discount rate of 15 %, the net present value was estimated to be Php 330 million or US$82.5 million. The benefit-cost ratio (BCR) is 33, while the estimated internal rate of return is more than 3,000 %.

To check the validity of estimates, the results were compared to other BCA studies for EWS (Table 17.16). The BCR estimate seems to be conservative, compared to the two case studies in the table, particularly in Bangladesh (with a BCR of 558) and Thailand (BCR of 176), but the study estimate exceeds that of Sri Lanka's with a low BCR of 0.93. The reason for this low estimate, however, as the authors cited, is the infrequent occurrence of floods in Sri Lanka.

Table 17.16 Summary of benefit-cost ratio estimates from various case studies (Subbiah et al. 2008)

Case study	BCR
Sri Lanka, May 2003 floods case study	0.93
Bangladesh, 2007 flood case study	558
Thailand, 2007 flood case study	176

Conclusions and Recommendations

Based on the foregoing analysis, a technology-based early warning system was found to be an economically sound adaptation option for the Sta. Cruz River Watershed. The project produces large social net benefits, with a net present value of Php 330 million or US$82.5 million and a benefit-cost ratio (BCR) reaching as much as 33.

In the analysis, benefits were measured using the *contingent valuation method* (CVM) and not through a market-based valuation technique (for instance, *damage cost avoided*). It would be interesting for future studies to compare benefit estimates of EWS using both techniques. A challenge that researchers would face in doing such an analysis, however, would be determining the magnitude of damages avoided that can be attributed to EWS.

References

Arrow K, Solow R, Portney P, Leamer E, Radner R, Schuman H (1993) Report of the NOAA panel on contingent valuation. Fed Regist 58(10):4601–4614

Bateman IJ, Carson RT, Day B, Hanemann M, Hanleys N, Hett T, Jones-Lee M, Loomes G, Mourato S, Ozdemiroglu E, Pearce D, Sugden R, Swanson J (2002) Economic valuation with stated preference techniques: a manual. Edward Elgar, Cheltenham

Basha E, Russ D (2007) Design of early warning flood detection systems for developing countries. Manuscript submitted to Massachusetts Institute of Technology, Cambridge, USA

Boardman AE, Greenberg DH, Vining AR, Weimer DL (2006) Cost-benefit analysis: concepts and practice, 4th edn. Prentice Hall, Upper Saddle River.

Carson RT (2000) Contingent valuation: a user's guide. Environ Sci Technol 34:1413–1418

Dillman D (2000) Procedures for conducting government-sponsored establishment surveys: comparisons of the total design method (TDM), a traditional cost compensation model, and tailored design. Retrieved October 20, 2012, from http://www.sesrc.wsu.edu/dillman/papers/2000/proceduresforconductinggovernment.pdf.

Haab TC, McConnell KE (2002) Valuing environmental and natural resources: the econometrics of non-market valuation. Edward Elgar, Cheltenham

Hanemann M (1994) Valuing the environment through contingent valuation. J Econ Perspect 8 (4):19–44

Leary N (1999) A framework for benefit-cost analysis on adaptation to climate change and climate variability. Mitigation and adaptation study on global climate change. Mitig Adapt Strat Global Change 4:307–318

Margulis S, Bucher A, Corderi D, Narain U, Page H, Pandey K, Linh Phu TT, Bachofen C, Mearns R, Blankespoor B, Dasgupta S, Murray S, Cushion E, Gronnevet L, Cretegny L, Ghosh P, Laplante B, Leony L, Schneider R, Ward P, Wheeler D (2008) The economics of adaptation to climate change: methodology report. The World Bank, Washington, DC

Pagulayan RC (2012) PAGASA and its flood early warning services. Paper presented at the First Philippine International River Summit, The Centennial Resort Hotel and Convention Center, Alta Tierra Village, Jaro, Iloilo City, 30 May–1 June 2012

Roa A (2012) Philippines cited for mobile phone use. Philippine Daily Inquirer, July 17. Retrieved October 20, 2012, from http://technology.inquirer.net/14162/philippines-cited-for-mobile-phone-use

Subbiah AR, Bildan L, Narasimhan R (2008) Background paper on assessment of the economics of early warning systems for disaster risk reduction, submitted to the Joint World Bank - UN Project on the Economics of Disaster Risk Reduction. Retrieved October 20, 2012, from https://www.gfdrr.org/sites/gfdrr.org/files/New%20Folder/Subbiah_EWS.pdf

Index

A
Adaptation, 6, 29, 30, 326–328, 330, 331, 334, 335, 337, 339, 341, 342, 347, 350, 363, 365, 366, 370, 376, 384, 386
Adaptation option, 330, 370
Adaptation strategy, 342
Adaptive, 9, 235
Agarwal, A.K., 52
Agarwal, D., 52
Agriculture, 4, 5, 8, 34, 66, 111, 144, 180, 187, 188, 190, 193, 194, 198, 233, 234, 237, 238, 242, 243, 246, 250, 252, 258, 259, 261, 300, 304–307, 309, 311, 312, 321, 347, 353, 372, 380
Air pollution, 5
Air temperature, 372
Amir, I., 309
Analytical techniques, 8
Anielski, M., 279, 280, 282
Anthropogenic, 326
Aquaculture, 8, 16, 40, 158, 161, 166, 167, 173, 352, 353, 357, 361, 363, 365, 366, 371
Arias, J.K.B., 16, 17, 34–37, 325–342, 369–386
Arrow, K., 377
Asis, C.A., 93–110
Assessments, 12, 14, 17, 26, 42, 43, 212, 303, 308, 341
Attributes, 15, 17, 36, 38, 39, 209, 210, 296
Awareness programs, 210, 211, 221

B
Ballaran, V.G., 369–386
Bangkok, 5, 9, 90, 111, 343
Bann, C., 15, 141
Baseline, 17, 98, 100, 101, 103, 105, 109, 140, 181, 182, 257, 279, 356, 363, 366, 384
Bateman, I.J., 374
BAU. *See* Business as usual (BAU)
Baylatry, M., 16, 20, 22, 227–250
Bazzani, G.M., 309
BCA. *See* Benefit-cost analysis (BCA)
BCR. *See* Benefit-cost ratio (BCR)
Behavioral economics, 6
Benefit-cost analysis (BCA), 11, 12, 28, 32, 42
Benefit-cost ratio (BCR), 17, 18, 20, 152, 154, 194, 234, 373, 385, 386
Bennett, J., 38
Bergland, O., 284
Bhadrani, T., 34
Biodiesel, 23, 26, 34, 49–67, 70, 72–76, 78–81, 85–91
Biodiversity, 4, 9, 114, 135, 141, 142, 153, 188, 207, 214, 222, 228, 246, 253, 258, 263, 264, 266, 268, 270, 272, 273, 278, 281, 284–287, 289, 290, 292
Biofuels, 21, 23, 65, 67, 69–71, 74, 77–80, 82, 88–91, 106
Blamey, R., 38
Boardman, A.E., 16, 349, 373
Breeding grounds, 300
Brown, R., 19
Bryman, A., 287
Bumrungsri, S., 275–296
Business as usual (BAU), 17, 328, 330, 334–342

D. James, H.A. Francisco (eds.), *Cost-Benefit Studies of Natural Resource Management in Southeast Asia*, DOI 10.1007/978-981-287-393-4

C

Callaway, J., 114, 125
Cambodia, 3, 4, 6, 8, 15, 111, 137–139, 155, 156, 229, 252, 253, 273
Campbell, H., 19
Canakci, M., 73
Capital costs, 53, 262, 265, 312, 313
Carbon emission, 53, 56, 60–62, 64, 66, 114, 116
Carbon price, 114–117, 123–127, 133, 134
Carbon stocks, 89, 234, 243, 246
Cash flow, 21, 55, 360, 362, 363
CBA. *See* Cost-benefit analysis (CBA)
CEA. *See* Cost-effectiveness analysis (CEA)
Certainty, 28, 29, 380
Cheng, J., 102
Chen, W.W., 280, 282
China, 4, 6, 8, 9, 23, 26, 42, 49–51, 53, 61, 65–67, 90, 91, 111, 156, 280, 296, 297, 300, 301, 308–310, 321, 322, 327, 348
Chladna, Z., 125
Choice modelling, 38, 39
Clean development mechanism (CDM), 115
Climate change, 5, 6, 9, 30, 33, 41, 65, 69, 78, 90, 109, 110, 114, 135, 282, 326, 328, 346–348, 366, 370
Coastal erosion, 29, 202, 204, 205, 326–332, 334, 339, 341–343
Coastal resources, 5, 6
Coastline, 327–329, 337, 342, 346
Colgan, C., 330
Consequences, 90, 95, 109, 114, 158, 159, 169, 171, 232, 279, 326, 370
Conservation, 9, 94, 135, 141, 142, 150, 152–155, 202, 212, 215, 219, 222, 229, 233–235, 240–242, 246, 250, 253, 261, 262, 264, 268, 277–280, 291, 293, 300, 301, 307, 309–311, 320, 321
Conservative, 98, 234, 379, 383, 385
Constraints, 27, 40, 95, 164, 171, 174, 302
Construction costs, 352
Consumers' surplus, 13, 14, 34, 36, 37
Contingent, 38, 222, 285, 342, 374, 386
Contingent valuation. *See* Valuation
Coping policy, 347
Coral reefs, 5, 328
Corton, T.M.J.B., 96–98, 102
Cost-benefit analysis (CBA), 7, 11, 139, 155, 160, 270, 278, 297, 325, 327, 337, 347, 363–365
Cost-effectiveness, 70, 78, 83, 85, 86, 88, 95, 99, 109, 302, 303, 312–314, 316–318
Cost-effectiveness analysis (CEA), 23, 74, 90, 98, 104, 302, 313, 315, 316
Cost of travel, 36
Costs, 14, 21, 82, 83, 86, 100, 111, 141, 142, 144, 145, 147, 148, 150, 151, 168, 218, 242, 243, 246, 263, 267–269, 288, 289, 291, 313–315, 317, 334, 335, 340, 360
Costs and benefits, 17, 22, 27, 51, 70, 77, 78, 139, 182, 188, 199, 206, 208, 231, 234–236, 253–255, 264, 288, 291, 330, 337, 351
Costs of production, 14, 16, 57, 74, 82, 242
Coxhead, I., 4
Crude oil, 49, 58
Cultural values, 287
Curry, S., 188
Curtis, I.A., 280

D

Dai, D., 52
Damage costs, 16, 26, 34, 38, 40, 79–82, 284, 342
Danh, V.T., 16, 26, 29, 32, 34, 35, 345–366
Dasgupta, A.K., 12
Davidson, M., 100
Degradation, 4, 6, 202, 203, 233, 235, 252, 263, 266, 267, 272, 278, 305
Delineate, 327, 331
Delivand, M., 102, 106
Demand, 4, 5, 13, 14, 36, 37, 69, 90, 106, 108–110, 114, 156, 159, 171, 173, 181, 190, 202, 204, 216–221, 241, 258, 284, 304, 305, 309, 310, 312, 321, 329
Demand functions, 13
Depletes, 6
Depreciation, 21, 54, 303, 329
Diaz-Balteiro, L., 114, 125
Dichotomous choice, 37, 375, 379
Dike heightening costs, 358, 366
Diminishing rate, 13
Discount rates, 17, 20, 21, 25–27, 29, 53, 55, 61, 62, 77, 83–88, 100, 114–116, 121–123, 125, 127, 129, 130, 132–134, 141, 142, 147, 150, 152, 153, 155, 194, 232, 249, 262–264, 267–273, 292, 293, 295, 312–317, 331, 336, 338, 339, 341, 342, 356, 359, 360, 363–366, 385
Distribution, 6, 17, 22, 27, 30–33, 52, 54, 59, 77–80, 83, 85, 90, 140, 195, 197, 231, 235, 241, 282, 304, 316, 328, 349, 352, 355, 356, 361, 366, 376
Dixon, J.A., 12

Doan, 114
Dorado, M.A., 325–342
Dorado, R.A., 325–342, 369–386
Dredging, 162, 217–221, 326

E

Eagle, A.J., 94
Early warning systems, 9, 370, 371, 373, 374, 376, 378, 381–383, 386, 387
Economic damage, 355, 366
Economic efficiency, 22, 27, 42
Economic growth, 4–6, 228, 310
Economics, 4–9, 11–13, 15–24, 26–29, 33, 34, 36, 37, 39, 41–43, 50, 52–54, 56, 57, 61, 62, 64–67, 70, 82, 89–91, 99, 103, 104, 108–110, 134, 139, 141, 142, 144, 147, 152, 155, 156, 158, 160, 174, 180, 182, 183, 187–189, 191, 194, 195, 197–199, 202–205, 209, 214–217, 228, 229, 231, 232, 235, 238, 243, 249, 250, 254, 255, 262, 270, 277–281, 292, 294, 297, 301–304, 309, 310, 314, 319–321, 326, 327, 329–331, 341, 347–349, 352, 353, 355, 357, 361, 364–367, 370, 371, 374, 375, 378, 383, 387
Economy and Environment Program for Southeast Asia (EEPSEA), 6–8, 19, 20, 44, 46, 273, 321
Ecosystems, 5, 15, 36, 41, 95, 111, 235, 250, 276–280, 296, 297, 300
Ecosystem services, 9, 15, 36, 38, 276–281, 285–288, 292, 294, 296, 297
Ecotourism, 9, 20, 22, 228–234, 236, 238, 240, 241, 243, 246, 249, 250
EEPSEA. *See* Economy and Environment Program for Southeast Asia (EEPSEA)
Effectiveness, 23, 74, 82, 84, 85, 87, 88, 90, 98, 100, 104, 110, 299, 302, 309, 312–316, 318, 319, 365, 371, 379
Efficiency, 11, 12, 17–19, 22, 27, 42, 51, 54, 65, 66, 70, 73, 75, 83, 84, 87, 88, 91, 106, 111, 152, 182, 270, 279, 302, 306, 308, 309, 312, 371
Englin, J., 114, 125
Environmental benefits, 12, 16, 24, 35, 141, 183, 194, 195, 249, 253
Environmental costs and benefits, 12, 183
Environmental economics, 6, 222
Environmental impacts, 9, 16, 24, 35, 55, 95, 96, 110, 160, 179–181, 183, 184, 187, 189, 194, 198, 204, 208, 219, 228, 231, 233, 246, 280
Environmental issues, 8, 189, 202, 212, 216
Environmental problems, 6, 159, 202, 220, 254
Environmental protection, 23, 35, 40, 49, 173, 174, 203, 213, 233
Environmental resources, 4–6
Environmental services, 36, 134, 153, 329, 330, 332, 342
Environmental values, 12, 15, 24, 36, 37, 39, 42, 183, 194, 255, 279, 280
Equity issues, 22
Erosion, 29, 186, 202, 205, 207, 234, 255, 258, 260, 279, 326, 330–332, 339, 341–343
Evaluations, 8, 12, 16
Excessive fishing, 5
Excessive harvesting, 5
Expenditures, 22, 35, 59, 238
Experimental data, 96
Explicit cost, 54
External, 16, 34, 53, 61, 62, 64, 65, 70, 71, 78–83, 85, 86, 90, 91, 100, 161, 173, 205, 206, 234, 341
Externalities, 16, 21, 37, 78, 91, 111, 174, 183, 207, 231, 235, 341
Eyidogan, M., 73

F

Faustmann, M., 115
Faustmann rotation, 114, 120–122, 124, 126–128
Feasibility(ies), 52–58, 61–65, 104, 109, 110, 169, 172, 180, 194, 198, 204, 295, 302, 310, 326, 327, 341, 342, 366, 370, 371
Financial analysis, 20, 21, 55, 57–59, 182, 370
Fish, 5, 21, 23, 34, 158–162, 164, 166–169, 171–175, 207, 219, 229, 328
Fisher, A.C., 24, 28
Fisher, F.M., 309
Fisheries, 5, 8, 27, 34, 40, 188, 207, 214, 215
Flinn, J.C., 107
Flood control, 11, 35, 208, 284, 304
Flood zone, 284
Forest, 4, 6, 9, 15, 16, 26, 34, 38, 113–116, 118–124, 127, 129, 133–136, 138–142, 150, 152, 153, 155, 212, 222, 229, 230, 233–235, 237–240, 242, 243, 246, 249, 252–270, 272, 273, 276–278, 280–283, 285–292, 293, 294, 296, 297
Forest management, 6, 9, 252
Forgone income, 35, 246
Forson, F.K., 51
Francisco, H.A., 3–9, 32
Freeman, A.M., 12
Free-riding, 383
Fu, Q., 309

G

Gadde, B., 97, 106
GHG emissions, 4, 60, 66, 70, 71, 78–83, 86, 94, 96, 98, 100–102, 104, 106, 109
Gittinger, J.P., 12
Global warming, 56, 78, 100, 108
Gnansounou, E., 73, 74
Gómez-Limón, J.A., 309
Goods and services, 12, 15, 16, 21, 25, 234, 276, 296, 297
Growth rates, 24, 171
Guitart, B., 281
Gunaratne, L.H.P., 34, 37, 38, 42, 201–221
Guo, S., 309
Guo, Z., 281, 282
Gutrich, J., 114

H

Hanemann, M., 375
Hanoi, 5, 89–91, 119, 134–136, 367
Hansen, K., 141
Hazard, 279, 326, 327, 341, 370
Hensher, D., 38
Hicks, J.R., 11, 13
Highest NPV, 18, 32
Hillen, M.M., 351, 352, 358, 359, 366
Ho Chi Minh, 5
Hou, K., 15, 17, 20, 34, 251–273
Howarth, R.B., 114
Hufschmidt, M.A., 12
Human activity, 16, 17, 36, 41, 230, 234

I

Im, M., 251–273
Incremental benefits, 17, 82
Incremental costs, 17
Indonesia, 3–6, 45, 326
Industrialization, 5, 65, 300
Infrastructure, 142, 147, 202, 207, 208, 214, 215, 228, 240–242, 284, 306, 309, 312, 313, 326, 329, 341, 348, 352, 366
Ingall, R., 202
Interest rate, 19, 21
Internal economic rate of return (IERR), 19
Internal rate of return (IRR), 17–20, 152, 154, 198, 234, 373
Inundated, 190, 326, 331, 346, 380
Investment, 17, 19, 21, 28, 31, 62, 100, 114, 141, 150, 155, 161, 168, 171, 172, 174, 218, 231, 243, 262, 263, 265, 267–269, 313, 317, 331, 342, 347, 351, 384
IRR. *See* Internal rate of return (IRR)
Irrigation system, 304, 311–314, 317, 319, 320

J

Jakarta, 5
James, D., 3–9, 11–44
Jatropha curcas, 50, 66, 91
Javier, E.F., 93–110
Jiang, J.S., 138
Jian, W., 23, 42, 299–320
Jim, C.Y., 280, 282

K

Kaldor, N., 11
Keam, H., 251–273
Keleş, A.S.H., 282
Khun, K., 139
Kortelainen, M., 280
Kovacevic, V., 70, 77
Krutilla, J.V., 24, 28
Kuala Lumpur, 5
Kunyu, N., 299–320
Kuosmanen, T., 280
Kyophilavong, P., 233

L

Lake, F., 330
Land area, 3, 4, 29, 138, 159, 168, 172, 202, 252, 258, 260, 329–332, 348, 355, 356, 366, 372
Land conversion, 139, 140
Land reclamation, 5
Lang, V.T., 16, 21, 23, 34, 157–175
Lao PDR, 3, 6, 8, 20, 156, 227–230, 240, 250
Launio, C.C., 16, 38, 93–110
Leary, N., 374
Lechon, 76
Le Thanh, L., 16, 21, 23, 38, 69–88
Li, C., 59
Lic, V., 137–156
Likelihood, 29, 107, 211, 349, 355, 376, 384
Little, I.M.D., 11, 12
Liu, X.L., 309
Logging, 4, 138, 235, 240
Loss, 4, 5, 32, 35, 65, 94, 100, 114, 166, 173, 183, 191, 192, 198, 202, 204–208, 214, 216, 233, 234, 243, 246, 254, 258, 260, 276–278, 282, 294, 311, 326, 330, 332, 334, 335, 341, 342, 348, 352–354, 361, 366, 371
Louviere, J., 38
Lucking, R., 188

M

Maintenance costs, 147, 262, 265, 266, 282, 330, 334, 342, 356, 358, 359, 366
Mai, V.C., 351, 352, 358, 359, 366
Malaysia, 4–6, 326, 343
Malivarn, S., 227–250
Manalili, R.G., 93–110
Mangrove forests, 346
Mangroves, 5, 9, 328, 343
Mao, C., 309
Marginal cost, 14, 62
Marglin, S., 12
Margulis, 373
Mariano, V.P., 107
Market-based technique, 34, 206, 386
Market equilibrium, 14, 34
Market mechanism, 13, 291
Market price, 13, 14, 306
Marshallian demand functions, 13
Meas, S., 251–273
Mejias, P., 309
Mendoza, M.E.T, 369–386
Metro manila, 5
Mining, 9, 16, 21, 27, 34, 38, 42, 179, 182, 184, 186, 187, 189, 193, 202–222, 230, 326, 328
Mirrlees, J.A., 12
Mitigation, 6, 23, 69, 94, 106, 110, 191, 198, 331, 336, 343
Monetary incentives, 20
Moons, E., 280
Myanmar, 3, 4, 6

N

Nanayakkara, S.M.A., 205
Nash, C.A., 12
Natural, 4, 6–9, 12, 15, 17, 21, 27, 36, 38, 41, 44, 50, 102, 138, 140, 159, 174, 182, 183, 186, 192, 198, 199, 205, 209, 213, 215, 228, 229, 233, 237, 238, 240–242, 250, 254, 259, 261, 276, 277, 279, 281, 285, 287, 291, 292, 294, 296, 304, 326, 330, 336, 347, 348, 352, 358, 365, 370
Natural resources, 4, 6, 7, 233, 261
Navrud, S., 284
Negative impacts, 96, 108, 187, 191, 195, 196, 232, 234, 356
Negative NPV, 338
Net benefits, 13, 14, 18, 19, 24, 28, 30, 141, 150, 152–155, 235, 246, 255, 262, 270–272, 284, 292, 338, 386
Net income, 191, 315, 333, 334
Net present value (NPV), 17–20, 24, 26, 29–32, 62, 63, 77, 115, 121–124, 126–133, 152, 154, 183, 194, 232, 234, 235, 243, 246, 271, 291, 292, 295, 331, 338, 339, 342, 363, 366
Nghiem, T.H.N., 116
Nguyen, 114
Nguyen, N.B., 114
Nhung, N.T.H., 16, 21, 26, 27, 34, 113–134
Non-market values, 24, 40
Non-use values, 15, 37
NPV. *See* Net present value (NPV)

O

Olewiler, N., 279, 280
Open access, 220, 262, 264, 272
Open ended questions, 37
Operational costs, 312
Opportunity cost, 16, 21, 24, 26, 62, 189, 207, 232, 246
Optimal rotation, 27, 114, 121, 122, 125, 127, 130, 132–134
Options, 9, 12, 16–20, 23, 24, 26–30, 32, 42, 60, 94, 96, 98–101, 103, 104, 106, 109, 110, 140, 150, 152, 153, 158–160, 162, 169–175, 203, 205, 212–214, 216, 233, 255, 261, 262, 270–272, 278, 287, 293, 301–303, 310–312, 316–321, 326, 327, 331, 335, 338, 341, 342, 349–351, 356, 359–366, 370
Outputs, 17, 53, 59, 214, 233, 234
Ozsezen, A.N., 73

P

Pareto, 13, 22
Payment card, 37, 379
Pay-off, 29, 30
Pearce, D.W., 12, 15
Philippines, 3–6, 8, 16, 38, 46, 94, 96, 98, 102, 106, 109–111, 222, 325–327, 331, 343, 370, 371, 373, 384, 386
Planted forests, 113, 114, 117, 134, 135
Policy analysis, 7, 278, 279
Policy initiatives, 8
Pollution, 5, 6, 8, 9, 16, 23, 34, 38, 39, 89, 95, 108, 158–162, 164, 167, 171, 173, 174, 186, 204, 214, 276

Population, 3–6, 10, 17, 39, 81, 96, 119, 141, 155, 159, 166, 173, 181, 202, 220, 230, 233, 237, 252, 253, 258, 260, 262, 264, 266, 268, 277, 285, 302, 327, 329, 333, 349, 372, 373, 377, 384, 385
Positive NPV, 18
Pouliquen, L.Y., 31
Poverty, 4, 6, 250, 273
Predo, C.D., 11–44
Preferences, 12, 31, 33, 38, 39, 203, 209, 221
Present values, 18, 19, 24, 25, 27, 29, 32, 55–57, 77, 114, 115, 118, 125, 133, 141, 142, 147, 150, 153, 183, 195, 232, 243, 246, 255, 262–264, 266–272, 294, 302, 331, 337, 339, 359, 363–366, 385
Producer surplus, 330
Production chains, 58
Project, 8, 12, 16–20, 22, 23, 26, 28, 29, 31, 32, 35, 53, 55, 61, 77–79, 91, 109–111, 127, 134, 141, 150, 153, 155, 169, 174, 180–183, 187–199, 216–218, 231–236, 240, 241, 243, 246, 249, 250, 253, 254, 256, 257, 262, 264–266, 271, 278, 281, 304, 313, 314, 319, 327, 330–332, 347, 352, 356, 365, 370–373, 378, 379, 382, 383, 386
Project appraisal, 12, 31
Property value, 36, 335
Protected areas, 9, 222, 233, 250
Public interest, 11, 12, 20, 213

Q

Quantitative risk, 30

R

Ranganathan, J., 279
Ray, A., 12
Recreational value, 330, 332, 333
Region, 3–7, 66, 135, 172, 228, 234, 240, 304, 309, 347, 353, 355
Remote areas, 254
Replacement cost, 35, 207, 282
Research grants, 6
Reservoir, 24, 180, 181, 187, 188, 190–195, 197, 198, 308, 312, 313
Resource depletion, 6
Resource management, 6, 8, 12, 15, 181, 213, 279, 307, 308, 320
Revealed preference, 209, 222
Rice cultivation, 98, 306
Rice straw, 38, 94–111
Rice straw burning, 94, 95, 108
Riesgo, L., 309
Risk, 25, 28–33, 35, 37, 52, 106, 108, 110, 326–333, 347–350, 352, 355, 356, 366, 370, 376, 382–384, 387
Rodriguez, L.C.E., 114, 125, 281
Rola, A.C., 4
Roongtawanreongsri, S., 15, 34, 36, 38, 275–296
Ros, C., 251–273
Rubber plantations, 20, 21, 138–140, 142, 144, 150, 152–156, 277, 283, 285, 287, 288, 291, 294

S

Salinity, 29, 220, 348, 352–356, 361–366
Sand mining. *See* Mining
Sawangchote, P., 275–296
Scitovsky, 11
Sea dikes, 9, 346–353, 355, 356, 359, 363–366
Sedimentation, 5
Sensitivity analysis, 17, 19, 20, 26, 28, 29, 43, 54, 74, 117, 121, 123, 129, 130, 153, 227, 235, 255, 272, 273, 331, 337, 341, 345, 356, 365
Shadow prices/pricing, 12, 21, 54
Shoreline erosion, 5
Shushan, L., 299–320
Singapore, 3, 4, 10, 44, 45, 134, 135, 156, 273, 321, 343
Singhal, 97
Siringan, F., 328
Siriwardana, K.A.D.S.W, 217
Small-scale dike, 351, 363, 366
Social benefits, 20–22, 207
Soil erosion, 139, 206, 233, 234, 243, 246, 254, 255, 258, 260, 277, 294
Solid waste, 5, 94, 108
Southeast Asia, 3, 6, 9, 10, 134, 135, 229, 326
Spawning grounds, 5
Spillover effects, 22
Stated preference, 34, 38, 209, 280
Stenger, A., 279
Storm event, 29, 349, 351
Stratified random sampling, 373, 377
Sugden, R., 12
Suksaroj, C., 275–296
Supply curve, 14
Surcharge, 378
Sustainability, 6, 9, 42, 50, 211, 232, 273
Sustainable, 6, 56, 88, 91, 111, 134, 228, 229, 231, 234, 252, 279, 280, 301, 310, 321
Su, Y.X., 309

T

Taxes and subsidies, 12, 114, 135
Technical support, 6, 318
Thailand, 3, 6, 8, 38, 91, 102, 106, 110, 111, 276, 284, 326, 343, 385, 386
Theoretical framework, 374
Thomson, T.A., 125
Thoradeniya, B., 16, 37, 43
Thoradeniya, W.B.M., 181
Threat, 5, 234
Threatened, 38, 162, 179, 184, 219, 277, 300, 329, 330, 332, 341, 342, 371
Threshold values, 24, 28
Timber extraction, 4
Timber price, 115, 120, 129–134
Timilsina, G., 102
Tong, C., 280, 282
Top, N., 137–156
Total benefits, 13, 264, 266, 268, 269, 272, 273, 336, 365
Total cost, 14, 57–59, 61, 62, 103, 144, 302, 336, 351
Tradeoff, 23, 42, 43
Transesterification, 51, 58, 60, 64, 67
Transparent, 43, 371
Travel cost method (TCM), 36, 190, 191
Truc, N.T.T., 97, 157–175
Turner, R.K., 15
Typhoons, 342, 370

U

Uncertainty(ies), 20, 25, 28–30, 32, 33, 41, 81, 109, 115, 212, 235, 321, 341, 355, 358, 363, 364, 380, 382
Unemployment, 12, 327, 334
Urbanization, 5, 6, 276, 277, 300, 326, 349
Use and non-use values, 15
Utility, 12, 13, 209, 374

V

Valuation, 7, 12, 35, 36, 38, 40, 43, 55, 183, 189, 190, 192, 193, 207, 222, 234, 280, 281, 284, 285, 296, 333, 347, 373, 386
van Essen, H., 100
van Kooten, G.C., 114, 121, 122, 125
Varela-Ortega, C., 309
Vietnam, 3, 6, 8, 16, 23, 26, 69, 70, 79–82, 88–91, 111, 113–115, 117–119, 133–136, 158, 173, 175, 326, 343, 346, 347, 351, 352, 355, 360, 361, 367
Vinh, K.Q., 157–175
Viravouth, P., 227–250
Vo, D.H., 117, 118
Vulnerability, 326–330, 370, 373
Vulnerable, 9, 184, 204, 210, 213–215, 220, 221, 329, 341

W

Wage differential, 37
Wang, P., 61
Wang, R., 309
Wang, R.S., 138
Wang, Y.L., 309
Wang, Z., 16, 21, 26, 34, 49–66
Warford, J., 302
Waste disposal, 184, 187, 188, 191, 192
Wastewater, 158–162, 164, 166–169, 171–174, 188
Water demands, 300, 304, 305, 320, 321
Water pollution, 161
Water quality, 5, 37, 158, 160, 161, 165, 166, 171–173, 180, 192, 199
Water resource allocation, 309, 320
Water resources, 5, 9, 167, 300, 303, 304, 306–310, 315, 321, 322
Watershed, 139, 255, 282, 303, 310, 311, 371, 372
Water-stressed rivers, 179
Water supply, 5, 172, 174, 180, 181, 184, 186–189, 191, 192, 194, 196, 199, 205, 239, 240, 276, 277, 281, 282, 289, 290, 292, 301, 304, 306, 309–312, 320, 321
Wei, Y.P., 309
Welfare, 11, 13, 20, 22, 27, 30, 194, 262, 276, 292, 316, 334, 342, 370, 374
Wellbeing, 11–16, 22, 42, 43, 209, 235, 384
Wen, J.J., 305
Wesseler, J., 70, 77
Wetlands, 9, 23, 162, 164, 168, 171–174, 206, 229, 230, 237, 241, 297, 300–302, 304–312, 314, 318–321, 326
Wildlife, 142, 191, 229, 230, 241, 255, 256, 277, 287, 297, 300
Williams, A., 12
Willig, R., 13
Willingness to pay, 13, 22, 35–37, 285, 309, 371, 373–377, 381–384
Wilson, S., 279, 280, 282
Worldfish, 6

X

Xia, G.L., 305
Xiaoxia, W., 299–320

Xi, J., 282
Xing, B., 61

Y
Yem, D., 15, 17, 20, 21, 34, 137–156
Yolasiğmaz, H.A., 282

Z
Zhang, M., 282
Zhao, Z., 309
Zhou, H., 51
Zhou, X.H., 309

MIX
Papier aus verantwortungsvollen Quellen
Paper from responsible sources
FSC® C105338

If you have any concerns about our products,
you can contact us on
ProductSafety@springernature.com

In case Publisher is established outside the EU,
the EU authorized representative is:
Springer Nature Customer Service Center GmbH
Europaplatz 3, 69115 Heidelberg, Germany

Printed by Libri Plureos GmbH
in Hamburg, Germany